Ultra-wideband Positioning Systems

Accurate determination of the location of wireless devices forms the basis of many new and interesting applications. Ultra-wideband (UWB) signals enable such positioning, especially in short-range wireless networks. This text provides a detailed account of UWB positioning systems, offering comprehensive treatment of signal and receiver design, time of arrival estimation techniques, theoretical performance bounds, ranging algorithms, and protocols. Beginning with a discussion of the potential applications of wireless positioning, and investigating UWB signals for such applications, later chapters go on to establish a signal processing framework for analyzing UWB ranging and positioning systems. The recent IEEE 802.15.4a standard related to UWB is also studied in detail. Each chapter contains examples, problems, and MATLAB® exercises to help readers grasp key concepts. This is an ideal text for graduate students and researchers in electrical and computer engineering, and for practitioners in the communications industry, particularly those in wireless communications. Further resources are available at www.cambridge.org/9780521187831.

ZAFER ŞAHİNOĞLU is currently working at the Mitsubishi Electric Research Laboratories as a Principal Technical Member, and is a Senior Member of the IEEE. He received his Ph.D. in Electrical Engineering from the New Jersey Institute of Technology in 2001, receiving their Hashimoto Prize the following year.

SİNAN GEZİCİ is an Assistant Professor in the Department of Electrical and Electronics Engineering at Bilkent University, Ankara, Turkey. He received his Ph.D. in Electrical Engineering from Princeton University in 2006, and is also a member of the IEEE.

İSMAİL GÜVENÇ is a Research Engineer at DoCoMo USA Communications Laboratories, Palo Alto, CA, and is a member of the IEEE. He received his Ph.D. in Electrical Engineering from the University of South Florida in 2006, receiving their Outstanding Dissertation Award in the following year.

Ultra-wideband Positioning Systems

Theoretical Limits, Ranging Algorithms, and Protocols

ZAFER ŞAHİNOĞLU, SİNAN GEZİCİ,
AND İSMAİL GÜVENÇ

CAMBRIDGE
UNIVERSITY PRESS

University Printing House, Cambridge CB2 8BS, United Kingdom

Cambridge University Press is part of the University of Cambridge.

It furthers the University's mission by disseminating knowledge in the pursuit of education, learning and research at the highest international levels of excellence.

www.cambridge.org
Information on this title: www.cambridge.org/9780521873093

© Cambridge University Press 2008

This publication is in copyright. Subject to statutory exception and to the provisions of relevant collective licensing agreements, no reproduction of any part may take place without the written permission of Cambridge University Press.

First published 2008
First paperback edition 2011

A catalogue record for this publication is available from the British Library

Library of Congress Cataloguing in Publication data
Sahinoglu, Zafer, 1971–
Ultra-wideband positioning systems : theoretical limits, ranging algorithms, and protocols/Zafer Sahinoglu, Sinan Gezici, and Ismail G|veng.
  p. cm.
Includes bibliographical references and index.
ISBN 978-0-521-87309-3
1. Ultra-wideband devices.   I. Gezici, Sinan, 1979–   II. G|veng, Ismail.   III. Title.
TK7872.M25S36 2008
621.384–dc22

2008025081

ISBN 978-0-521-87309-3 Hardback
ISBN 978-0-521-18783-1 Paperback

Additional resources for this publication at www.cambridge.org/9780521187831

Cambridge University Press has no responsibility for the persistence or accuracy of URLs for external or third-party internet websites referred to in this publication, and does not guarantee that any content on such websites is, or will remain, accurate or appropriate.

Dedicated to
my parents, Hakki and Saziye Sahinoglu, my brother Fatih and my sister Filiz, whose support I have always had from thousands of miles away.

Zafer Sahinoglu

Dedicated to
my parents, Ergül and Muammer Gezici, my sister Sevinc, my brother-in-law M. Bekir and my dear niece Sila.

Sinan Gezici

Dedicated to
my wife Zeynep, my daughter Beyza and my parents, Mehmet and Hatice Guvenc, and my brother Oguz.

Ismail Guvenc

Contents

Preface		*page* ix
1	**Introduction**	1
	1.1 Trends in location-aware applications	2
	1.2 Taxonomy of localization systems	6
	1.3 Ranging and localization with UWB	10
	1.4 Problems	19
2	**Ultra-wideband signals**	20
	2.1 Definition of UWB	20
	2.2 International regulations for UWB signals	24
	2.3 Emerging UWB standards	32
	2.4 Problems	42
3	**Ultra-wideband channel models**	44
	3.1 UWB versus narrowband	45
	3.2 UWB channel characterization	49
	3.3 UWB channel measurement campaigns	53
	3.4 Problems	61
4	**Position estimation techniques**	63
	4.1 Measurement categories	64
	4.2 Position estimation	74
	4.3 Position tracking	92
	4.4 Problems	97
5	**Time-based ranging via UWB radios**	101
	5.1 Time-based positioning	101
	5.2 Error sources in time-based ranging	103
	5.3 Time-based ranging	108
	5.4 Fundamental limits for time-based ranging	117

	5.5	Maximum likelihood-based ranging techniques	125
	5.6	Low-complexity UWB ranging techniques	133
	5.7	Problems	145

6 Ranging protocols — 148

	6.1	Layered protocols	149
	6.2	Time-based ranging protocols	150
	6.3	Ranging in IEEE 802.15.4a standard	158
	6.4	Problems	180

7 Special topics in ranging — 181

	7.1	Interference mitigation	182
	7.2	Coded payload modulation	196
	7.3	Private ranging	198
	7.4	Problems	202

8 Practical considerations for UWB system design — 203

	8.1	Signal design for ranging	203
	8.2	Link budget calculations	210
	8.3	Hardware issues	214
	8.4	Problems	225

9 Recent developments and future research directions — 227

	9.1	Development of accurate ranging/positioning algorithms	227
	9.2	Training-based systems and exploiting the side information	229
	9.3	NLOS mitigation	231
	9.4	Multiple accessing and interference mitigation	232
	9.5	Cognitive ranging and localization	233
	9.6	Anchor placement	235
	9.7	UWB radar in health-care	236
	9.8	UWB for simultaneous localization and mapping	237
	9.9	Secure ranging and localization	238
	9.10	Concluding remarks	240

References   241
Index   265

Preface

Ability to locate assets and people will be driving not only emerging location-based services, but also mobile advertising, and safety and security applications. Cellular subscribers are increasingly using their handsets already as mapping and navigation tools. Location-aware vehicle-to-vehicle communication networks are being researched widely to increase traffic safety and efficiency. Asset management in warehouses, and equipment and personnel localization/tracking in hospitals are among other location-based applications that address vast markets. It is a fact that application space for localization technologies is very diverse, and performance requirements of such applications vary to a great extent.

The Global Positioning System (GPS) requires communication with at least four GPS satellites, and offers location accuracy of several meters. It is used mainly for outdoor location-based applications, because its accuracy can degrade significantly in indoor scenarios. Wireless local area network (WLAN) technology has recently become a candidate technology for indoor localization, but the location accuracy it offers is poor, and also high power consumption of WLAN terminals is an issue for power-sensitive mobile applications. Ultra-wideband technologies (UWB) promise to overcome power consumption and accuracy limitations of both GPS and WLAN, and are more suitable for indoor location-based applications.

The Federal Communications Commission (FCC) and European Commission (EC) regulate certain frequency bands for UWB systems. These have prompted worldwide research and development efforts on UWB. Another consequence was development of international wireless communication standards that adopt UWB technology such as IEEE 802.15.4a WPAN and IEEE 802.15.3c WPAN.

The writing of this book was prompted by the fact that UWB is the most promising technology for indoor localization and tracking. As of today there is no book with particular focus on theoretical and practical evaluation of the capabilities of various UWB localization systems. The book is written for graduate-level students and practicing engineers. Prior knowledge in probability, linear algebra, digital signal processing, and signal detection and estimation is assumed.

The scope of the book is not limited to time-based UWB ranging systems, because in addition to signal design and time of arrival estimation, most location systems should adopt a ranging protocol and perform certain position estimation and tracking techniques. For completeness of the course, in depth coverage from signal design to position solving and tracking techniques is given. Each chapter includes examples and problems to accelerate readers' understanding. Programming exercises allow readers to simulate various techniques in UWB systems and help them see impacts of various design parameters.

Although the main focus of all chapters is on UWB systems, Chapters 1, 4, and 9 are not limited to UWB. Current trends for location-aware applications and taxonomy of localization systems are given in the first chapter. Position estimation and tracking techniques, which are applicable to any location system, are discussed in Chapter 4. Recent developments and future research directions form the main topic of Chapter 9.

UWB-specific treatment starts with Chapter 2, in which various UWB signal waveforms are studied, international regulations for UWB signal emissions are presented, and various UWB standards are discussed. UWB channel models arising from channel measurements conducted for 2–10 GHz, below 1 GHz and 57–66 GHz frequency band regions are overviewed in Chapter 3. Also, differences between narrowband and UWB channels are highlighted in this chapter. Treatment of time based ranging via UWB radios is given in Chapter 5. Its content includes discussion of potential error sources and quantification of fundamental performance limits via Cramer–Rao and Ziv–Zakai lower bounds. Chapter 6 is devoted to the discussion of various ranging protocols, and their pros and cons. The ranging aspect of the recently published IEEE 802.15.4a UWB WPAN standard is studied in detail, including preamble and start of frame delimiter design, timing counter management, and clock frequency offset mitigation. Narrowband and multiuser interference mitigation techniques, ranging privacy mechanisms and the state-of-the-art coded payload modulation technique are the special topics covered in Chapter 7. Practical considerations for UWB system design are given in Chapter 8, including signal design under practical constraints, link budget analysis, and specific hardware issues.

Solutions for the problems at the end of each chapter and Matlab simulation scripts can be found by visiting the website for this book, which is currently at www.cambridge.org/9780521873093. The most up-to-date errata sheet and references to additional material can also be found at the same site.

We would like to thank experts in the field, who have reviewed and commented on the draft of the manuscript. Their inputs greatly helped us improve the presentation. Special thanks to Andreas F. Molisch from Mitsubishi Electric Research Labs for his suggestions about the channel modeling chapter, Davide Dardari from University of Bologna for his thorough review of Chapter 5, Henk A. Wymeersch from Massachusetts Institute of Technology and Qin Wang from Harvard University for their inputs in general and for helping organize Chapter 6 in particular, Yihong Qi from AMD for her inputs on Chapters 4 and 7, Rainer Hach from Nanotron Inc. for his review of Chapter 6, Chia Chin Chong from NTT DoCoMo Labs and Fikret Altinkilic from Syracuse University for their suggestions on Chapter 3, Philip Orlik from Mitsubishi Electric Research Labs for reviewing and providing suggestions and comments on Chapters 2 and 4, and furthermore Fujio Watanabe from NTT DoCoMo Labs, Huseyin Arslan from University of South Florida and Volkan Efe from Motorola for providing comments on various chapters.

We also thank many colleagues in Mitsubishi Electric Research Labs, namely Jinyun Zhang, Kent Wittenburg, Fatih Porikli, Giovanni Vannucci, Richard Waters, Joseph Katz,

Darren Leigh, Huifang Sun, Masashi Saito, and Chunjie Duan. We are indebted to many amazing researchers with whom we closely interacted in the IEEE 802.15.4a standard and through other collaboration activities. These researchers are Patrick W. Kinney, Vern Brethour, Jay Bain, John Lampe, Ismail Lakkis, Michael McLaughlin, Francois Chin, Shahriar Emami, Ryuji Kohno, Yves Paul Nakache, Bin Zhen, Lars Menzer, Patricia Martigne, Huan Bang Li, Richard Roberts, Laurent Ouvry, Arnaud Tonnerre, Benjamin Rolfe, Moe Win, Huilin Xu, Hasari Celebi, Amer Catovic, Hiroshi Inamura, Yasuhiro Naoi, Hisashi Kobayashi, and H. Vincent Poor. Finally, we acknowledge the great support of our editor, Phil Meyler, at Cambridge University Press, and thank our families and beloved ones for being patient during the writing of this book.

1 Introduction

Wireless communications are becoming an integral part of our daily lives. Satellite communications, cellular networks, wireless local area networks (WLANs), and wireless sensor networks (WSNs) are only a few of the wireless technologies that we use every day. They make our daily lives easier by keeping us connected anywhere, anytime.

Since more and more devices are going wireless every day, it is essential that future wireless technologies can coexist with each other. Ultra-wideband (UWB) is a promising solution to this problem which became popular after the Federal Communications Commission (FCC) in the USA allowed the unlicensed use of UWB devices in February 2002 subject to emission constraints. Due to its unlicensed operation and low-power transmission, UWB can coexist with other wireless devices, and its low-cost, low-power transceiver circuitry makes it a good candidate for short- to medium-range wireless systems such as WSNs and wireless personal area networks (WPANs).

One of the most promising aspects of UWB radios are their potential for high-precision localization. Due to their large bandwidths, UWB receivers can resolve individual multipath components (MPCs); therefore, they are capable of accurately estimating the arrival time of the first signal path. This implies that the distance between a wireless transmitter and a receiver can be accurately determined, yielding high localization accuracy.

Such unique aspects of UWB make it an attractive technology for diverse communications, ranging, and radar applications such as robotics, emergency support, intelligent ambient sensing, health-care, asset tracking, and medical imaging (see Fig. 1.1). Potential of UWB technology for future wireless communication networks was also recognized by the IEEE, which adopted UWB in the IEEE 802.15.4a WPAN standard for the creation of a physical layer for short-range and low data rate communications and for precise localization.

Various aspects of UWB ranging and localization systems are discussed in the subsequent chapters. In this chapter, first, general trends in location-aware applications are reviewed. Then, a taxonomy of localization systems is presented. This is followed by a discussion on UWB localization applications and available UWB localization technologies.

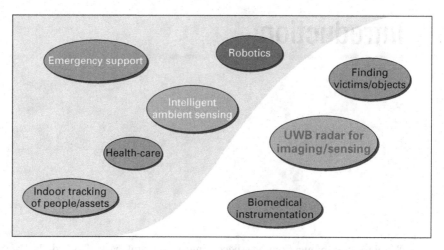

Fig. 1.1. Diverse business opportunities for UWB communications, ranging, and radar.

1.1 Trends in location-aware applications

1.1.1 Location-based applications and services

As wireless devices are becoming more and more integrated into our daily lives, they are also getting more and more intelligent. In other words, today's wireless devices are becoming more context aware. In [1], *context awareness* (or *context sensitivity*) is defined as the ability of the mobile device to be aware of the user's surrounding physical environment and state. Another related definition in [2] states that a system is *context aware* if it uses contexts[1] to provide relevant information and services to the user, where relevancy depends on the user's tasks. These services may be based on the context information such as time, location, temperature, speed, orientation, biometrics, audio/video recordings, etc.

Between these different variables that define a context, location and time are probably the two most important inputs that define a specific situation [3], and *location awareness* can be considered as a special and important form of context awareness. Localization serves as an enabling technology that makes numerous context-aware services and applications possible.

A location-aware wireless device may use the location information in different ways for different technologies. For example in cellular networks, location information can be used for Emergency-911 (E-911) services, location-sensitive billing, fraud detection, resource management, and intelligent transportation systems [4]. Location of the doctors/patients in hospitals, injured skiers on mountains, or fire-fighters and victims inside a building are a few examples on how location information can be used to save lives in emergency situations. While knowing your position will be handy to find the closest

---

[1] Context is defined as any information that can be used to characterize the situation of an entity [2].

printer in a wireless local area network, it may as well be used to locate your friend in a university campus. In warehouses, laboratories, and hospitals, location of portable and in-demand equipment may be needed [5]. A customer may receive location-based advertisements in a shopping mall, or personal digital assistants with location support can be used for guided tours in museums. Smart homes/offices/highways that exploit the location information are a few other examples of how our daily lives can be made easier [6].

The services provided to the user based on location information are commonly referred as location-based services (LBS). Some definitions of the LBS available in the literature are as follows [3].

- *"Network-based services that integrate a derived estimate of a mobile device's location or position with other information so as to provide added value to the user."*
- *"Recent concept that denotes applications integrating geographic location (i.e. spatial coordinates) with the notion of service. Examples of such applications include emergency services, car navigation systems, tourist tour planning, or yellow maps (combining of yellow pages and maps) information delivery."*
- *"Will allow mobile users to receive personalized and lifestyle-oriented services relative to their geographic location."*
- *"Most location-based services will include two major actions: (1) Obtaining the location of a user, and (2) Utilizing this information to provide a service."*

The classification and characteristics of different LBS are presented in [3], and summarized in Table 1.1, which also includes the business models ("C:" Customer, "B:" Business, "G": Government, "W": Workforce), location update specifications (Pull: Information is provided after a request by the mobile, Push: Network delivers the information to the mobile based on an event or trigger condition), and typical accuracy requirements.

The LBS are traditionally considered for cellular networks. This may be correlated with the fact that cellular technology has long been an integral part of our daily lives and has a wide consumer acceptance. Hence, killer applications are relatively obvious.

On the other hand, technologies such as WLANs and WSNs have only recently become widely deployed. As they become more integrated to daily practices, the killer applications for these technologies will become more obvious. As a matter of fact, we have already started seeing new LBS using these technologies. For example, LOKI software [7], developed by Skyhook Wireless Inc., uses the WiFi network to pinpoint a mobile user's location and provide services such as finding the closest restaurant to the user's location. Applications such as guided tours in museums, location-based advertisements, people/inventory tracking, etc. are possible through similar technologies. While such LBS are not widely deployed today, they are expected to become more common with the advances in different relevant technologies, wide consumer acceptance, and decreases in device costs.

The accuracy and precision requirements of location-based applications are highly dependent on the application characteristics. Accuracies on the order of tens of meters

Table 1.1. Classification and characteristics of location-aware applications and location-based services (After [3]).

Category	Description	Examples	Business Model	Pull versus Push	Accuracy
Location-based information services	User requests information related to its location	Local weather forecast, navigation, local maps, local bus schedules	C2B	Pull	< 1 km
Points of interest	The mobile user looks up stationary objects or facilities in the near area	Restaurant, hotel, etc. finder services, service look-up (e.g. printers)	C2B	Pull	< 1 km
Discovering other users	The mobile user looks up other users in the nearby area	Games, friend-finder, flirt-finder	C2C	Pull	< 200 m
Tracking services	A (likely stationary) user looks up the location of a mobile person or object	Fleet management, tracking children, patients, doctors, tracking assets	C2C, B2B, B2W	Tracking entity pulls	< 200 m
Assistance services	A service center receives the location of a mobile caller who needs assistance	Emergency calls, breakdown services	C2G, C2B	Push	< 20 m
Messaging and announcement services	Mobile users receive a message from another user broadcast to a certain area	Local advertisements, messages to nearby friends	C2C, B2C	Push	< 1km
Trigger services	A mobile user receives a trigger when entering a certain location	Location-based reminders, traffic warnings, weather warnings	C2C, C2B, B2C	Push	< 500 m
Location-based billing	A user is charged according to his/her location	Toll billing, home zones	B2C, B2B	Push towards billing side	< 500 m

Table 1.2. Accuracy requirements of potential localization applications (After [8]).

Applications	Accuracy
Automated handling	0.5 cm
Route-guidance for blind	1 cm
In-building survey	1 cm
Tool positioning	1 cm
In-building robot guidance	8 cm
Formation flying	10 cm
Recreation and toys	10 cm
Urban canyon (off-road)	30 cm
Urban canyon (marine)	50 cm
Incidence tracking/guidance	80 cm
Urban canyon (other)	80 cm
Exhibit commentary	1 m
Goods and item tracking	1 m
Hazard warnings	1 m
Pedestrian route guidance	1 m
In-building tracking (other)	1 m
In-building worker tracking	1 m
Urban canyon (rail)	1 m
Precision landing	1 m
Access control	3 m
Location-based services	3 m
Public services tracking	3 m
Docking	5 m
Parolee tracking	10 m
Local information	30 m
Train / air / bus information	30 m
Advertising	100 m

might be satisfactory for applications such as location-based handover in cellular networks. On the other hand, a meter of positioning error may mean a life-or-death situation for a fire-fighter depending on whether he is on the correct side of a building wall or not. In [8], tentative accuracy requirements of various localization applications are depicted, which are tabulated in Table 1.2. These show that the required accuracy of the location estimate can range from less than a centimeter to over tens of meters. Note that accuracy is only one aspect of the overall system; factors such as cost, range, and complexity are other issues to be considered, and no single localization system fits to all applications.

There are numerous localization technologies currently available which have different ranges, accuracy levels, costs, and complexities. While some of these technologies date back to World War II, significant improvements in localization technologies have been observed, particularly over the last few decades. Some of the important current localization technologies are classified and their key characteristics are summarized in Table 1.4 at the end of this chapter.

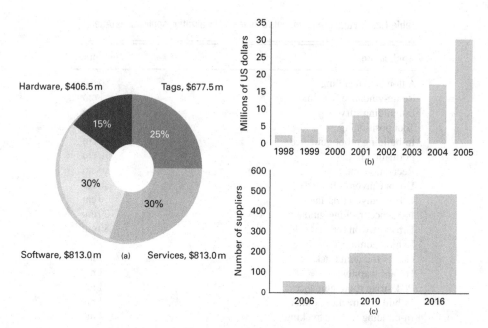

Fig. 1.2. (a) Share of spendings on RTLS in 2016 in millions of US dollars. (b) Global market on RTLS in millions of US dollars, from 1998 to 2005. (c) Trend in number of significant suppliers into parts of the RTLS value chain in 2006, 2010 and 2016 (After [9]).

1.1.2 Trends in real-time location systems

Due to its importance, there is a significant interest in the industry on real-time location systems (RTLS). In Fig. 1.2, estimated market share of spendings on RTLS in 2016 is depicted [9], which shows that software and services will dominate the total share rather than the hardware. Global market on RTLS and trend in number of significant suppliers is also seen to be exponentially increasing, indicating the importance and significance of the technology. While the number of significant suppliers into parts is 50 in 2006, it is expected to be around 200 in 2010, and around 500 in 2016.

According to IDTechEx forecast for RTLS [9], the global RTLS market will increase to 2710 million US dollars in 2016, while it is only 70 million US dollars in 2006. According to the same report, the major applications of RTLS in 2016 will be in military (44%, US$1.2 billion), health-care (30%, US$0.8 billion), logistics and other (26%, US $0.7 billion, including manufacturing, prison/parole service, and postal/courier sectors, etc.); however, there will be increasing interest from other sectors such as leisure, retail, and agricultural.

1.2 Taxonomy of localization systems

Different classifications of localization technologies have been previously presented in the literature [2, 10, 11]. In the following, some of the important classifications are briefly overviewed.

1.2.1 Signaling scheme

A fundamental classification is based on the signaling scheme that a localization technology uses. *Radio frequency (RF)* is probably the most commonly used signaling scheme for localization purposes. This is because RF signals can penetrate through obstacles and can propagate to long distances.

Infrared signals are low power and inexpensive; however, they cannot penetrate through obstructions (as opposed to RF), and they are susceptible against sunlight. Therefore, one usually has to install infrared sensors all over the indoor environment to pick up the signals from a transmitter.

Optical signals also require line-of-sight (LOS) conditions, are affected by sunlight, and require low power. They provide high accuracies and are typically more appropriate for short ranges (e.g. around 10 m).

Another inexpensive signaling alternative is *ultrasound* signals, which provide high accuracies in the short range. An advantage of acoustic signals is that the sound travels slowly. Hence, slow clocks are sufficient, and high accuracy can be achieved inexpensively in LOS conditions. On the other hand, acoustic emitters are power hungry, and they do not work well in non-line-of-sight (NLOS) scenarios.

1.2.2 RF signaling waveforms

Among different RF technologies, *ultra-wideband (UWB)*, *code division multiple-access (CDMA)*, and *orthogonal frequency division multiplexing (OFDM)* are a few of the RF technologies that may be considered for localization. Depending on their accuracy and range requirements, different versions of these technologies are used in various wireless systems such as *cellular systems*, *WLANs*, WPANs, *radio frequency identification (RFID) systems*, and *WSNs*.

1.2.3 Position-related parameters

Localization systems can employ various parameters/information obtained from received signal(s), such as the *time-of-arrival (TOA)*, *time difference of arrival (TDOA)*, *angle of arrival (AOA)*, and *received signal strength (RSS)*. Hybrid approaches that use combinations of the above are also possible. These different approaches are discussed in detail in Chapter 4.

1.2.4 Data fusion and localization methods

Different metrics of a received signal can be processed in various ways for obtaining a location estimate. The simplest way of estimating the target's location is the *cell ID* localization, where the target's position is approximated to be the location of the serving reference node (RN)[2], and the positioning accuracy is limited to the cell size. Alternatively, with *proximity detection*, the distance to a particular RN can be estimated.

In *triangulation*-based systems, the intersection(s) of the *at least* two lines (obtained e.g. from AOA information) from *at least* two RNs is/are used to estimate the terminal location.

On the other hand, in *trilateration* systems, at least three RNs are required for two-dimensional (2-D) localization, and at least four RNs are required for three-dimensional (3-D) localization. The intersections of the circles (or hyperbolas) obtained from TOA/RSS information (or TDOA information) are used to estimate the terminal location.

Fingerprint-based or *pattern-matching* localization technologies compare real-time measurements with a location database to infer the terminal's location. The positioning accuracy is limited to the granularity of the training locations, and an off-line calibration stage is required, which may need to be repeated if the propagation characteristics of the environment change.

1.2.5 Location estimation unit

Depending on where the localization is performed, localization technologies can be classified as *handset-based* (location aware), or *network-based* (location support) systems. In handset-based localization, the target receives signals from the RNs, and calculates its own location (also called *self-positioning*). It is more commonly used in military or public safety applications, such as a fireman trying to find his way out of a building. Global Positioning Systems (GPS) systems also fall under the same category.

In network-based localization, the RNs forward the received signal information, such as TOA, AOA, and RSS, to a central processing unit, where the target's location is estimated (also called *remote-positioning*). Note that privacy issues may be a big concern in this type of localization system, since the target may not always wish to be tracked by the network. In such a case, the target may estimate its own location, and may choose not to report its location to the central server.

1.2.6 Indoor versus outdoor localization

Due to significant differences in the propagation characteristics of the environments, it is common to classify the localization systems as *indoor* and *outdoor* localization systems.

A typical example of an outdoor localization system is the GPS. It uses TDOA information from four or more of 24 satellites around the world to estimate target's position with an accuracy between 1 and 5 m. It performs poorly indoors since buildings block GPS signals. Another widely used outdoor localization system is the E911 service in cellular networks.

---

[2] A reference node may be a base station (BS) in cellular networks, an access point (AP) in WLANs, or an anchor node (AN) in WSNs.

Although GPS and E911 systems can provide location information outdoors, they are not designed for the indoor environments, where unique technical challenges exist and accuracy requirements are typically much higher. Indoor localization systems may require a completely different infrastructure installed within buildings (e.g. active badges [10]), or they may rely on the existing communications infrastructure such as wireless LANs (e.g. RADAR by Microsoft Research [21]).

1.2.7 Active versus passive localization

In *active* localization systems, the network sends specific signals to estimate the location of a target. *Passive* localization systems, on the other hand, use incumbent signals received from the mobile. In other words, as opposed to an active system, a passive system does not transmit any signal for location estimation purposes.

1.2.8 Centralized versus distributed localization

In *centralized* localization systems, position-related information (such as the TOA, AOA, RSS, etc.) is forwarded to a data fusion center, where the target location is estimated. The terminals that use a *distributed* localization system determine their location jointly by communicating with each other.

1.2.9 Software-based versus hardware-based localization

Software-based localization systems can be implemented by using the existing infrastructure, and there is no need for deploying extra hardware for localization purposes. As an example, Ekahau positioning engine [22] uses the existing WiFi infrastructure; it uses signal processing algorithms to estimate (and track) the target location from the RSS metrics obtained from different access points. *Hardware-based* localization systems need installation of extra hardware, such as in the case of SpotOn technology [23].

1.2.10 Relative coordinate versus absolute coordinate localization

Absolute location is the actual physical coordinate of a target with respect to a global reference; e.g. expressed as 24° 35′ 53.2″ North, 10° 45′ 11.5″ East. On the other hand, relative location is the position of a target with respect to a local reference within the network.

Another related term is the *semantic* (or *symbolic*) location, which is much easier to interpret by the targets. Absolute location, as discussed before, may be good enough for a missile but it is not much use for a taxi driver [1]. Examples of semantic location are "Topkapi Palace, Istanbul, Turkey", or "Stanford University Campus, Palo Alto, CA". Semantic locations have the advantage that they can be easily used as search keys in traditional databases [3].

1.2.11 Range-based versus range-free localization

If a localization system depends on the distances (or angles) between the nodes (e.g. the target and the base station), such a localization system is referred to as a *range-based* system. The distances are typically estimated using TOA, TDOA, and RSS metrics (or AOA for direction of arrival estimation).

On the other hand, there are localization systems which do not require estimation of absolute distances. Such approaches typically fall under two categories [2]: (1) techniques that rely on high density of anchors, such as the centroid algorithm, which calculates the position estimate to be the average location of all the *connected* anchors, and (2) hop counting techniques such as the DV-hop algorithm [24].

1.2.12 Accuracy versus precision

How well a certain localization technology performs is commonly measured with its *accuracy*, which is defined as how far the estimated location of the target is away from its actual location (e.g. 1 m accuracy). It is also desired that a certain localization accuracy is achieved with high probability; *precision* defines the percentage that a certain accuracy (or better) is achieved (e.g. 95% precision).[3]

1.3 Ranging and localization with UWB

1.3.1 Applications of UWB localization

As discussed earlier, there is not one localization technology that fits to all applications. For example, whereas GPS is an excellent technology for many scenarios and typically has fine precision outdoors, it fails to yield the desired accuracies in indoor environments due to multipath effects and blocked LOS. In addition, GPS devices are usually too expensive for many applications.

UWB is an excellent signaling choice for high accuracy localization in short to medium distances due to its high time resolution and inexpensive circuitry. It is also considered to be the unique signaling choice for short-range, low-data rate communications such as in WSNs. Some of the key applications for low-rate UWB communication and ranging systems are summarized in Fig. 1.3.

The low-rate UWB was standardized in 2007 under the IEEE 802.15.4a. The potential of UWB for high precision ranging, its possible applications, and implementation issues were extensively discussed and documented during the standardization process. Some key UWB localization applications as well as their range/accuracy requirements are tabulated in Table 1.3 based on [25]. The UWB technology is an excellent match that makes these exciting applications possible with sub-meter accuracies at distances smaller than 300 m.

---

[3] Some other performance measures in localization systems were listed in [2] as calibration requirements, responsiveness, self-organization, cost, power consumption, and scalability.

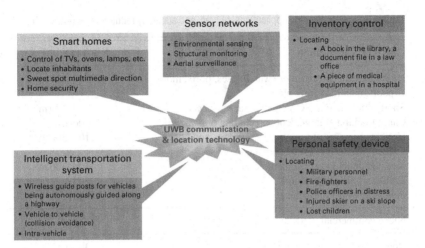

Fig. 1.3. Applications and business opportunities for low-rate UWB.

Probably the most suitable technology where UWB may be used as a physical layer signaling scheme is WSNs. Together with the advances in RF and MEMS IC technologies, wireless sensors are becoming cheaper, smaller, and more capable. We are probably living the last years in which furniture, buildings, cars, streets, highways, etc. are not dominated by WSNs. Localization of sensors in WSNs is important for a number of reasons, including: (1) in order for sensor data (e.g. temperature, humidity, and light intensity, etc.) to be meaningful, it is essential that the sensor's location is known, (2) some geographic routing algorithms can be enhanced if the location information is available, (3) location itself can be the data to be sensed, especially in logistics management [26].

Below, some of the recent applications of WSNs are briefly oveviewed.[4] The motivation is to make it clear that localization is a key component in many of the WSN applications.

- In the Great Duck Island project, 150 sensing nodes are deployed throughout the island, to collect and relay data such as temperature, pressure, and humidity, etc. to a central device. Then, these data are made available through Internet using a satellite link [28].
- In the ZebraNet project, WSNs are used to study the behavior of zebras, where special GPS equipped collar devices are attached to the zebras [29].
- In order to monitor the volcano activity in Ecuador, WSNs are used in the areas where human presence is discouraged [30].
- In agricultural monitoring applications (such as the wireless vineyard project), data are collected using WSNs and processed to make decisions, such as detecting parasites to automatically choose the right insecticide, or watering and fertilization only wherever and whenever necessary [26, 31].

---

[4] For some other applications and a detailed discussion on the design space of WSNs, the reader is referred to [27].

Table 1.3. Key localization applications, ranges, and accuracy requirements (After [25]).

Core RTLS applications	Range	Accuracy
High value inventory items (warehouses, ports, motor pools, manufacturing plants)	100–300 m	30–300 cm
Sports tracking (NASCAR, horse races, soccer)	100–300 m	10–30 cm
Cargo tracking at large depots including port facilities	300 m	300 cm
Vehicles for large automobile dealerships and heavy equipment rental establishments	100–300 m	300 cm
Key personnel in office/plant facility	100–300 m	15 cm
Children in large amusement parks	300 m	300 cm
Pet/cattle/wild-life tracking	300 m	15–150 cm
Niche commercial markets	**Range**	**Accuracy**
Robotic mowing and farming	300 m	30 cm
Supermarket carts (matching customers with advertised products)	100–300 m	30 cm
Vehicle caravan/personal radios/family radio service	300 m	300 cm
Military applications	**Range**	**Accuracy**
Military training facilities	300 m	30 cm
Military search and rescue: lost pilot, man overboard, coast guard rescue operations	300 m	300 cm
Army small tactical unit friendly forces situational awareness – rural and urban	300 m	30 cm
Civil government/safety applications	**Range**	**Accuracy**
Tracking guards and prisoners	300 m	30 cm
Tracking firefighters and emergency responders	300 m	30 cm
Anti-collision system: aircraft/ground vehicles	300 m	30 cm
Tracking miners	300 m	30 cm
Aircraft landing systems	300 m	30 cm
Detecting avalanche victims	300 m	30 cm
Locating RF noise and interference sources	300 m	30 cm
Extension to LoJack vehicle theft recovery system	300 m	300 cm

- Avalanche victims can be rescued by the help of WSNs [32]. The people at risk (skiers, hikers, etc.) carry wireless sensors with an oximeter (to measure oxygen level in blood), oxygen sensor (to detect air pockets around victim) and accelerometers (to detect orientation of victim), which are communicated to the PDAs of a rescue team.
- A prototype network of meteorological and hydrological sensors has been deployed in Yosemite National Park to monitor natural climate fluctuations, global warming, and the growing needs of water consumers [33].
- WSNs were used to monitor 44 days in the life of a 70-m tall redwood tree, at a density of every 5 min in time and every 2 m in space, where each sensor

reported the air temperature, relative humidity, and photosynthetically active solar radiation [34].
- A virtual fence application was presented in [35], where an acoustic stimulus is given to animals which cross a virtual fence line. It can be dynamically shifted based on the movement data of the animals, improving the utilization of feed-lots and reducing overheads for installing and moving physical fences.
- Example military applications: counter-sniper systems (detect and locate shooters as well as the trajectory of bullets) [36], self-healing land-mines (ensure that a certain geographical area remains covered with land-mines; if an enemy tampers with a mine, an intact mine hops into the breach using a rocket thruster) [27, 37], tracking of military vehicles (e.g. tanks) using sensors dropped from an unmanned aerial vehicle (UAV) [27], and UAV flock control [38].
- Example medical and commercial applications: damage detection in civil structures (such as smart structures actively responding to earthquakes and making buildings safer [26]), continuous medical monitoring [39], elder care [40], aware home [41], smart kindergarten [42], condition-based maintenance of the equipment [26], and active visitor guidance systems [43].

These examples prove that localization may be needed as a key enabling component for numerous WSN applications, and UWB is an excellent fit for communications and localization for WSNs.

1.3.2 Available UWB localization technologies

There are already a number of UWB ranging and positioning devices in the market. Together with the completion of the IEEE 802.15.4a standard, standard-compliant UWB localization technologies are also being announced. Some of the available UWB localization technologies and their key characteristics are overviewed below and listed in Table 1.4.

Sapphire DART
The Sapphire DART system from Multispectral Solutions, Inc. (MSSI) is an active RFID and RTLS system (see Fig. 1.4) and has the following characteristics [44].

- Tag read ranges in excess of 200 m. (650 feet) line-of-sight, and better than 50 m (160 feet) indoors through multiple obstructions.
- Real-time location (not just active RFID).
- Battery life of up to 10 years, even at one tag transmission per second.
- Real-time location accuracies better than 30 cm (10 cm with averaging).
- Immunity to ISM and WiFi interference, and multipath effects.
- Microminiature tag sizes (e.g. 0.5 × 1.0 × 0.25 inches and 10 g).
- Tags certified UL1604 for use in hazardous locations.

Table 1.4. Current location technologies.

Technology	Location method	Accuracy	Remarks
GPS	Localization using time-of-flight information from four or more of 24 satellites	1–5 m (95–99%)	Expensive (US$100 receivers), does not work indoors
Loran	TDOA	Better than 0.25 nautical miles (460 m) within published areas	Loran-A developed during World War II, and Loran-C around 1950s. Land-based system operating at 90–110 kHz. More robust to jamming than GPS. Mostly used by mariners.
Enhanced Loran (eLoran)	Incorporates signals from all stations in range	8–20 m	Being installed in US in 2004, a variation is used in north-west Europe.
ARGOS [52]	Makes use of Doppler effect. Six NOAA satellites are currently in service with ARGOS instruments.	150–1000 m	May locate any platform equipped with a suitable transmitter, anywhere in the world. Average daily power consumption as low as a few milliamps, miniaturized models can be as compact as a small matchbox, weighing as little as 15 g.
Polaris Wireless [53]	Uses multipath signatures for localization	Well within 100 m	Fingerprint information compared with a database.
A-GPS (Assisted GPS)	Modified handsets that use a GPS receiver	Around 10 m	Works well in rural and suburban areas with unobstructed sky view. Specialized network server to assist in location estimation.
WLANs			
Ekahau [22]	RSS-based pattern matching, usage of Bayesian inference methods	Up to 1 m	No extra cost over existing wireless LAN structure, extensive utilities
Microsoft RADAR [21]	RSS-based pattern matching	3–4.3 m (50%)	Scalability problems, no extra cost over existing wireless LAN structure
Wireless Andrew [10]	Closest AP	802.11cell size	Poor accuracy
Aeroscout [54]	TDOA and RSSI	1–5 m with TDOA	Uses Wi-Fi-based Active RFID tags.

Table 1.4. (cont.)

Technology	Location method	Accuracy	Remarks
PanGo [55]	RF fingerprint technique	Room level accuracy (< 3 m)	Asset tracking system for companies. includes an integrated rules-based notification application that sends event-triggered alerts to users based on asset location, presence/absence duration and status.
LOKI (Skyhoow Wireless, Inc.) [7]	Closest AP	802.11 cell size	Installed as a free software. Used for locating the closest restaurant, etc. through a search engine.
WSNs			
Active Badges [56]	Infra-red-based proximity of wearable badges to predeployed sensors	Room size	Installation costs, cheap tags and sensors, sunlight and fluorescent interference, limited IR range
Active Bats [57]	Ultrasound time-of-flight lateration	9 cm (95%)	Ceiling sensor installation costs
Cricket [58]	RSS and ultrasound-based localization	4 x 4 feet regions	US$10 beacons and receivers, installation costs
SpotON [23]	RSS-based ad-hoc lateration	Depends on cluster size	US$30 per tag, inaccuracy of RSS metric
UWB			
Ubisense [45]	TDOA and AOA	30 cm in 3-D	Maximum tag-sensor distances greater than 50 m
PAL650 (MSSI) [47]	TDOA	Up to 1 foot	The world's first FCC-certified UWB-based active RFID tracking system.

Table 1.4. (cont.)

Technology	Location method	Accuracy	Remarks
Sapphire DART UWB (MSSI) [44]	TDOA	Better than 30 cm (10 cm with averaging)	Tag read ranges in excess of 200 m.
Aetherwire [49]	Time-based	As low as centimeter accuracy	30 m range (good penetration). Can hop through the network to cover kilometer distances.
Others			
Easy Living [10]	Localization using camera vision	Variable	Requires three cameras per room and processing power
PinPoint 3D-iD [59]	RF-based TOA	1–3 m	Expensive hardware, installation costs, 802.11 interference
Motion Star [10]	Scene analysis	1 mm	Very expensive hardware, precise installation required
ActiveTag RFID (AXCESS, Inc.) [60]		Of the order of meters	Active RFID positioning system for asset and personnel tracking. Capability to link asset IDs to employee IDs. A tag activator wakes up the tag to transmit up to 100 m.
Spot (Inner Wireless, Inc.) [61]		Room-level	Uses ZigBee technology (IEEE 802.15.4 standard). Designed specifically for hospitals and health-care organizations.
Indoor GPS (Metris, Inc.) [62]	AOA	Less than 1 mm (static), around 1 mm (dynamic)	Laser positioning system for indoors. Transmission range expandable from 2 to 300 m.

1.3 Ranging and localization with UWB

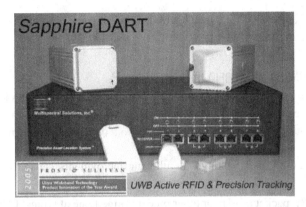

Fig. 1.4. The Sapphire Dart localization system from MSSI (used with permission) [44].

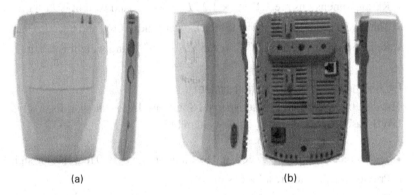

Fig. 1.5. (a) A Ubisense tag is a small tag worn by a person or attached to an asset allowing it to be accurately located within an indoor environment. Tags have two programmable buttons, two LEDs, and a programmable buzzer. (b) The Ubisense sensor receives UWB pulses from Ubisense tags which are then used to determine exact location based on time difference of arrival (TDOA) and angle of arrival (AOA). Sensors have an array of four UWB receivers enabling angle to be calculated with a high degree of accuracy (used with permission) [45].

Ubisense

Ubisense delivers a precise, real-time location system utilizing UWB technology (see Fig. 1.5) and has the following characteristics [45].

- Up to 12 inch / 30 cm 3-D accuracy even within a complex indoor environment.
- Maximum tag-sensor distance: greater than 150 feet (50 m)
- Can track each tag several times a second.
- Dynamically manages the update rates of individual tags so that fast-moving tags will be located more frequently than stationary or slow-moving ones, simultaneously increasing system performance and battery lifetime.
- Monitors real-time spatial interactions involving people and objects. For analysis, Ubisense provides historic reporting and playback of a user-defined time period.
- Uses a cellular sensor and processing architecture to achieve exceptional scalability using low-cost, off-the-shelf servers and Ethernet networks.

- Can scale to sites up to 3 000 000 feet/300 000 m and to track tens of thousands of Ubisense Tags in real time.

Time Domain PulseON350 active RFID tracking system

The PulseON350 active RFID tracking system by Time Domain Corporation is capable of providing presence detection, 1-D, 2-D, and 3-D location with sub-foot accuracy in best case environments, and less than 3 feet accuracy in a typical environment [46].

- TDOA-based positioning and tracking with full scale deployments, capable of tracking thousands of tags.
- RF tag transmits RF packet at 1 Hz or other predetermined rate at typical LOS ranges of 75 feet.
- RF tag dimensions are $1.1'' W \times 1.2'' D \times 0.4'' T$ and weight is 11.8 g.
- The reader receives tag transmissions, determines TOA, and decodes data.
- Reader dimensions are $8.5'' W \times 6.5'' D \times 1.5'' T$ and weight is 816 g.

PAL650 precision asset location system

The PAL 650 system from Multispectral Solutions, Inc. is the world's first FCC-certified, UWB-based, active RFID tracking system [47].

- Tag-to-receiver ranges exceed 300 feet indoors (over 600 feet line-of-sight).
- Provide localization resolutions of better than 1 foot.
- Permits tag operation (at a 1 update per second rate) of up to 4 years on a single 3.0 V lithium cell.
- Operating frequency: 6.2 GHz.
- Uses time differences-of-arrival.
- Positioning data made available through an IP socket interface for use in client applications.

Other UWB localization devices and technologies

In addition to the earlier products discussed above, there are also other UWB ranging and localization technologies from other companies and research centers. Thales Research and Technology reported better than 30 cm accuracy with both pulsed-UWB and frequency hopped DS-UWB [48]. Aetherwire Inc. claims on the order of centimeter accuracy at distances up to kilometer range [49].

In February 2007, IMEC announced its first ever published IEEE 802.15.4a standard compliant transmitter which can transmit at all bands between 3 and 10 GHz [50]. Another recent announcement by Fujitsu in January 2007 states that they achieved 17 cm accuracy with UWB localizers using one way ranging in LOS situations [51]. With the completion of the IEEE 802.15.4a standard, numerous other products from different vendors are also expected.

1.4 Problems

(1) Explain the differences between the following terms.
   (a) Context awareness vs. location awareness.
   (b) Accuracy vs. precision.
   (c) Active localization vs. passive localization.
   (d) Relative coordinate vs. absolute coordinate.
   (e) Trilateration vs. triangulation.

(2) List five applications where UWB technology can be used for sub-meter ranging and localization.

(3) Discuss why UWB is more suitable for wireless sensor network type of technologies rather than other technologies such as cellular networks.

(4) Explain why GPS is not a suitable technology for indoor localization. Name three alternative technologies for indoor localization, and elaborate on why they are better alternatives compared to the GPS.

2 Ultra-wideband signals

Commonly, an ultra-wideband (UWB) signal is defined to be a signal with a fractional bandwidth of larger than 20% or an absolute bandwidth of at least 500 MHz. The main feature of UWB signals is that they occupy a much wider frequency band than conventional signals; hence, they need to share the existing spectrum with incumbent systems. Therefore, certain regulations are imposed on systems transmitting UWB signals. In this chapter, after a detailed description of UWB signals, various regulatory rules on UWB systems in different parts of the world are investigated. Then, emerging UWB standards for wireless personal area network (WPAN) applications are studied.

2.1 Definition of UWB

Although Guglielmo Marconi's spark gap radio transmitters were sending UWB signals across the Atlantic Ocean in 1901, the rigorous investigation of UWB systems was stimulated by the studies on impulse response characterization of microwave networks in the 1960s [63, 64]. Instead of the conventional swept-frequency response characterization, a linear-time-invariant (LTI) system was characterized by its response to an impulse in the time domain. After employing impulses to characterize behavior of various systems, it was also realized that such impulses could also be used in radar and communications systems [65]. The first UWB communications patent was issued in 1973 to Gerald F. Ross on transmission and reception of baseband pulse signals [66].

Early names for UWB technology include *baseband*, *carrier-free*, *non-sinusoidal* and *impulse*. The term UWB was coined by the US Department of Defense in the late 1980s. A UWB signal is characterized by its very large bandwidth compared to the conventional narrowband systems. Namely, a signal is called UWB if it has an absolute bandwidth of at least 500 MHz, or a fractional (relative) bandwidth larger than 0.2.[1] The absolute bandwidth is calculated as the difference between the upper frequency f_H of the -10 dB emission point and the lower frequency f_L of the -10 dB emission point; i.e.

$$B = f_\mathrm{H} - f_\mathrm{L}, \tag{2.1}$$

---

[1] This definition is in accordance with the definition of the US FCC [67].

which is also called $-10\,\text{dB}$ bandwidth (Fig. 2.1). On the other hand, the fractional bandwidth is defined as

$$B_{\text{frac}} = \frac{B}{f_c}, \qquad (2.2)$$

where f_c is the center frequency and is given by

$$f_c = \frac{f_H + f_L}{2}. \qquad (2.3)$$

From (2.1) and (2.3), the fractional bandwidth B_{frac} in (2.2) can be expressed as

$$B_{\text{frac}} = \frac{2(f_H - f_L)}{f_H + f_L}. \qquad (2.4)$$

According to the US FCC [67], a UWB system with f_c larger than 2.5 GHz must have an absolute bandwidth larger than 500 MHz, and a UWB system with f_c smaller than 2.5 GHz must have a fractional bandwidth larger than 0.2 (Fig. 2.1).

Due to their large bandwidth, UWB systems are characterized by very short duration waveforms, usually on the order of a nanosecond. Commonly, a UWB system transmits ultra-short pulses with a low duty cycle. In other words, the ratio between the pulse transmission instant and the average time between two consecutive transmissions is usually kept small. However, for UWB communications systems, both low duty cycle schemes and continuous transmissions can be considered.

A type of UWB communications system that transmits UWB pulses with a low duty cycle is called *impulse radio (IR)* [68]. In an IR UWB system, a number of pulses are transmitted per information symbol and information is usually conveyed by the positions or the polarities of the pulses, as shown in Fig. 2.2. Each pulse resides in an interval called a "frame", and the positions of the pulses in the frames are determined according to a *time-hopping (TH)* code in order to reduce the probability of collisions with pulses of other UWB systems in the environment. For example, in Fig. 2.2, three information bits are being transmitted, and each bit consists of two pulses (or, two frames). The TH code for the first bit is given by {2, 1}, which means that the pulse in the first frame is shifted by $2T_c$ seconds and the one in the second frame is shifted by T_c seconds, where T_c represents the chip interval.

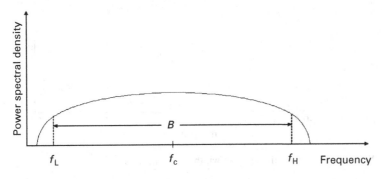

Fig. 2.1. A UWB signal is defined to have an absolute bandwidth $B \geq 500$ MHz, or a fractional bandwidth $B_{\text{frac}} = B/f_c > 0.2$.

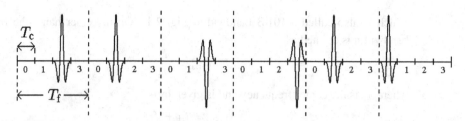

Fig. 2.2. An IR UWB signal, in which two pulses are transmitted per information symbol, and information is conveyed by the polarities of the pulses (BPSK). Hence, +1, −1 and +1 are being transmitted in this example. Note that each pulse resides in an interval of T_f seconds, called a "frame", and the positions of the pulses in different frames are determined by a TH code, which is {2, 1, 2, 3, 1, 0} in this example.

Some common UWB pulse shapes include derivatives of the Gaussian pulse [69], pulses based on modified Hermite polynomials [70] and wavelet pulses [71, 72]. For example, the second derivative of the Gaussian pulse is expressed as

$$\omega(t) = A\left(1 - \frac{4\pi t^2}{\zeta^2}\right) e^{-2\pi t^2/\zeta^2}, \qquad (2.5)$$

where $A > 0$ and ζ are parameters that determine the energy and the width of the pulse, respectively.[2] In Fig. 2.3, a unit energy pulse with the width of around 1 ns is plotted according to (2.5) ($\zeta = 0.4$ ns). As another example, Fig. 2.4 illustrates UWB pulses

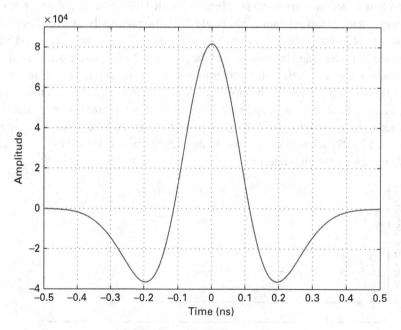

Fig. 2.3. The UWB pulse in (2.5) with the pulse width of around 1 ns.

---

[2] The pulse width is roughly equal to 2.5ζ.

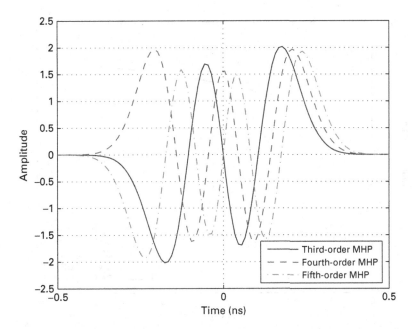

Fig. 2.4. UWB pulses based on modified Hermite polynomials (©2006 IEEE) [70].

based on third-, fourth-, and fifth-order modified Hermite polynomials (MHPs). For UWB systems operating under regulatory constraints, it is important to employ optimal pulse shapes in order to utilize available bandwidth and power. Therefore, optimal and suboptimal waveform design techniques are studied extensively to generate UWB pulses with optimal spectral properties [73–75].

In addition to UWB systems with low duty cycles, it is also possible to realize UWB systems with continuous transmissions. For example, a DS-CDMA system with a very short chip interval can be used as a UWB communications system [76]. Alternatively, transmission and reception of very short duration OFDM symbols can be considered as an OFDM UWB scheme [77].

For both low duty cycle and continuously-transmitting UWB systems, the common property of very large bandwidth brings many advantages for positioning, communications, and radar applications. The main advantages can be summarized as follows:

- penetration through obstacles
- high ranging, hence positioning, accuracy
- high-speed data communications
- low cost and low power implementation.

The penetration capability of a UWB signal is a result of its large frequency spectrum that includes low frequencies as well as high frequencies. This large spectrum also results in high time resolution, which improves the ranging accuracy, as will be studied in Chapter 4.

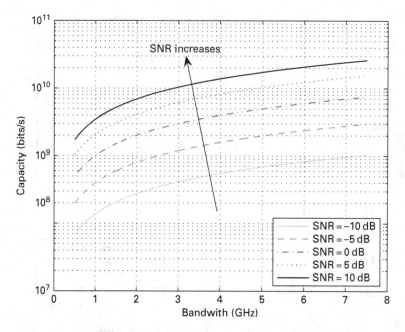

Fig. 2.5. Capacity versus bandwidth curves for UWB systems over AWGN channels.

The suitability of UWB signals for high-speed data communications can be observed from the Shannon capacity formula. For an AWGN channel with bandwidth of B Hz, the maximum data rate that can be acquired is given by

$$C = B \log_2(1 + \text{SNR}) \quad \text{(bits/second)}, \tag{2.6}$$

where SNR is the signal-to-noise ratio of the system. In other words, as the bandwidth of the system increases, it becomes possible to transmit more information from the transmitter to the receiver, as shown in Fig. 2.5. Also note that when the bandwidth is very large, signal power can be kept low to increase the battery life of the system and to minimize the interference to the other systems in the same frequency spectrum.

Moreover, a UWB system can be operated in baseband, meaning that UWB pulses can be transmitted without a sine-wave carrier ("carrier-free"). In that case, the system does not require IF processing, which facilitates low cost implementations.

2.2 International regulations for UWB signals

As stated in the previous section, UWB signals have unique properties that prove to be very useful for communications, ranging and radar applications. However, since UWB signals occupy a very large portion in the spectrum, they need to coexist with the incumbent systems without causing significant interference. For example, frequency allocation of some wireless systems is shown in Fig. 2.6. If UWB signals were allowed to transmit over the range of frequencies of these systems without any restrictions, all these

2.2 International regulations for UWB signals

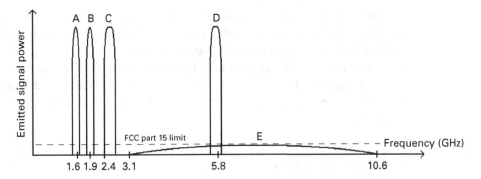

Fig. 2.6. Spectrum allocation of various wireless systems. A. Global positioning system (GPS) (1.56–1.61 GHz), B. Personal communication system (PCS) (1.85–1.99 GHz), C. Microwave ovens, cordless phones, bluetooth, IEEE 802.11b (2.4–2.48 GHz), D. IEEE 802.11a (5.725–5.825 GHz), E. UWB (3.1–10.6 GHz). Note that the bandwidths and power levels of various systems are not drawn to scale. The FCC Part 15 limit is also shown in the figure [78]. UWB systems are required to operate below the Part 15 limit (−41.3 dBm/MHz), which is the limit for unintentional radiators, such as televisions and computer monitors.

systems could be jammed by UWB emission. Therefore, a UWB transmitter must meet certain requirements in order not to cause any adverse effects on the functionality of other systems. Especially, GPS, which is used not only for commercial but also for military and homeland security purposes, should not experience performance degradation due to UWB emission.

In order to benefit from advantages of UWB without degrading the performance of other systems, the Federal Communications Commission (FCC) in the USA started a specification definition system for UWB in 1998. Then, in February 2002, it announced its "First Report and Order", which allowed the limited use of UWB devices [67]. According to this regulation, UWB systems must transmit below certain power levels in order not to cause significant interference to the other systems in the same frequency spectrum. Specifically, the power spectral density must not exceed −41.3 dBm/MHz for frequency ranges from 3.1 to 10.6 GHz, and it must be even lower outside this band, depending on the specific application. In other words, the FCC spectral mask specifies a useful spectrum of 7.5 GHz for most UWB systems.

After the FCC legalized the use of UWB signals in the USA, a considerable amount of effort has been put into development and standardization of UWB systems. In what follows, the FCC regulations in the USA are investigated in detail, and then the regulatory efforts in other parts of the world are summarized.

2.2.1 FCC regulations

The FCC specifies a set of rules to control harmful interference from UWB devices. Mainly, it imposes certain power emission limits for various types of UWB systems. These emission limits are specified in terms of equivalent isotropically-radiated power (EIRP), which is defined as the product of the power supplied to an antenna and its gain in a given direction relative to an isotropic antenna. According to the FCC regulations, maximum EIRP in any direction should not exceed the Part 15 limit of -41.3 dBm,[3] which is the limit for unintentional radiators, such as television and computer monitors [78]. In addition, various systems must have even lower limits than that of Part 15 in some frequency bands depending on the specific application area. In this respect, the FCC limits can be studied for three different systems: communications, vehicular radar and imaging.

Communications systems

For communications systems, slightly different FCC limits are specified for indoor and outdoor systems, as shown in Figs. 2.7 and 2.8, respectively. Specifically, emissions of outdoor systems in the frequency band from 1.61 to 3.1 GHz should have an extra attenuation of 10 dB compared to those of indoor systems.

UWB devices for indoor systems are not allowed to be used outdoors, or to direct their radiation outside. Only peer-to-peer communication is permitted; i.e. each transmitter may only transmit to an associated receiver. Indoor UWB systems have many potential applications such as high-speed wireless personal area networks (WPANs) and wireless USB (wUSB).

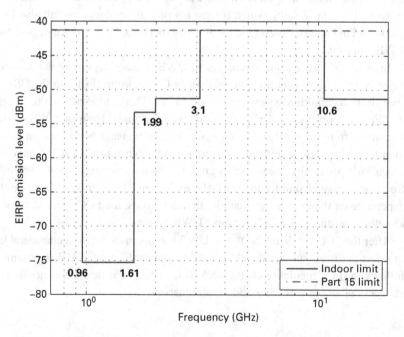

Fig. 2.7. FCC emission limits for indoor UWB systems.

---

[3] According to the FCC regulations, emissions (EIRPs) are to be measured using a resolution bandwidth of 1 MHz (except for 1.16–1.24 GHz and 1.56–1.61 GHz for which a resolution bandwidth of at least 1 kHz is required).

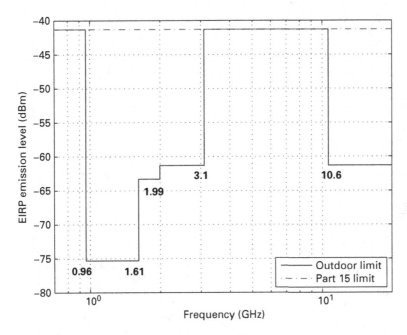

Fig. 2.8. FCC emission limits for outdoor UWB systems.

For outdoor systems, the FCC requires that UWB devices operate without a fixed infrastructure and only communicate with their associated receivers. Applications of outdoor UWB communications include object positioning and tracking, which make use of precise ranging capability of UWB signals.

Vehicular radar systems

Vehicular radar systems can operate at the Part 15 limit between 22 and 29 GHz, as shown in Fig. 2.9. Furthermore, the center frequency of such systems is required to be higher than 24.075 GHz, and an additional 25 dB attenuation is required for the 23.6–24 GHz band if the elevation angle is greater than 30° above the horizon, which is required to prevent any interference to passive sensing systems operating on low Earth-orbiting satellites [67]. Finally, vehicular radar systems can operate only when the vehicle is operating; that is, when the engine is running.

Imaging systems

Under the imaging systems category, ground-penetrating radars (GPRs), wall imaging, medical imaging, through-wall imaging and surveillance systems can be considered. Table 2.1 summarizes the FCC emission limits for those systems, and Fig. 2.10 plots the limits for medical imaging systems (the same plot applies to GPR and wall imaging systems).

The FCC requires that operation of various imaging systems (GPR, imaging, and medical) must be coordinated, and dates and locations of operation must be reported. Also, use of the imaging systems stated in Table 2.1 requires licensing, and use of each

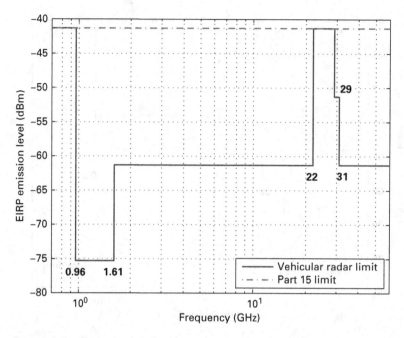

Fig. 2.9. FCC emission limits for vehicular radar systems.

Table 2.1. The FCC emission limits (EIRP in dBm) for UWB imaging systems.

Frequency range (GHz)	0.96–1.61	1.61–1.99	1.99–3.1	3.1–10.6	Above 10.6
GPR, wall imaging, medical imaging	−65.3	−53.3	−51.3	−41.3	−51.3
Through-wall imaging (low-frequency)	−65.3	−53.3	−51.3	−51.3	−51.3
Surveillance systems	−53.3	−51.3	−41.3	−41.3	−51.3

system is limited to certain organizations. For example, surveillance systems can be operated only by public safety, manufacturing, petroleum and power licensees [79].

Although the FCC's emission limits differ for various types of UWB systems, there are also a number of other FCC regulations that are common for all UWB systems [67, 79].

- The frequency f_M at which the highest power is emitted must be within the $-10\,\text{dB}$ absolute signal bandwidth.
- Peak emissions within a 50 MHz bandwidth around f_M may not exceed 0 dBm EIRP.
- Emissions below 0.96 GHz are limited by the FCC Part 15 limit of $-41.3\,\text{dBm/MHz}$ for unintentional radiators.
- Operation on aircraft, ship or satellite is not permitted.

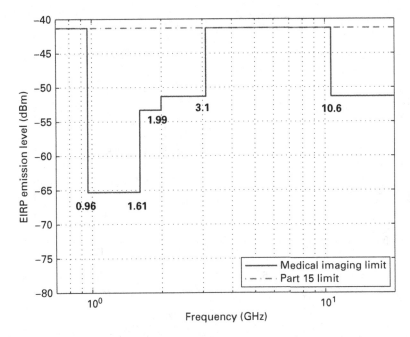

Fig. 2.10. FCC emission limits for medical imaging systems.

2.2.2 Other regulatory efforts

After UWB was authorized by the FCC in the USA in 2002, regulatory efforts have been underway in Europe and Asia to approve the use of UWB devices under certain restrictions. For compatibility issues, it would be beneficial to have the same regulations worldwide. However, the emission limits have some differences in different parts of the world. Therefore, UWB devices should have sufficient flexibility to operate worldwide, or be designed according to the worst-case scenario.

In the following, the regulatory efforts in Europe and Japan are summarized. Both Japan and Europe have recently allowed the use of UWB systems although modifications to initial regulations are expected in the near future.

Europe

In Europe, the Electronic Communications Committee (ECC) of the European Conference of Postal and Telecommunications Administrations (CEPT) undertook technical studies for UWB regulations. The studies and recommendations of the ECC were considered by the Radio Spectrum Committee (RSC) of the European Commission (EC), which made the final decision (at the beginning of 2007) for UWB regulations that are valid in the member countries [80].

The spectrum mask imposed by the EC is as shown in Fig. 2.11 for UWB systems that do not employ appropriate interference mitigation techniques. Mainly, such UWB systems can transmit at −41.3 dBm/MHz over the 6–8.5 GHz band. This limit is valid also for the 4.2–4.8 GHz band until the end of 2010. Starting from 2011, the EIRP will be limited to −70 dBm/MHz for that band. Note that the FCC in the USA allows an EIRP

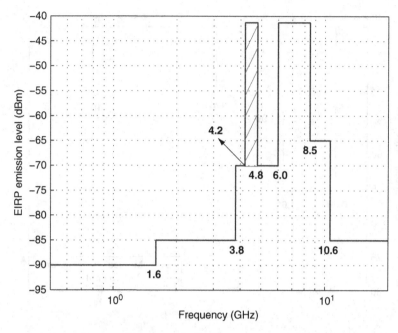

Fig. 2.11. ECC emission limits without appropriate mitigation techniques. UWB systems can transmit −41.3 dBm/MHz in the 4.2–4.8 GHz band until the end of 2010. After then, the limit of −70 dBm/MHz will be imposed on that band.

of −41.3 dBm/MHz over a wider frequency band, 3.1–10.6 GHz (cf. Fig. 2.7). In other words, the EC regulations are more strict than the FCC regulations.

For UWB systems employing appropriate interference mitigation techniques, the EC limits are as shown in Fig. 2.12. Namely, such systems can transmit at −41.3 dBm/MHz in the 3.4–4.8 GHz band provided that they have low duty cycle transmissions [80]. This low duty cycle requirement is specified in terms of T_{on} and T_{off}, which are defined as the duration of a burst (irrespective of the pulses contained in the burst) and the duration between two consecutive bursts, respectively [81]. The EC requires that the maximum duration of a burst should not exceed 5 ms; i.e. $T_{on} \leq 5$ ms, and that the total off-time per second should be larger than 950 ms whereas the total on-time should not exceed 5% per second and 0.5% per hour.

Although the current regulations[4] of the EC are quite strict, an amendment at the beginning of 2008 is expected to relax the restrictions and take into account other interference mitigation techniques such as detect-and-avoid.

Japan
In Japan, the Ministry of Internal Affairs and Communications (MIC) authorized the regulations for indoor UWB devices in March 2006. However, these initial regulations

---

[4] The regulations specified in Figs. 2.11 and 2.12 are for indoor UWB devices, and for outdoor UWB devices that are not attached to a fixed installation, a fixed infrastructure, a fixed outdoor antenna, or an automotive or railway vehicle. UWB applications for automotive short-range radars are regulated by [82] and [83].

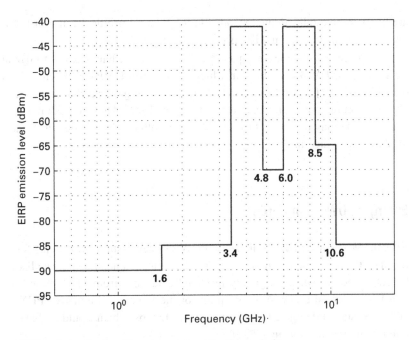

Fig. 2.12. ECC emission limits with appropriate mitigation techniques.

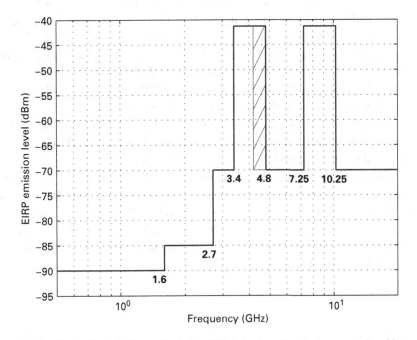

Fig. 2.13. MIC emission limits in Japan for indoor UWB devices. For the frequency band between 3.4 and 4.8 GHz, interference mitigation techniques such as LDC or DAA are required (The shaded band of 4.2–4.8 GHz is excepted from this requirement until the end of 2008.).

are likely to be modified and extended in the future. Specifically, use of UWB devices outdoors and inside automobiles is being investigated.

The current regulations in Japan specify two usable bands 3.4–4.8 GHz and 7.25–10.25 GHz for UWB operation, in which −41.3 dBm EIRP can be emitted (measured over 1 MHz) [84]. For the first band, interference mitigation techniques are required, or else the average power must be reduced to −70 dBm.[5] Also the peak power measured over 50 MHz may not exceed 0 dBm EIRP for both bands. The details of the spectrum mask are illustrated in Fig. 2.13.

2.3 Emerging UWB standards

After the FCC allowed the limited use of UWB systems, standardization efforts were initiated by the IEEE for a high-speed PHY enhancement amendment to the IEEE 802.15.3 WPAN standard, which originally provided data rates from 11 Mbps up to 55 Mbps. The IEEE formed a new task group 3a for the high rate alternative PHY. The IEEE 802.15.3a standard targeted data rates up to 480 Mbps, which would enable imaging and multimedia applications in WPANs.

The IEEE 802.15.3a task group evaluated a number of PHY proposals from various companies, and ended up with two proposals. The first one was based on multiband OFDM (MB-OFDM) UWB [77], supported by the WiMedia Alliance [85], and the other was DS-UWB [76], supported by the UWB Forum [86]. However, a final decision on which technology to use in the standard could not be reached and the task group was dissolved at the beginning of 2006. Although the IEEE 802.15.3a standard could not be finalized, the industry groups in favor of MB-OFDM UWB and DS-UWB systems have been working on pushing their products into the marketplace.

After the IEEE 802.15.3a task group was dissolved, the WiMedia Alliance had Ecma International[6] approve their WPAN standard based on MB-OFDM UWB technology. The details of this standard are investigated in Section 2.3.1.

In addition to high-rate WPAN applications, UWB signals have also been considered for low-rate WPANs that focus on low power and low complexity devices. The IEEE formed the task group 4a (TG4a) in March 2004 for an amendment to the IEEE 802.15.4 standard for an alternative PHY. The IEEE 802.15.4a provides high-precision ranging/location capability, high aggregate throughput and ultra-low-power consumption. This standard is studied in Section 2.3.2.

---

[5] For the 600 MHz band of 4.2–4.8 GHz, interference mitigation techniques are not required until the end of 2008.

[6] Ecma International is an industry association founded in 1961 and works on the standardization of information and communication technology and consumer electronics (http://www.ecma-international.org).

2.3.1 Ecma standards on multiband OFDM UWB

At the end of 2005, Ecma International approved two standards for UWB technology based on MB-OFDM approach, which are ECMA-368, high-rate UWB PHY and MAC standard, and ECMA-369, MAC-PHY interface for ECMA-368 [87, 88]. Mainly, these Ecma standards specify a basis for high-speed and short-range WPANs, utilizing all or part of the spectrum between 3.1 and 10.6 GHz with data rates of up to 480 Mbps.

Operating band frequencies and time-frequency codes

According to the Ecma standards, the frequency band 3.1–10.6 GHz is divided into 14 bands, with a 528 MHz spacing between consecutive center frequencies. In other words, the center frequency for the nth band, $f_c^{(n)}$, is given by

$$f_c^{(n)} = 2.904 + 0.528n \quad \text{(GHz)}, \tag{2.7}$$

for $n = 1, \ldots, 14$. These 14 frequency bands are also classified into five band groups as shown in Fig. 2.14.

The transmitted signal at any given time occupies one of the 14 bands in Fig. 2.14. Depending on the time-frequency code (TFC) at the transmitter, data can be interleaved over a number of bands, which is called time-frequency interleaving (TFI), or it can be transmitted over a single band, which is called fixed-frequency interleaving (FFI). For example, in Fig. 2.15, the first three bands are used to transmit information symbols. The first symbol is transmitted over band-1, the second is transmitted over band-2, the third is transmitted over band-3 and this structure is repeated thereafter, which corresponds to a TFC of $\{1, 2, 3, 1, 2, 3\}$. Note that TFI, rather than FFI, is employed here since data is transmitted over three different bands.

In the Ecma standard, a total of seven TFCs are defined for the first band group as shown in Table 2.2. Similarly, seven TFCs are defined for band group 2, band group 3 and band group 4. For band group 5, two FFI codes, namely $\{13, 13, 13, 13, 13, 13\}$ and $\{14, 14, 14, 14, 14, 14\}$, are specified. Overall, 30 channels are specified in the standard.

Transmitter structure

A multiband OFDM UWB transmitter according to the Ecma standard is illustrated in Fig. 2.16. Information bits to be transmitted are first scrambled, and then encoded using a convolutional encoder, which implements a type of forward error correction (FEC) scheme. A convolutional encoder encodes the input bits by passing them through a linear finite state machine, where the number of states determines the *constraint length* of the code, and the ratio between the number of output bits and the number of input bits defines the *rate* of the code. For the Ecma standard, a convolutional encoder with rate 1/3 and constraint length 7 is employed. By using this encoder, various code rates can also be obtained by employing a technique called *puncturing*, which is a method for omitting some of the encoded bits at the output of the encoder, hence for increasing the coding rate. For example, by omitting 7 bits from each 15 encoded output bits of the rate 1/3 convolutional encoder, the rate can be increased to 5/8. According to the standard,

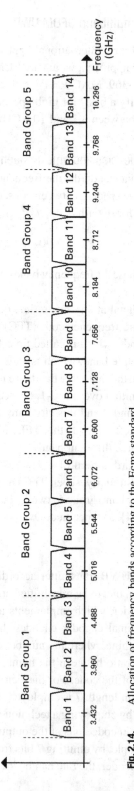

Fig. 2.14. Allocation of frequency bands according to the Ecma standard.

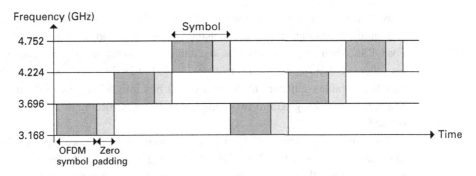

Fig. 2.15. Time-frequency allocation for a system using the first three bands with a TFC of {1, 2, 3, 1, 2, 3}.

Table 2.2. TFCs for band group 1.

TFC-1	1	2	3	1	2	3
TFC-2	1	3	2	1	3	2
TFC-3	1	1	2	2	3	3
TFC-4	1	1	3	3	2	2
TFC-5	1	1	1	1	1	1
TFC-6	2	2	2	2	2	2
TFC-7	3	3	3	3	3	3

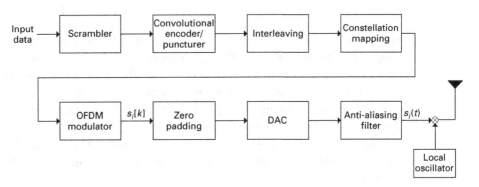

Fig. 2.16. Basic blocks of an multiband OFDM UWB transmitter according to the Ecma standard.

a coding rate of 1/3, 1/2, 5/8 or 3/4 can be used in the system corresponding to various data rate options.

After the convolutional encoding, the coded bits are interleaved, which is a process that spreads bits over a series of symbols so as to provide robustness against burst errors. The Ecma standard defines both inter-symbol and intra-symbol interleaving. For the inter-symbol interleaving, bits are permuted over six symbols, whereas for the intra-symbol interleaving arrangement, bits inside symbols are changed according to certain structures.

After the interleaving, the bits are mapped onto a complex constellation. For data rates of 53.3, 80, 106.7, 160 and 200 Mbps, the binary data is mapped to a QPSK constellation, whereas for data rates of 320, 400 and 480 Mbps, the binary data is mapped to a multi-dimensional constellation using a dual-carrier modulation (DCM) technique. For QPSK, each pair of binary bits, b_{2i} and b_{2i+1}, is mapped to a complex number as $\frac{1}{\sqrt{2}}(2b_{2i} - 1 + j(2b_{2i+1} - 1))$ for $i = 0, 1, \ldots$. For DCM, each 200 bits are converted into 100 complex numbers by grouping 200 bits into 50 groups of 4 bits, and then mapping each 4-bit group to two complex numbers according to a certain pattern, as defined in [87].

The complex numbers obtained by constellation mapping are input to the OFDM modulator as shown in Fig. 2.16, and zero padding is applied to the output of the OFDM modulator. After that, the discrete signal is converted to a continuous-time waveform by a digital-to-analog converter (DAC) and an anti-aliasing filter. Finally, depending on the TFC, a local oscillator is used to set the center frequency of the signal, which is then transmitted through the antenna as shown in Fig. 2.16.

The details of the signal structures between the OFDM modulator and the antenna are explained in the following section.

Signal model

According to the Ecma standard, each transmitted packet is expressed as

$$s_{tx}(t) = \text{Re}\left\{\sum_{i=0}^{N_s} s_i(t - iT_s) \exp\left(j2\pi f_c^{(q(i))} t\right)\right\}, \tag{2.8}$$

where T_s is the symbol length, N_s is the number of symbols in the packet, $s_i(t)$ is the complex baseband signal representation for the ith symbol, $f_c^{(n)}$ is the center frequency for the nth frequency band as given in (2.3), and $q(i)$ is a function that maps the ith symbol to the appropriate frequency band according to the TFC at the transmitter. For example, for the TFC in Fig. 2.15, $q(i) = \text{mod}\{i, 3\} + 1$ can be used, where $\text{mod}\{x, y\}$ denotes the remainder of the division of x by y.

Since each packet consists of a synchronization preamble, a header and a PHY service data unit (PSDU),[7] the symbol $s_i(t)$ in (2.8) is defined according to the symbol index as follows:

$$s_i(t) = \begin{cases} s_{\text{sync},i}(t), & 0 \leq i < N_{\text{sync}} \\ s_{\text{hdr},i-N_{\text{sync}}}(t), & N_{\text{sync}} \leq i < N_{\text{sync}} + N_{\text{hdr}}, \\ s_{\text{frame},i-N_{\text{sync}}-N_{\text{hdr}}}(t), & N_{\text{sync}} + N_{\text{hdr}} \leq i < N_s \end{cases} \tag{2.9}$$

where N_{sync} and N_{hdr} are the number of symbols in the synchronization preamble and header sections of the packet, respectively. In the following, the detailed descriptions of the signal structures are described only for the header and the PSDU. Interested readers are referred to [87] for the detailed description of the synchronization signals.

---

[7] The PSDU is formed by concatenating the frame payload with the frame check sequence, tail bits and pad bits, which are inserted in order to align the data stream on the boundary of the symbol interleaver [87].

2.3 Emerging UWB standards

Consider the discrete signal $s_i[k]$, which is obtained by taking the Inverse discrete Fourier transform (IDFT) of the complex modulated data:

$$s_i[k] = \frac{1}{\sqrt{N_{\text{FFT}}}} \sum_{l=-61}^{61} b_{i,l} \exp(j2\pi l k / N_{\text{FFT}}), \quad (2.10)$$

for $k = 0, 1, \ldots, N_{\text{FFT}} - 1$ and $i = N_{\text{sync}}, \ldots, N_s - 1$, where $b_{i,l}$ is the complex information at the lth subcarrier of the ith symbol, and N_{FFT} is the size of the IDFT. Note that $s_i[k]$ in (2.10) is an OFDM symbol, which effectively divides the frequency spectrum (528 MHz) into overlapping but orthogonal sub-bands by using N_{FFT} subcarriers and transmits information symbols $(b_{i,l})$ at each subcarrier [89].

The Ecma standard specifies that the total number of subcarriers N_{FFT} is 128, and out of 128 subcarriers, 122 are used in the system, as can be noted by the limits of the summation in (2.10) (the subcarrier corresponding to the DC component is also set to zero; i.e. $b_{i,0} = 0$). The subcarriers are classified into data subcarriers, pilot subcarriers and guard subcarriers. According to the standard, there are 100 data subcarriers, which are used to carry information, whereas there exist 12 pilot subcarriers, which transmit known data for the purposes of signal parameter estimation at the receiver. Furthermore, there are 10 guard subcarriers, five on each side of the OFDM symbol, which carry the same information as the outermost data subcarriers.

In order to mitigate the effects of multipaths and to provide a time window to allow the transmitter and the receiver sufficient time to switch between the different bands, zero-padding is applied to $s_i[k]$ after the IDFT operation, and $s_{\text{frame},i}[k]$ and $s_{\text{hdr},i}[k]$ are obtained as

$$s_{\text{hdr},i}[k] = \begin{cases} s_i[k], & k = 0, 1, \ldots, N_{\text{FFT}} - 1, \\ 0, & k = N_{\text{FFT}}, \ldots, N_{\text{FFT}} + N_{\text{ZPS}} - 1, \end{cases} \quad (2.11)$$

for $i = N_{\text{sync}}, \ldots, N_{\text{sync}} + N_{\text{hdr}} - 1$, and

$$s_{\text{frame},i}[k] = \begin{cases} s_i[k], & k = 0, 1, \ldots, N_{\text{FFT}} - 1, \\ 0, & k = N_{\text{FFT}}, \ldots, N_{\text{FFT}} + N_{\text{ZPS}} - 1, \end{cases} \quad (2.12)$$

for $i = N_{\text{sync}} + N_{\text{hdr}}, \ldots, N_s - 1$, where N_{ZPS} represents the number of samples in the zero-padded suffix.

Then, from the discrete-time signals $s_{\text{hdr},i}[k]$ and $s_{\text{frame},i}[k]$, the continuous-time symbols $s_i(t)$ are obtained by digital-to-analog conversion and filtering, as shown in Fig. 2.16.

System parameters

In this section, some of the system parameters in the Ecma standard are summarized. Table 2.3 lists the data modes supported by the standard, which ranges from 53.3 to 480 Mbps. Note that various data rates are obtained by adjusting the rate of convolutional encoder, and/or by using spreading in the frequency and/or time domain. Time-domain spreading (TDS) involves transmitting the same information across two consecutive OFDM symbols, whereas frequency-domain spreading (FDS) involves transmitting the same information on two separate subcarriers within an OFDM symbol.

Table 2.3. Various data options and corresponding parameters for the Ecma International standard for MB-OFDM UWB transmitters.

Data rate (Mbps)	Modulation	Coding rate	FDS factor	TDS factor
53.3	QPSK	1/3	2	2
80	QPSK	1/2	2	2
106.7	QPSK	1/3	1	2
160	QPSK	1/2	1	2
200	QPSK	5/8	1	2
320	DCM	1/2	1	1
400	DCM	5/8	1	1
480	DCM	3/4	1	1

Table 2.4. Systems parameters for the MB-OFDM UWB transmitter according to the Ecma International standard.

Parameter	Definition	Value
N_{FFT}	Total number of subcarriers (FFT size)	128
N_T	Total number of subcarriers used	122
N_D	Number of data subcarriers	100
N_P	Number of pilot subcarriers	12
N_G	Number of guard subcarriers	10
N_{ZPS}	Number of samples in zero-padded suffix	37
T_s	Symbol interval	312.5 ns
T_{FFT}	IFFT and FFT period	242.42 ns
T_{ZP}	Zero-padding duration	70.08 ns
T_{switch}	Time to switch between bands	9.47 ns

In Table 2.4, some of the important system parameters are listed. Since each symbol is transmitted over 312.5 ns, and 100 data subcarriers are transmitted per symbol, a total of 3.2×10^8 subcarriers are transmitted per second. As each subcarrier carries two bits of information (for both QPSK and DCM), the raw data rate is obtained as 640 Mbps. Then, according to the rate R of the convolutional encoder, and TDS and FDS factors, the data rate can be calculated as

$$\text{Data rate} = \frac{\text{Raw data rate} \times R}{N_{TDS} \times N_{FDS}}, \quad (2.13)$$

where N_{TDS} and N_{FDS} are the TDS and FDS factors, respectively. Note that the various data rate options in Table 2.3 can be verified by (2.13).

Ranging and location awareness

Ranging is an optional capability in the Ecma standard. If a device implements ranging, it should have an accuracy of 60 cm or better.

Ranging in the Ecma standard is based on estimation of propagation delay between a pair of devices. This propagation delay estimation should be performed with respect to a reference point in the preamble, called the *ranging reference point*. During the reception (transmission) of a ranging packet, the value of a ranging counter corresponding to the ranging reference point is used to obtain the packet reception (transmission) time. Also, the system should provide processing delays corresponding to the transmission or reception of a packet. By using the packet transmission and reception times, and the corresponding processing delays, two devices can perform a two-way protocol for ranging estimation. The principles of two-way ranging are studied in Chapter 6.

2.3.2 IEEE 802.15.4a standard

In 2004, the IEEE 802.15 low-rate alternative PHY task group (TG4a) was formed to design an alternate PHY specification for the already existing IEEE 802.15.4 standard for WPANs [90]. The main purpose of the TG4a was to provide communications and high-precision ranging with low-power and low-cost devices. The TG4a's efforts resulted in the IEEE 802.15.4a standard in 2007. With additional features provided by the 15.4a amendment, the IEEE 802.15.4 standard now facilitates new applications and market opportunities.

The IEEE 802.15.4a specifies two optional signaling formats based on IR-UWB[8] and chirp spread spectrum (CSS). The IR-UWB option can use 250–750 MHz, 3.244–4.742 GHz, or 5.944–10.234 GHz bands; whereas the CSS uses the 2.4–2.4835 GHz band. For the IR-UWB there is an optional ranging capability, whereas the CSS signals can only be used for communications purposes. Since the focus of this book is on ranging algorithms, only the IR-UWB option of the IEEE 802.15.4a standard is studied in this section. The interested reader is referred to [91] for a detailed description of the CSS signaling employed in the IEEE 802.15.4a.

Channel allocations

As specified above, a UWB device can transmit in one or more of the following bands according to the IEEE 802.15.4 standard:

(i) Sub-GHz: 250–750 MHz
(ii) Low band: 3.244–4.742 GHz
(iii) High band: 5.944–10.234 GHz.

Over these three bands, 16 channels are supported for the UWB PHY: one in the sub-GHz band, four in the low band and 11 in the high band. These channels and their center frequencies and bandwidths are listed in Table 2.5, along with the specification of mandatory channels in each band. Specifically, a UWB device that implements the low band (high band) should support channel 3 (channel 9), whereas the remaining channels in the band are optional.

---

[8] The UWB option in the IEEE 802.15.4a standard does not employ a conventional IR-UWB signal. Instead bursts of pulses are transmitted in different burst intervals and information is carried by the positions and the polarities of the bursts, as will be investigated in the subsection entitled "Transmitter structure and signal model."

Table 2.5. UWB channels for the IEEE 802.15.4a standard.

Channel No.	Center freq. (MHz)	Bandwidth (MHz)	UWB band	Mandatory
0	499.2	499.2	Sub-GHz	Yes
1	3494.4	499.2	Low band	No
2	3993.6	499.2	Low band	No
3	4492.8	499.2	Low band	Yes
4	3993.6	1331.2	Low band	No
5	6489.6	499.2	High band	No
6	6988.8	499.2	High band	No
7	6489.6	1081.6	High band	No
8	7488.0	499.2	High band	No
9	7987.2	499.2	High band	Yes
10	8486.4	499.2	High band	No
11	7987.2	1331.2	High band	No
12	8985.6	499.2	High band	No
13	9484.8	499.2	High band	No
14	9984.0	499.2	High band	No
15	9484.8	1354.97	High band	No

Transmitter structure and signal model

The main components of an IR-UWB transmitter according to the standard are illustrated in Fig. 2.17. The information bits are first encoded by a Reed–Solomon (RS) encoder, which is a type of block error-correcting code that works by over-sampling a generator polynomial constructed from the input data [92]. The RS encoder takes a block of 330 bits at a time, and adds 48 parity bits according to a generator polynomial specified in the standard. So, the RS encoder has a rate of around 0.87. Then, the encoded bits from the RS encoder are encoded by a convolutional encoder with a rate of 1/2.

Each pair of encoded bits is carried by one UWB symbol. A UWB symbol structure is shown in Fig. 2.18, where the symbol duration T_{sym} is divided into two intervals, denoted as T_{BPM}. At each symbol interval, one burst of UWB pulses is transmitted, and the location of the burst in either the first or the second interval indicates one bit of information. In other words, if the burst resides in the first half of the symbol, a "0" is transmitted; if the burst is in the second half of the symbol, a "1" is transmitted. This is called burst position modulation (BPM). In addition, the polarity of the burst carries another bit of information, corresponding to binary phase shift keying (BPSK). Overall, BPM–BPSK modulation is used to carry two bits of information per symbol.

Also note from Fig. 2.18 that the burst can be transmitted in one of the possible intervals, each with length T_{burst}, in the first or third quarter of the symbol. The position

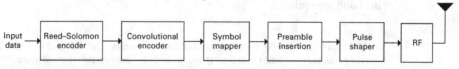

Fig. 2.17. Basic blocks of an IR-UWB transmitter according to the IEEE 802.15.4a standard (After [91]).

2.3 Emerging UWB standards

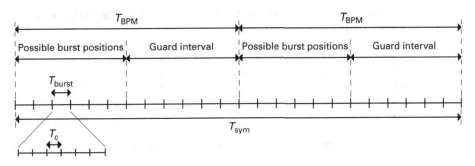

Fig. 2.18. UWB symbol structure according to the IEEE 802.15.4a standard (After [91]).

of the burst can be determined by a burst-hopping sequence, which provides robustness against multi-user interference.

After the symbol mapper in Fig. 2.17, a preamble is added prior to the header of each packet, which is used for timing acquisition, coarse and fine frequency recovery, packet and frame synchronization, channel estimation, and leading edge signal tracking for ranging. After that, bits are transmitted by means of UWB pulses, using the pulse shaper, the RF components and the antenna, as shown in Fig. 2.17.

The transmitted signal for the ith symbol can be expressed as

$$s_i(t) = (1 - 2b_{i,1}) \sum_{n=0}^{N_{\text{cpb}}-1} \left(1 - 2s_{n+iN_{\text{cpb}}}\right) \omega \left(t - b_{i,0} T_{\text{BPM}} - \tilde{h}_i T_{\text{burst}} - nT_{\text{c}} - iT_{\text{sym}}\right), \tag{2.14}$$

where N_{cpb} is the number of chips per burst, i.e. $T_{\text{burst}} = N_{\text{cpb}} T_{\text{c}}$, with T_{c} denoting the chip interval, $\omega(t)$ is the UWB pulse waveform, $\{s_{n+iN_{\text{cpb}}}\}_{n=0}^{N_{\text{cpb}}-1}$ is the binary spreading sequence, and $\tilde{h}_i \in \{0, 1, \ldots, N_{\text{burst}}/4 - 1\}$ is the burst-hopping position for the ith symbol, where $N_{\text{burst}} = T_{\text{sym}}/T_{\text{burst}}$. Note that the limitation of the burst-hopping position to a quarter of the number of bursts per symbol provides a guard interval in the symbol as shown in Fig. 2.18.

The information bits carried by the ith symbol are denoted by $b_{i,0}$ and $b_{i,1}$, where $b_{i,0} \in \{0, 1\}$ is the BPM information determining the position of the burst, and $b_{i,1} \in \{0, 1\}$ is encoded into the burst polarity for BPSK modulation.

System parameters

The UWB PHY of the IEEE 802.15.4a standard supports various data rates through the use of variable-length bursts. The bit rates supported by a given channel are $\{0.11, 0.85, 1.7, 6.81, 27.24\}$ Mbps. In addition, the channels can transmit pulses with various mean pulse repetition frequency (PRF) options, which are 3.90, 15.6 and 62.4 MHz. As an example, in Table 2.6, the parameters are listed for a mean PRF of 62.4 MHz. Note that by changing the number of chips per burst, N_{cpb}, and keeping the number of bursts per symbol, N_{burst}, fixed, various symbol lengths (in terms of number of chips per symbol, N_{c}, in Table 2.6), hence various data rates are obtained.

Table 2.6. System parameters for UWB PHY of the IEEE 802.15.4a standard for a mean PRF of 62.4 MHz, where CC refers to convolutional coding.

RS rate	CC rate	N_{burst}	N_{cpb}	N_c	Bit rate (Mbps)
0.87	0.5	8	512	4096	0.11
0.87	0.5	8	64	512	0.85
0.87	0.5	8	8	64	6.81
0.87	0.5	8	2	16	27.24

Ranging and location awareness

The IEEE 802.15.4a standard supports optional ranging capability for the UWB PHY option. The ranging estimation is obtained from time-delay estimates. In order to obtain time-delay estimates, packet preambles are used, and timing parameters are exchanged between two devices according to certain two-way protocols. The details of the ranging algorithms for the IEEE 802.15.4a standard will be studied in Chapter 6.

2.4 Problems

(1) What are the advantages of UWB signals in positioning applications?

(2) (a) Calculate the Fourier transform of the UWB pulse in (2.5). Hint: The Fourier transform of e^{-at^2} is $\sqrt{\pi/\partial}.e^{-\pi^2 f^2/a}$.

(b) Find the maximum value of the Fourier transform magnitude, and the frequency at which the maximum value is attained. Comment on the relation between that frequency and ζ.

(3) (a) Calculate the average power spectral density of the following signal

$$s(t) = \sum_{i=-\infty}^{\infty} a_i \omega(t - iT_f), \qquad (2.15)$$

where $\omega(t)$ is a UWB pulse, T_f is the pulse repetition interval, which is larger than the pulse width, and a_i is a randomization sequence uniformly distributed on $\{-1, +1\}$.

(b) For the UWB pulse in (2.5) with $\zeta = 0.2$ ns, and for $T_f = 100$ ns, calculate the maximum value of A in (2.5) such that the average power spectral density of $s(t)$ never exceeds -41.3 dBm/MHz.

(4) (*programming exercise*) The fifth derivative of the Gaussian pulse is expressed as follows

$$\omega(t) = A\left(-\frac{t^5}{\sigma^4} + \frac{10t^3}{\sigma^2} - 15t\right)\frac{e^{-t^2/(2\sigma^2)}}{\sqrt{2\pi}\,\sigma^7}, \qquad (2.16)$$

where $A > 0$ and $\sigma > 0$ are the parameters to adjust the energy and the width of the pulse.

(a) Plot the pulse in (2.16) for $\sigma = 10^{-11}$, $\sigma = 5.10^{-11}$ and $\sigma = 10^{-10}$. For each case, choose A such that the pulse is normalized to unit energy.

(b) Obtain the Fourier transform of the pulse, and plot the Fourier transform magnitude for the scenarios considered in part (a). Observe the relations between σ, the bandwidth, and the center frequency of the pulse spectrum. Hint: The approach in problem 2 can be followed to obtain the Fourier transform of the Fifth derivative of the Gaussian pulse.

(d) Assume a low duty cycle UWB signal with average power spectral density $k|\Phi(f)|^2$, where $k = 10^7 \text{s}^{-1}$ and $\Phi(f)$ is the Fourier transform of the UWB pulse $w(t)$.

For such a signal, choose appropriate values for A and σ such that the signal spectrum fits tightly to the FCC mask for indoor communication systems. In other words, find the optimal A and σ to achieve the maximum signal power under the FCC regulations. Plot both the average power spectrum of the signal for the calculated values and the FCC mask.

3 Ultra-wideband channel models

Wireless channel models carry significant importance for gaining insight into designing physical layer systems and selecting certain system parameters. For instance, in an IR-UWB system, a design engineer might need to know how much apart to transmit two sequential pulses in order to avoid inter-frame interference at the receiver, or how likely the first arriving signal component contains the highest energy among all signal components for accurate ranging. Answers to such questions can be obtained either directly from channel measurements conducted in an environment of interest, or from statistical models derived from channel measurement campaigns.

There are various channel modeling techniques (e.g. ray tracing and statistical modeling) [93–96] and channel sounding methods (e.g. time-domain vs. frequency domain) [97, 98], which have been studied extensively in the literature. The focus of this chapter is not those well-known channel modeling techniques, but mainly the UWB channel models recently proposed and their interpretations for positioning applications.

Many UWB channel modeling campaigns have been performed within the past few years, mainly due to emerging UWB standards (e.g. multiband OFDM-UWB, IEEE 802.15.4a, and IEEE 802.15.3c) [96, 97, 99–101]. Although channel statistics and models of various frequency bands are publicly available, many of those do not explicitly include ranging-related statistics. Therefore, one of the aims of this chapter is to investigate UWB channel models from a range estimation perspective.

Designing a wireless system typically involves the steps illustrated in Fig. 3.1. First, application requirements need to be explored. Low attenuation at low frequencies makes through-the-wall communications and tracking applications attractive, but it is difficult to adopt sub-GHz UWB systems due to coexistence issues with existing narrowband systems. Therefore, UWB systems above 3 GHz can find implementation opportunities more easily. Inline with international regulations and restrictions, a frequency band plan should be tailored. For instance, the FCC does not allow the use of UWB for toys and games. Therefore, UWB would not be an option if the goal is to produce location capability for toys. Once the band plan is in place, channel measurement campaigns can be conducted to help set technical design criteria such as data rate, achievable link distance, and maximum supportable mobility. Regulations also impact design criteria. An emission level constraint within a particular frequency band can limit communication range for a given data rate. The last two steps involve completion of system design, and implementation and testing.

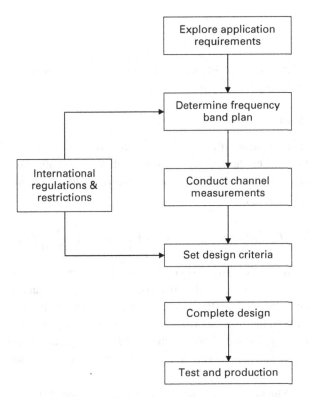

Fig. 3.1. Steps in designing a wireless system.

This chapter focuses on UWB channel models. It first explains differences between UWB and narrowband systems, and studies the effects of those differences in respective channel models. Then, UWB channel characteristics are studied for various frequency bands of operation, which are 0.1–1 GHz, 2–10 GHz, and 57–66 GHz. In addition, key statistics that need particular consideration in development of a ranging system are addressed.

3.1 UWB versus narrowband

Large bandwidths of UWB systems result in significant differences in channel characterization compared to that for narrowband systems.[1] For narrowband systems, the properties of the objects in a given environment, such as their reflection and scattering properties, can be considered as constant with respect to frequency due to the small frequency band of interest. However, for UWB systems, the frequency dependence of material properties as well as that of transmit and receive antennas become significant. In this section, the frequency dependence of propagation is investigated, and the main differences in channel characterization for UWB and narrowband systems are explained.

---

[1] Both narrowband and wideband systems can be considered as "narrowband" as compared to UWB systems.

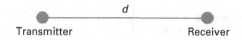

Fig. 3.2. A transmitter and a receiver spaced at a distance of d. It is assumed that the transmitted signal arrives at the receiver only through the direct path.

3.1.1 Free-space propagation

First, an oversimplified propagation model is considered as shown in Fig. 3.2, where the transmitted signal reaches the receiver only through a direct path. For this free-space propagation scenario, it is well known that the received power at frequency f and distance d can be expressed as [102]

$$P_{\text{rx}}(d, f) = P_{\text{tx}} G_{\text{tx}}(f) \eta_{\text{tx-ant}}(f) G_{\text{rx}}(f) \eta_{\text{rx-ant}}(f) \left(\frac{c}{4\pi f d} \right)^2, \quad (3.1)$$

where P_{tx} is the transmit power, G_{tx} and G_{rx} are the antenna gains for the transmit and receive antennas, respectively, c is the speed of light, and $\eta_{\text{tx-ant}}$ and $\eta_{\text{rx-ant}}$ are, respectively, the efficiencies for the transmit and receive antennas. From (3.1), it is observed that if the gains of the antennas vary considerably with frequency, the received power level can change significantly, too. In other words, unlike narrowband systems, for which the gains can be considered as frequency independent, the gains can vary over the frequency range of a UWB system.

Another source of frequency dependency comes into play through the antenna efficiency terms in (3.1). One factor that affects the antenna efficiency is the impedance bandwidth of an antenna, which specifies a frequency band over which the signal loss is not very significant. For UWB antennas, it is quite challenging to limit this signal loss to low and fixed levels over a wide frequency band. Therefore, antenna efficiency is also commonly a frequency-dependent term.[2]

3.1.2 Propagation in a realistic environment

In a realistic environment, there are other objects, in addition to the transmitter and the receiver, that affect the propagation characteristics. For example, the transmitted signal can arrive at the receiver by reflecting from an object in the environment in addition to directly reaching the receiver. Therefore, the properties of the objects in a given environment are also important in determining the response of a channel.

In Fig. 3.3, the signal path from the transmitter to the receiver via a reflection from an object is illustrated. If the transmitted power is denoted as P_{tx}, the received power at frequency f can be expressed as [103]

$$P_{\text{rx}}(d_1, d_2, f) = P_{\text{tx}} G_{\text{tx}}(f) \eta_{\text{tx-ant}}(f) G_{\text{rx}}(f) \eta_{\text{rx-ant}}(f) \frac{c^2 \sigma_{\text{rcs}}(f)}{4\pi (4\pi f d_1 d_2)^2}, \quad (3.2)$$

---

[2] UWB antennas will be investigated in detail in Chapter 8.

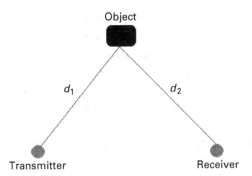

Fig. 3.3. The signal from the transmitter reflects from the object and reaches the receiver.

where d_1 (d_2) is the distance between the transmitter (receiver) and the object, and σ_{rcs} is the radar cross-section (RCS) of the object. The RCS can be considered as a fictional surface area that intercepts the incident wave and scatters the energy isotropically in space. Generally, it is a complex function of wave frequency, polarization, aspect angle and the shape of the object. In the radar literature, the RCS is considered to be a random variable; and the radar equation (3.2) is used only to estimate its mean power. The interested reader is referred to [103] for a detailed table of common statistical models for RCSs of various scatterers.

In order to illustrate frequency dependent effects in propagation, consider a simple sphere of radius r as the object in Fig. 3.3. The RCS for such an object is given by πr^2 for $r \gg \tilde{\lambda}$, where $\tilde{\lambda}$ represents the wavelength, and it is directly proportional to $r^6/\tilde{\lambda}^4$ for $r \ll \tilde{\lambda}$ [104]. In this case, the RCS is independent of the frequency for high frequencies (i.e. for $f \gg c/r$), but increases with frequency for low frequencies (i.e. for $f \ll c/r$). In fact, for geometric shapes other than spheres, the RCS generally increases with frequency [103, 105]. In other words, RCS is a frequency-dependent parameter in general, which can cause significant changes in the received signal power over a range of frequencies.

In addition to reflection, transmission through objects in a given environment is another factor that affects the channel characteristics. This is especially important when there is an object between the transmitter and the receiver. The amount of transmission through a dielectric layer of width d_{layer} is determined by the transmission coefficient, which is given by [102, 106]

$$C_T = \frac{C_{T,1} C_{T,2} e^{-j\varphi(f)}}{1 + C_{R,1} C_{R,2} e^{-2j\varphi(f)}}, \qquad (3.3)$$

where C_T and C_R represent, respectively, the transmission and reflection coefficients (index 1 for air, and index 2 for the layer material), and $\varphi(f)$ is the *electrical length* of the dielectric as seen by waves that are at an angle ψ with respect to the layer, which is given by

$$\varphi(f) = \frac{2\pi f}{c} \sqrt{\varepsilon}\, d_{\text{layer}} \cos \psi, \qquad (3.4)$$

with ε denoting the relative dielectric constant of the layer material. From (3.3) and (3.4), frequency-dependent characteristics of the transmission mechanism can be observed.

In general, dielectric properties of objects impact both transmission and reflection coefficients (e.g. (3.3) and (3.4)). For most materials, the dielectric characteristics vary with the frequency. As an example, the dielectric constant of a brick wall is plotted in Fig. 3.4, which increases monotonically in the 1.31–7.01 GHz band [107].

Since materials can respond to different frequency components in different ways, UWB signals can have both good reflection and transmission properties for various objects at the same time. This is because a UWB signal consists of many frequency components, some of which can reflect well from some objects, while some others can transmit well through them. Therefore, UWB signals can be used in various scenarios, such as in through-the-wall applications [108, 109].

In addition to reflection and transmission, two other propagation characteristics that show significant dependence on frequency are diffraction at the edge of a screen or wedge, and scattering on rough surfaces [102]. All in all, the objects in a given environment have frequency-dependent effects on the propagation, which should be considered in a UWB system.

Having gained some insight into how UWB channels differ from narrowband ones, in the next section, UWB channel models derived from various measurement campaigns are presented and parameters that characterize UWB channels are discussed.

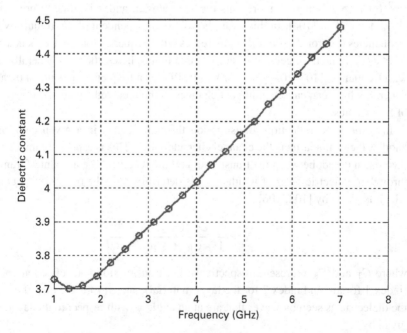

Fig. 3.4. Dielectric constant of a brick wall (thickness: 8.71 cm, length: 19.8 cm, and height: 5.82 cm) [107].

3.2 UWB channel characterization

There are two common methods for characterization of UWB channels. In the first method, an environment with materials of known electromagnetic characteristics is considered, and it is assumed that complete geometric information of the environment is available. Then, one can generate the propagation characteristics of the environment by using an electromagnetic simulation tool with ray-tracing techniques [93, 94]. This approach is called *deterministic modeling*. Although it has the advantage of offering a good representation of the propagation environment, one major drawback of the deterministic modeling is that it is site-specific. Furthermore, gathering accurate site-related information might be quite cumbersome. If the site geometry changes, the corresponding model might easily become obsolete. Results from various deterministic modeling-based propagation characterizations are reported in detail in [110].

A more common way for channel modeling is to derive statistical models from actual channel measurements. This so-called *statistical-modeling* approach is less complex than deterministic modeling. The key channel parameters that need accurate modeling are path loss, shadowing, power delay profile, and small-scale fading. In this section, statistical characterization of UWB channels is studied by reviewing these key channel parameters [100].

3.2.1 Path loss

Path loss (PL) is defined as the ratio of the received signal power P_{rx} to the transmitted signal power P_{tx}, and it is a frequency-dependent parameter for UWB systems as discussed in the previous section.

For a narrowband system, the PL at distance d can be defined as

$$PL(d) = \frac{E\{P_{rx}(d, f_c)\}}{P_{tx}}, \quad (3.5)$$

where f_c is the center frequency, and the expectation is taken over a sufficiently large area to average out shadowing and small-scale fading [100].

On the other hand, for UWB systems, a frequency-dependent PL can be defined as

$$PL(d, f) = E\left\{\int_{f-0.5\Delta f}^{f+0.5\Delta f} |H(d, \tilde{f})|^2 d\tilde{f}\right\}, \quad (3.6)$$

where $H(d, f)$ is the transfer function (including the effects of the antennas), and Δf is a sufficiently small interval over which material properties, such as dielectric constants, can be considered constant [100]. For a given UWB system, the expression in (3.6) should be integrated over the frequency range of the system in order to obtain the total PL.

For simplicity, distance and frequency dependencies can be treated independently, and the PL formula can be expressed as [100]

$$PL(d, f) = PL(d)PL(f), \quad (3.7)$$

where $PL(f) \propto f^{-2\kappa}$ [111] and $PL(d) \propto d^{-n}$, with κ and n denoting the frequency decaying factor and the PL exponent, respectively.[3] The distance dependence is commonly expressed in dB as

$$PL(d) = PL_0 + 10n \log_{10}\left(\frac{d}{d_0}\right), \qquad (3.8)$$

where d_0 is a reference distance (e.g. 1 m) and PL_0 is the PL at the reference distance.

Both the PL exponent and frequency decaying factor depend on the environment. For example, larger PL exponents are observed in NLOS situations compared to LOS ones.

The PL expressions considered in this section include the effects of the transmit and receive antennas, as well. In Section 8.2, a more flexible model will be developed when performing link budget calculations by taking into account both the spectral regulations and the explicit effects of different antenna efficiencies.

3.2.2 Shadowing

Shadowing, also referred to as large-scale fading, is defined as the slow variation of the local mean signal power around the PL. This variation is basically due to changes in the surrounding environment. Shadowing is commonly modeled as a log-normal distribution [100]. Thus, with inclusion of shadowing, the PL in dB can be expressed as

$$PL(d) = PL_0 + 10n \log_{10}(d/d_0) + S, \quad d > d_0 \qquad (3.9)$$

where S is a Gaussian-distributed random variable with zero-mean and standard deviation σ_{sh}.

3.2.3 Power delay profile

Power delay profile gives received power level $P(\tau)$ with respect to a reference timeline τ typically initialized to zero upon arrival of the first signal component as illustrated in Fig. 3.5. In a UWB channel, multipath components (MPCs) arrive in multiple clusters at various attenuation levels and delays. Therefore, the received signal becomes widely dispersed with respect to the transmitted signal. The power delay profile is an indicator of the degree of this dispersion. If the channel impulse response is $h(t)$, the power delay profile of the channel can be computed as the local spatial average of $|h(t)|^2$.

The complex baseband impulse response of UWB channels is given by [100, 113]

$$h(t) = \sum_{k=0}^{K}\sum_{l=0}^{L_k} \alpha_{k,l} e^{j\phi_{k,l}} \delta(t - T_k - \tau_{k,l}), \qquad (3.10)$$

where K is the number of clusters, L_k is the number of rays (MPCs) in the kth cluster, $\alpha_{k,l}$ is the channel coefficient of the lth ray in the kth cluster, T_k is the delay of the kth cluster, and $\tau_{k,l}$ is the delay of the lth ray with respect to the arrival time of the kth cluster. The phases $\phi_{k,l}$ are uniformly distributed within $[0, 2\pi]$.

---

[3] The frequency dependency of the PL is claimed to be $\sqrt{PL(f)} = e^{-f\kappa}$ in [112].

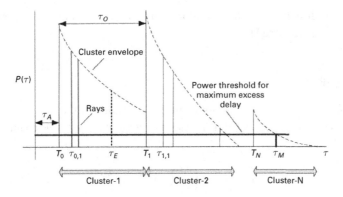

Fig. 3.5. Illustration of power delay profile and characteristic parameters.

In this channel model, the number of clusters is an important parameter, which is modeled as Poisson distributed in [114]; i.e.

$$p_K(x) = \frac{\mu_K^x e^{-\mu_K}}{x!}, \tag{3.11}$$

where μ_K represents the mean number of clusters.

Another important channel characterization is related to the statistics of the cluster and ray arrival times. In [100], cluster arrival times are modeled by a Poisson process and ray arrival times as a mixture of two Poisson processes. In other words, for the cluster arrival times,

$$p(T_k|T_{k-1}) = \Lambda_k e^{\Lambda_k(T_k - T_{k-1})}, \quad k > 0, \tag{3.12}$$

where Λ_k is the cluster arrival rate; and for the ray arrival times,

$$p(\tau_{k,l}|\tau_{k,l-1}) = \beta_{\text{mix}}\lambda_1 e^{-\lambda_1(\tau_{k,l} - \tau_{k,l-1})} \\ + (1 - \beta_{\text{mix}})\lambda_2 e^{-\lambda_2(\tau_{k,l} - \tau_{k,l-1})}, \quad l > 0, \tag{3.13}$$

where β_{mix} is the mixture probability, and λ_1 and λ_2 are the ray arrival rates.

In addition to the statistics of cluster and ray arrivals, the distribution of the cluster and ray powers should be determined in order to obtain the power delay profile. For UWB channels, the power delay profile is exponential within each cluster, and also the mean energy of the clusters follows an exponential decay. For example, for the 2–10 GHz band [100],

$$E\{|\alpha_{k,l}|^2\} = \frac{\Omega_k e^{-\tau_{k,l}/\gamma_k}}{\gamma_k[(1 - \beta_{\text{mix}})\lambda_1 + \beta_{\text{mix}}\lambda_2 + 1]}, \tag{3.14}$$

where Ω_k is the mean energy of the kth cluster and γ_k is the intra-cluster decay time constant, which linearly increases with T_k. The mean cluster energy Ω_k is also modeled as exponentially decaying, i.e.

$$\Omega_k = M_{\text{clus}} e^{-T_k/\Gamma}, \tag{3.15}$$

where Γ is the cluster decay time constant, and M_{clus} is commonly modeled as a log-normal random variable [100].[4]

3.2.4 Small-scale fading

Small-scale fading refers to the variations in the amplitude of the channel coefficient $|\alpha_{k,l}|$, which is commonly modeled as Nakagami distributed for UWB channels [100]. Mathematically,

$$p_{|\alpha|}(x) = \frac{2}{\tilde{\Gamma}(m)} \left(\frac{m}{\tilde{\Omega}}\right)^m x^{2m-1} e^{-mx^2/\tilde{\Omega}}, \quad (3.16)$$

for $x \geq 0$, $m \geq 0.5$ and $\tilde{\Omega} \geq 0$, where m is the Nakagami m-factor, $\tilde{\Omega}$ is the mean power of the channel coefficient, and $\tilde{\Gamma}(m)$ is the gamma function given by

$$\tilde{\Gamma}(z) = \int_0^\infty t^{z-1} e^{-t} dt \quad (3.17)$$

for a complex number z with a positive real part.

3.2.5 Temporal dispersion and auxiliary parameters

There are several auxiliary parameters related to the power delay profile, such as time of first arrival τ_A, mean excess delay τ_E, root-mean-square (RMS) delay spread τ_{rms}, maximum excess delay τ_M, and the peak-to-lead delay τ_{pld}. These parameters not only provide intuitive measures for certain channel properties, but can also provide guidelines for the design and evaluation of ranging algorithms.

Time of first arrival

Time of first arrival τ_A corresponds to the arrival time of the first signal component. Accurate estimation of τ_A carries great importance for ranging. In cases that the first arriving signal component is not the strongest arrival, the range estimation error may increase drastically. The response should be to implement algorithms that search for the first arriving signal component backwards from the strongest. The leading signal detection is addressed in Chapter 5 in detail.

Mean excess delay and RMS delay spread

The first moment of the power delay profile $P(\tau)$ is referred to as the mean excess delay, and the square root of the second central moment of the power delay profile as the RMS delay spread [115]; i.e.

---

[4] For some NLOS environments, the shape of the power delay profile can be different from the common exponential model [100].

$$\tau_E = \frac{\int \tau P(\tau) d\tau}{\int P(\tau) d\tau}, \qquad (3.18)$$

$$\tau_{rms} = \left[\frac{\int (\tau - \tau_E)^2 P(\tau) d\tau}{\int P(\tau) d\tau} \right]^{1/2}. \qquad (3.19)$$

RMS delay spread is a measure of multipath spread within a given channel, and is an important parameter for characterizing time dispersion. For example, in IR-UWB systems, pulses might need to be transmitted further apart in order to avoid inter-frame interference (IFI). Especially, the presence of IFI in the preamble of a communications packet can be detrimental to ranging performance. Mean excess delay and RMS delay spread can provide some insight into necessary pulse separations to avoid the IFI.

Maximum excess delay

The maximum excess delay τ_M is defined as the excess delay for which the power level falls below a threshold. Assume that the threshold is 10 dB below the peak power. Then the τ_M is referred to as the -10 dB maximum excess delay. Similar to the RMS delay spread, the maximum excess delay provides information about the multipath spread of the channel.

Peak-to-lead delay

The peak-to-lead delay τ_{pld} specifies the time interval between the first and the strongest MPCs. For the cases in which the first signal path is the strongest, $\tau_{pld} = 0$, which is very desirable for time-of-arrival (TOA), hence range, estimation algorithms. In channels that are likely to have a weaker first arrival, selection of the delay corresponding to the strongest MPC as the TOA estimate can result in large ranging errors. In such cases, after determining the delay of the strongest MPC, a *search-back* algorithm can be implemented to determine the delay of the first signal component [116]. The probability density function of τ_{pld} might be used to develop accurate search-back schemes. For instance, the length of the maximum search-back window can be determined from τ_{pld}. The sampling rate at the receiver) [117] is also another critical parameter that affects the statistics of the τ_{pld} and consequently the range estimation performance.

3.3 UWB channel measurement campaigns

In the IEEE 802.15.4a standard, channel measurement campaigns were conducted mainly in two frequency bands: 0.1–1 GHz (sub-GHz) and 2–10 GHz [114]. There is also an emerging standard called IEEE 802.15.3c for the 60 GHz region. In this section, statistical channel parameters are given for each of these frequency bands. Antenna effects are excluded in all measurements, except for the outdoor environments.

Table 3.1. Residential environments.

	LOS	NLOS
PL_0 (dB)	43.9	48.7
n	1.79	4.58
κ	1.12 ± 0.12	1.53 ± 0.32
σ_{sh}	2.22	3.51
μ_K	3	3.5
Λ (1/ns)	0.047	0.12
λ_1 (1/ns) λ_2 (1/ns), β_{mix}	1.54, 0.15, 0.0095	1.77, 0.15, 0.0045
Γ (ns)	22.61	26.27

3.3.1 2–10 GHz band

Channel models in IEEE 802.15.4a are classified as follows:

- CM-1: Residential LOS
- CM-2: Residential NLOS
- CM-3: Office LOS
- CM-4: Office NLOS
- CM-5: Outdoor LOS
- CM-6: Outdoor NLOS
- CM-7: Industrial LOS
- CM-8: Industrial NLOS
- CM-9: Open outdoor environment NLOS (e.g. farm, snow-covered area).

In Table 3.1, some channel parameters for residential environments are listed based on measurements for 7–20 m ranges and frequencies up to 10 GHz (please refer to Section 3.2 for the definitions of the parameters). The PL exponent n drastically increases from 1.79 to 4.58 after the introduction of LOS obstruction. The mean number of clusters μ_K in LOS and NLOS cases is 3 and 3.5, respectively. Clusters decay faster in NLOS environments compared to the LOS ones.

Table 3.2 shows the channel parameters for office environments. The model relies on measurements for distances of 3–28 m and a frequency band of 2–8 GHz. The PL exponents are smaller than those in residential environments for both LOS and NLOS scenarios. The mean number of clusters is only 1 for the NLOS case. This is due to the fact that channel measurements exhibiting a power delay profile shape that do not follow a multicluster model were considered in modeling the NLOS cases (see [100]).

For outdoor environments, the model is derived from measurements for 5–17 m ranges and 3–6 GHz frequency band (see Table 3.3). The mean number of clusters is 13.6 and 10.6 for LOS and NLOS cases, respectively. The values of n and κ for the NLOS case are only coarse estimates.

The industrial environment channel model is representative of distances from 2 to 8 m. The NLOS is described by a single power delay profile shape, and there is no distinction into clusters [114]. This is mainly caused by both dense reflections from metal surfaces

Table 3.2. Office environments.

	LOS	NLOS
PL_0 (dB)	35.4	57.9
n	1.63	3.07
κ	0.03	0.71
σ_{sh}	1.9	3.9
μ_K	5.4	1
Λ (1/ns)	0.016	NA
λ_1 (1/ns), λ_2 (1/ns), β_{mix}	0.19, 2.97, 0.0184	NA
Γ (ns)	14.6	NA

Table 3.3. Outdoor environments.

	LOS	NLOS
PL_0, dB	45.6	73.0
n	1.76	2.5
κ	0.12	0.13
σ_{sh}	0.83	2
μ_K	13.6	10.5
Λ (1/ns)	0.0048	0.0243
λ_1 (1/ns) λ_2 (1/ns), β_{mix}	0.27, 2.41, 0.0078	0.15, 1.13, 0.0062
Γ (ns)	31.7	104.7

Table 3.4. Industrial environments.

	LOS	NLOS
PL_0 (dB)	56.7	56.7
n	1.2	2.15
κ	−1.103	−1.427
σ_{sh} (dB)	6	6
μ_K	4.75	1
Λ (1/ns)	0.0709	NA
λ_1 (1/ns), λ_2 (1/ns), β_{mix}	NA	NA
Γ (ns)	13.47	NA

and scatterers. Remember that κ indicates the frequency dependency of the path loss. Unlike other channel models, it takes negative values in the industrial environment case (see Table 3.4).[5]

---

[5] Please refer to [100] for detailed lists of channel parameters for all the environments including the open outdoor NLOS environment.

Analysis of ranging-related parameters

From the channel model described in Section 3.2, and the parameters considered in this section, statistics of various channel parameters that can be useful for ranging algorithms can be obtained. One important parameter for consideration in ranging system design is τ_{pld}. As stated earlier, it measures the delay between the first arriving signal component and the strongest one. Typically, receivers lock onto the strongest signal component for synchronization and acquisition. However, ranging requires further effort that is the detection of the leading signal component. *A-priori* knowledge of how much earlier than the strongest path the leading signal path should be searched for helps implementer set a proper search-back window size. If it is set too short, the leading path remains outside the window region. Alternatively, if it chosen to be too long, it becomes more likely to pick a noise peak as the signal, causing large ranging errors. The latter happens typically at low SNR levels.

In Fig. 3.6, the probability density functions of τ_{pld} are plotted for channels CM-1 through CM-8. The probability density functions are obtained by polynomial fitting with the actual histograms. The results are obtained from 2000 channel impulse response (CIR) realizations per model. The minimum value of τ_{pld} is 0 ns for all channels corresponding to the first component being the strongest one. Among all the channel models, CM-1 has

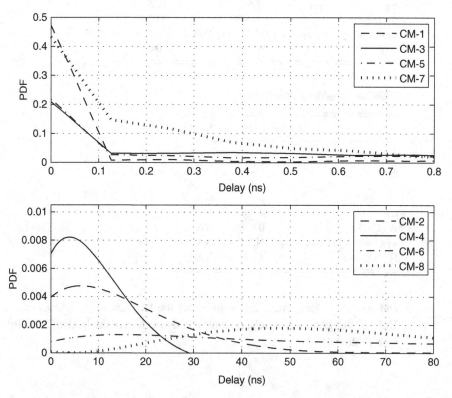

Fig. 3.6. Probability density functions (PDFs) of τ_{pld} based on CIRs of various UWB channels (After [118]).

Table 3.5. The log-normal parameters for the mean excess delay and the RMS delay spread of the IEEE 802.15.4a channels. (©2007 IEEE) [119].

Channel	Mean excess delay (τ_E)			RMS delay spread (τ_{rms})		
	μ [ns]	σ [ns]	K–S	μ [ns]	σ [ns]	K–S
CM-1	2.6685	0.4837	95.7%	2.7676	0.3129	94.8%
CM-2	3.3003	0.3843	95.8%	2.9278	0.1772	95.2%
CM-3	2.0993	0.3931	96.2%	2.2491	0.3597	96.2%
CM-4	2.7756	0.1770	95.3%	2.5665	0.1099	95.4%
CM-5	3.0864	0.4433	94.6%	3.3063	0.2838	94.6%
CM-6	4.6695	0.4185	94.9%	4.2967	0.3742	95.7%
CM-7	1.3845	0.9830	98.9%	1.9409	0.7305	93.9%
CM-8	4.7356	0.0225	94.7%	4.4872	0.0164	95.9%

the highest peak probability for τ_{pld}. More specifically, it has the highest probability that the first arriving path is the strongest path, which is very desirable in ranging applications. On the other hand, for CM-8, the probability of the first arriving path being the strongest one is nearly zero. In fact, the probability that $\tau_{pld} < 10$ ns is quite low for CM-8, which makes accurate range estimation quite challenging. A general observation from Fig. 3.6 is that for NLOS channels, τ_{pld} has a heavy-tailed distribution, which can potentially cause larger ranging errors.

Although mean excess delay and RMS delay spread are not explicit parameters in a range estimation process, they can still give some useful information about the structure of the channel, hence can provide some guidelines for the design of ranging algorithms. Analysis of the mean excess delay and the RMS delay spread statistics for the IEEE 802.15.4a channels shows that their histograms can be well modeled by a log-normal distribution given by

$$p(x) = \frac{1}{x\sqrt{2\pi}\sigma} \exp\left[-\frac{(\ln(x) - \mu)^2}{2\sigma^2}\right], \quad (3.20)$$

where μ is the mean and σ is the standard deviation of $\ln(x)$. This observation is further verified by using the Kolmogorov–Smirnov (K–S) hypothesis test with a 5% significance level. The mean and standard deviation of $\ln(\tau_E)$ and $\ln(\tau_{rms})$ as well as the K–S passing rates are tabulated in Table 3.5 [119].

Figure 3.7 shows the probability density functions of τ_E and τ_{rms} for channels CM-5, CM-6, CM-7 and CM-8. The probability density functions for LOS and NLOS scenarios are quite distinct in both outdoor and industrial environments. This distinction can prove to be a useful tool for adaptive ranging algorithms. For example, in a search-back algorithm, the searchback window size can be varied depending on whether a channel is LOS or NLOS.

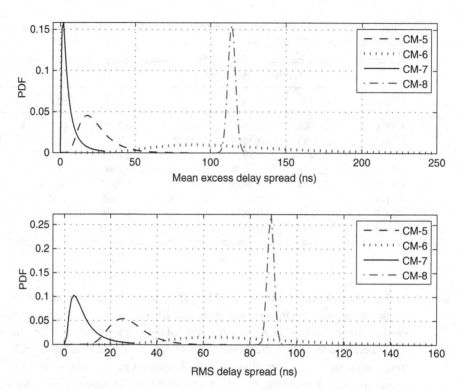

Fig. 3.7. The PDFs of the mean excess delay and RMS delay spread for CM-5–CM-8 channel models (©2007 IEEE) [119].

3.3.2 Below 1 GHz band

Only a few measurement campaigns are available for below 1 GHz (also called "sub-GHz") UWB channels. In [120], measurements for indoor office environments were conducted, using baseband UWB pulses. Based on those measurements and with a 2 ns delay resolution, a sub-GHz channel model is suggested in [121]. The channel for indoor office environment was modeled with an exponentially decaying single cluster. The decay time constant $\tilde{\gamma}$ is shown to be log-normal distributed. In addition, small-scale fading statistics fit the Nakagami distribution with a truncated Gaussian distributed m-parameter.

Later, the IEEE 802.15.4a standard developed a channel model for the frequency band between 100 and 960 MHz [114]. The standard basically adopts the model in [121]. The main difference is that in the IEEE 802.15.4a standard, the decay time constant is assumed to be deterministic with distance dependency such that

$$\tilde{\gamma} = 40\sqrt{d/10} \quad \text{(ns)}, \tag{3.21}$$

where d is the distance in meters.[6] However, in [122] a linear increase in the decay constant with distance is reported. Values of some channel parameters for the sub-GHz indoor office channel model are reported as $n = 2.4$, $\mu_K = 1$ and $m = 5$ in [114].

For the LOS case the MPCs are claimed to be correlated in [123]. From the average of 100 LOS channel measurements in a 3.7 m by 4.6 m room, the direct path energy is observed to be 2.3 dB higher than the total energy contained in the other MPCs. The RMS delay spread is approximated as

$$\tau_{\text{rms}} = 0.2d/c, \qquad (3.22)$$

where c is the speed of light.

In the NLOS case, an additional parameter is used to split total energy between the direct path and the others. A rule of thumb approximation for the RMS delay spread in nanoseconds for the NLOS case is given by

$$\tau_{\text{rms}} = 4.5\sqrt{d} \quad \text{(ns)}, \qquad (3.23)$$

where d is the distance in meters.

Unfortunately, due to insufficient measurement data to support these sub-GHz UWB channel models, it is difficult to derive general conclusions.

3.3.3 57–66 GHz band

The IEEE 802.15.3 Task Group 3c (TG3c) has been working on a millimeter-wave-based physical layer (PHY) for an amendment to the existing 802.15.3 WPAN standard. The millimeter-wave PHY will be operating in the millimeter-wave region of the spectrum allowed by the FCC regulations, including the 57–64 GHz unlicenced band.[7] The millimeter-wave WPAN is expected to provide data rates over 2 Gbps, and facilitate applications such as high speed internet access and streaming content download [125]. The typical communications ranges will be around 10 m.

The channel models related to this standard are derived from measurements conducted in several environments over the 57–66 GHz band [126–130]. Although the channel model adopted by the TG3c [101] resembles the channel characterization studied in Section 3.2, there are also a number of differences in certain statistical models.

Although PL for UWB systems is a frequency-dependent parameter, the channel model adopted by the TG3c considers a frequency-independent PL for simplicity [101]. Other than this simplification, the distance dependence of the PL and the shadowing statistics are similar to the generic model studied in Section 3.2. The PL exponent n ranges from 1.2 to 2 for LOS and from 1.97 to 10 for NLOS environments [131–133]. Shadowing is log-normal distributed, and its variance σ_{sh} is site specific. Especially, human movements can cause obstructions as high as 18–36 dB [134]. A summary of

---

[6] Also, the average power of the first bin does not obey the exponential decay of the average power delay profile corresponding to the other components [100].
[7] Please refer to FCC 47 CFR 15.255 for the related FCC regulations [124].

Table 3.6. Large-scale fading characteristics for 60 GHz.

Channel	Description	n	PL_0	σ_{sh}	−3 dB Beamwidth
CM-1	Residential LOS	1.53	75.1	1.5	Tx-72°, Rx-60°
CM-2	Residential NLOS	2.44	86.0	6.2	Tx-72°, Rx-60°
CM-3	Office LOS	1.16	84.6	5.4	Tx-Omni, Rx-30°
CM-4	Office NLOS	3.74	56.1	8.6	Tx-Omni, Rx-30°

large-scale fading characteristics is given in Table 3.6. These values are obtained from directional measurements, and antenna effects are included [101].

Similar to the IEEE 802.15.4a channel models, clustering of MPCs is observed in the millimeter wave channel. The mean number of clusters, μ_K, varies typically between 3 and 14. However, the number of clusters, which is modeled by a Poisson distribution in Section 3.2, does not follow a specific distribution in the 57–66 GHz band.

The channel impulse response takes into account both the spatial and temporal domains, and is given by

$$h(t, \phi) = \sum_{k=0}^{K} \sum_{l=0}^{L_k} \alpha_{k,l} \delta(t - T_k - \tau_{k,l}) \delta(\phi - \Theta_k - \vartheta_{k,l}), \qquad (3.24)$$

where K is the number of clusters, L_k is the number of rays in the kth cluster, $\alpha_{k,l}$ is the channel coefficient, T_k and Θ_k are, respectively, delay and mean AOA of the kth cluster, and $\alpha_{l,k}$, $\tau_{k,l}$ and $\vartheta_{k,l}$ are the complex channel amplitude, delay, and azimuth of the lth ray in the kth cluster. Note that this channel model is based on the assumption that the spatial and the temporal domains are independent.

In the presence of directional antennas and LOS situations, a strong LOS component is observed in the measurements [101]. Therefore, for such scenarios, the channel model can be extended to include a strict LOS component in addition to the clustered MPCs in (3.24); i.e.

$$h(t, \phi) = \alpha_{LOS} \delta(t, \phi) + \sum_{k=0}^{K} \sum_{l=0}^{L_k} \alpha_{k,l} \delta(t - T_k - \tau_{k,l}) \delta(\phi - \Theta_k - \vartheta_{k,l}), \qquad (3.25)$$

where $\alpha_{LOS} \delta(t, \phi)$ represents the strict LOS component.

Both the cluster and ray arrival times are modeled by Poisson distributions as

$$p(T_k | T_{k-1}) = \Lambda e^{-\Lambda (T_k - T_{k-1})}, \quad k > 0, \qquad (3.26)$$

$$p(\tau_{k,l} | \tau_{k,l-1}) = \lambda e^{-\lambda (\tau_{k,l} - \tau_{k,l-1})}, \quad l > 0, \qquad (3.27)$$

where Λ and λ are the cluster and ray arrival rates, respectively. Note that the ray arrivals are described by a mixture of two Poisson processes in (3.13), which is a more general model than that in (3.27).

Related to small-scale fading statistics, both cluster and ray amplitudes are modeled by log-normal distribution, which is another difference of the channel model for the

57–66 GHz band from the channel model studied in Section 3.2, which models the channel amplitudes by Nakagami distribution.

Even though the IEEE 802.15.3c standard targets communication ranges in excess of 10 m, channel measurements in the literature for 60 GHz cover a range from 2 to 5 m [101]. Channel models derived from such short-range measurements would not carry practical importance for location-aware applications, because range and location information is valuable for most applications typically at distances longer than 10 m. Therefore, a detailed analysis of the channel statistics for this band is omitted.

3.4 Problems

(1) In an indoor environment, how would the number of MPCs arriving at a receiver change if a UWB signal is employed instead of a narrowband signal at the same center frequency?

(2) Considering the frequency-dependent PL model in Section 3.2.1, calculate the ratio of the PL at $f = 4$ GHz and $d = 10$ m to the PL at $f = 5$ GHz and $d = 5$ m for both LOS and NLOS office environments.

(3) Consider a UWB channel profile with a single cluster and Poisson path arrivals. The conditional probability density function for the delay of the lth MPC given that of the $(l-1)$th one is expressed as

$$p(\tau_l|\tau_{l-1}) = \lambda\, e^{-\lambda(\tau_l - \tau_{l-1})}, \qquad (3.28)$$

for $l = 1, \ldots, L$.

(a) If the widths of the pulses received via different paths are all equal to T_p, calculate the probability that no pulses collide with the first pulse (i.e. the pulse via the 0th path).

(b) For the previous scenario, obtain the probability mass function of the number of pulses that collide with the first pulse.

(4) In a UWB channel, the peak of the first cluster is weaker than the peak of the second cluster, and the first path of the second cluster is the strongest MPC. Assume that a TOA estimation algorithm has first estimated the delay corresponding to that strongest MPC, and the aim is to estimate the delay of the first signal path in the first cluster, i.e. the TOA.

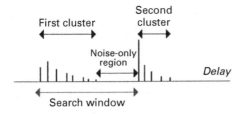

Fig. 3.8. Channel profile described in problem 4.

It is known that the first cluster is 15 ns long and that between the end of the first cluster and the next there happens to be a noise-only region. Assume that the length of the noise-only region is Gaussian distributed with zero mean and a standard deviation of 4 ns. What should be the length of the search window shown in Fig. 3.8 such that it contains the first path with 90% confidence?

(5) Consider a channel with a direct path and a single major reflection, the energy of which is 1.2 dB weaker than the direct path. Consider a range estimation algorithm that selects the strongest path as the direct path, and a ranging error occurs if the reflecting path is selected as the direct path. Assuming that the algorithm checks only those two MPCs to determine the direct path, and that the measured energy of each path is corrupted by independent Gaussian noise with zero mean and variance σ^2, calculate the ratio between the energy of the direct path and σ such that a ranging error occurs with 10% probability.

(6) (*programming exercise*) Consider a UWB channel with the delays for its first five MPCs being given by $\tau_0 = 0$, $\tau_1 = 1.2$ ns, $\tau_2 = 1.7$ ns, $\tau_3 = 3.2$ ns, and $\tau_4 = 5.1$ ns. Assume that these components belong to the same cluster, and that the average power of each path is given by

$$E\{|\alpha_l|^2\} = c_1 e^{-\tau_l/c_2}, \tag{3.29}$$

for $l = 0, 1, 2, 3, 4$. In addition, the amplitude of each MPC, $|\alpha_l|$, follows a Nakagami distribution with the Nakagami m-factor of 1.5.
  (a) For $c_1 = 1$ and $c_2 = 5$ ns, generate ten-thousand realizations for the amplitudes of those 5 MPCs.
  (b) From the channel realizations in part (a), calculate the probability that the first MPC becomes the strongest one.
  (c) Repeat part (b) for every other multipath component to be the strongest one, and calculate the mean square TOA estimation error for an algorithm that always selects the strongest component as the TOA estimate (assume no noise exists in the system).

4 Position estimation techniques

After the investigation of UWB signals and channel models in the previous chapters, this chapter focuses on position estimation techniques from a UWB perspective.

In order to estimate the position of a node (called the "target" node) in a wireless network, signals are exchanged between the target node and a number of reference nodes [135].[1] The position estimation can be performed directly from the signals traveling between the nodes, which is called *direct positioning* [137], or by a two-step approach in which certain parameters are extracted from the signals first, and then the position is estimated based on those signal parameters (Fig. 4.1) [136]. Although two-step approaches are suboptimal in general, their complexity is lower than the direct approach. Also, the performance of the two is usually very close for sufficiently high SNRs and/or signal bandwidths [137, 138]. Therefore, most practical systems adopt two-step approaches, which will be the main focus of this chapter.

In the first step of a two-step positioning technique, signal parameters, such as received signal strength (RSS) or time-of-arrival (TOA), are estimated. Various types of signal parameter measurements (estimation schemes) are studied in Section 4.1. Then, position

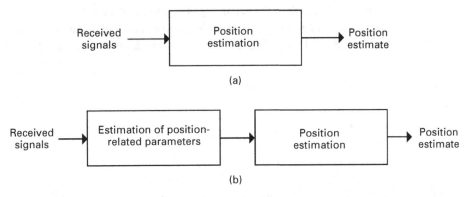

Fig. 4.1. (a) Direct positioning, (b) two-step positioning (with kind permission from Springer Science and Business Media) [135].

---

[1] In this book, we focus on *radiolocation*, which is the process of position estimation through the use of radio signals. Other techniques for position estimation/tracking include dead-reckoning and proximity systems [136].

estimation from signal parameters is considered in Section 4.2. In the case of constant monitoring of a node position, position tracking algorithms, such as Kalman filtering, can improve accuracy of position estimates. Tracking algorithms are investigated in Section 4.3.

4.1 Measurement categories

The first step to estimate the position of a target node in a wireless network involves measurement of a set of signal parameters, as shown in Fig. 4.1(b). Depending on accuracy requirements and constraints on transceiver design, various signal parameters can be employed. Commonly, a single parameter is estimated for each received signal, such as the arrival time of the signal. However, it is also possible to estimate multiple signal parameters in order to improve positioning accuracy.

4.1.1 Received signal strength (RSS)

RSS measurements provide information about the distance ("range") between two nodes based on certain channel characteristics. The main idea behind an RSS-based approach is that if the relation between distance and power loss is known, the RSS measurement at a node can be used to estimate the distance between that node and the transmitting node, assuming that the transmit power is known.

The distance between two nodes provides a circle of uncertainty[2] for the position of the target node, as shown in Fig. 4.2. However, due to inaccuracies in both RSS measurements and quantification of the distance versus path loss (PL) relation, distance estimates are subject to errors. Therefore, in reality, each RSS measurement defines an uncertainty area, such as the one in Fig. 4.3, instead of a circle.

As studied in Chapter 3, a UWB signal experiences multipath (small-scale) fading, shadowing and PL while traveling from one node to another. Ideally, average RSS

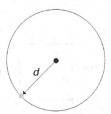

Fig. 4.2. The black node measures the RSS and determines the distance d between itself and the target node (gray node). In the absence of errors, the distance information defines a circle around the black node with a radius of d.

---

[2] Two-dimensional positioning is considered in this chapter for simplicity. Extensions to three-dimensional positioning easily follow.

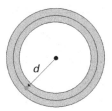

Fig. 4.3. The black node measures the RSS and determines the distance d with some uncertainty, which defines a ring around the black node with a center radius of d. Note that the uncertainty region can define more complicated areas than a ring depending on error statistics.

(equivalently, power) over a sufficiently long time interval would exclude the effects of multipath fading and shadowing, and would result in the following model[3]

$$\bar{P}(d) = P_0 - 10n \log_{10}(d/d_0), \quad (4.1)$$

where n is the PL exponent, $\bar{P}(d)$ is the average received power (dB) at a distance d and P_0 is the received power (dB) at a reference distance d_0.

For UWB systems, the multipath effects can be mitigated significantly by measuring the sum of the powers of multipath components (MPCs) [26]. In other words, if the integration interval T in the calculation of the average power,

$$P(d) = \frac{1}{T} \int_0^T |r(t)|^2 dt, \quad (4.2)$$

is long enough to include all the MPCs in the received signal $r(t)$, the small-scale fading effects can be mitigated.

However, the shadowing effects are usually present in the received power $P(d)$, which are modeled as log-normal random variables. In other words, the received power in dB can be modeled as a Gaussian random variable with mean $\bar{P}(d)$ given by (4.1) and variance σ_{sh}^2; i.e.

$$10 \log_{10} P(d) \sim \mathcal{N}\left(\bar{P}(d), \sigma_{sh}^2\right). \quad (4.3)$$

Note that this model can be used in both LOS and NLOS scenarios with an appropriate choice of channel-related parameters.

From the received power model in (4.3), the Cramer–Rao lower bound (CRLB)[4] for estimating the distance can be expressed as [139]

$$\sqrt{\text{Var}\{\hat{d}\}} \geq \frac{\ln 10}{10} \frac{\sigma_{sh}}{n} d, \quad (4.4)$$

where $\hat{d}$ represents an unbiased estimate of d. It is observed from (4.4) that the lower bound increases as the standard deviation of the shadowing increases, since RSS measurements vary more around the true average power in that case. Furthermore, a larger

---

[3] Note that there is also thermal noise in real systems, which is usually location independent. It is assumed that its effects can be mitigated sufficiently [139].
[4] Interested readers are referred to [140] for a detailed explanation of CRLBs.

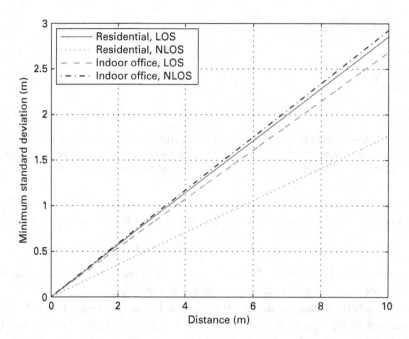

Fig. 4.4. Theoretical limits for distance estimation based on RSS measurements at different distances for various channel models.

Table 4.1. Channel parameters for the environments investigated in Fig. 4.4.

	n	σ_{sh}
Residential LOS	1.79	2.22
Residential NLOS	4.58	3.51
Indoor office LOS	1.63	1.90
Indoor office NLOS	3.07	3.90

PL exponent results in a better estimation accuracy, as the average power becomes more sensitive to distance for larger n. Finally, the distance dependence structure of (4.4) implies that the accuracy of RSS measurements deteriorates as the distance between the nodes increases.

In Fig. 4.4, the minimum standard deviations are plotted versus distance for various environments according to the IEEE 802.15.4a channel models studied in Section 3.3.1, for which the PL exponents and the standard deviations of the shadowing are given in Table 4.1. As observed from (4.4), the lower bound increases linearly with the distance; also, note that the NLOS residential environment has the lowest bound, since it has a significantly larger PL exponent than the other environments. In all the cases, the standard deviation of the error cannot be made smaller than 1 m for distances larger than 6 m. In other words, RSS measurements cannot provide very accurate range estimates for UWB systems.

4.1.2 Angle of arrival (AOA)

Unlike an RSS measurement that provides range information between two nodes, an AOA measurement provides information about the direction of an incoming signal, hence the angle between the two nodes, as shown in Fig. 4.5.

Commonly, antenna arrays are employed in order to measure the AOA of a signal.[5] The angle information is obtained at an antenna array by measuring the differences in arrival times of an incoming signal at different antenna elements. An example is illustrated in Fig. 4.6 for AOA estimation at a uniform linear array (ULA). When the distance between the transmitting and receiving nodes are sufficiently large, the incoming signal can be modeled as a planar wave-front. This results in $l \sin \psi / c$ seconds difference between the arrival times at consecutive array elements, where l is the inter-element spacing, ψ is the AOA and c represents the speed of light. Therefore, estimation of the time differences of arrivals provides angle information. More advanced array structures, such as uniform

Fig. 4.5. The reference node (black node) measures the AOA and determines the angle ψ between itself and the target node (gray node) (with kind permission from Springer Science and Business Media) [135].

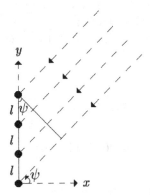

Fig. 4.6. Signal arrival at a ULA, and relation between arrival time differences and AOA.

---

[5] Another technique is to use the ratio of RSS measurements between at least two directional antennas located on a node [26].

circular arrays (UCAs) and rectangular lattices, operate on the same basic principle as the ULA; namely, estimation of time differences between array elements, the geometry of which is known by the receiver.

For a narrowband signal, time difference can be represented as a phase shift. Therefore, the combinations of the phase-shifted versions of received signals at array elements can be tested for various angles in order to estimate the direction of signal arrival [136]. However, for UWB systems, time-delayed versions of received signals should be considered since a time delay cannot be represented by a unique phase value for a UWB signal.

In order to obtain theoretical lower bounds on the achievable accuracy of AOA measurements, consider a ULA, as shown in Fig. 4.6, with N_a antenna elements. Let $r_i(t)$ denote the received signal at the ith element, which is expressed as[6]

$$r_i(t) = \alpha s(t - \tau_i) + n_i(t), \qquad (4.5)$$

for $i = 1, \ldots, N_a$, where $s(t)$ is the transmitted signal, α is the channel coefficient, τ_i is the delay for the signal arriving at the ith antenna element, and $n_i(t)$ is white Gaussian noise with zero mean and a spectral density of $\mathcal{N}_0/2$.

The delay τ_i can be expressed as

$$\tau_i \approx \frac{d}{c} + \frac{l_i \sin \psi}{c}, \qquad (4.6)$$

with

$$l_i = l\left(\frac{N_a + 1}{2} - i\right), \qquad (4.7)$$

for $i = 1, \ldots, N_a$, where d is the distance between the transmitter and the center of the antenna array at the receiver, and l is the inter-element spacing.

For independent noise at different antenna elements, the CRLB for estimating ψ is given by [141]

$$\mathrm{Var}\{\hat{\psi}\} \geq \frac{6c^2 \mathcal{N}_0}{\alpha^2 \tilde{E} N_a (N_a^2 - 1) l^2 \cos^2 \psi}, \qquad (4.8)$$

where $\tilde{E}$ represents the energy of the first derivative of $s(t)$ in (4.5); i.e.

$$\tilde{E} = \int_{-\infty}^{\infty} \left[s'(t)\right]^2 dt. \qquad (4.9)$$

---

[6] CRLBs for AOA estimation in multipath channels are studied in [141].

Applying the Parseval's relation[7] to (4.9), the bound in (4.8) can alternatively be expressed as

$$\sqrt{\text{Var}\{\hat{\psi}\}} \geq \frac{\sqrt{3}\,c}{\sqrt{2}\,\pi\sqrt{\text{SNR}}\,\beta\sqrt{N_a(N_a^2-1)}\,l\cos\psi}, \quad (4.10)$$

where $\text{SNR} = \alpha^2 E/\mathcal{N}_0$, with E denoting the energy of the signal $s(t)$, is the signal-to-noise (SNR) ratio for each element, and β is the effective bandwidth defined by

$$\beta = \left(\frac{1}{E}\int_{-\infty}^{\infty} f^2 |S(f)|^2 \mathrm{d}f\right)^{1/2}, \quad (4.11)$$

with $S(f)$ representing the Fourier transform of $s(t)$.

It is noted from (4.10) that an increase in the SNR, effective bandwidth, inter-element spacing or the number of antenna elements enhances the accuracy of AOA estimation. Therefore, the large bandwidth of UWB signals can facilitate accurate AOA measurements. It is also observed that a ULA cannot detect obtuse angles[8] as accurately as it can detect acute angles.[9]

In Fig. 4.7, the theoretical limits are plotted for a ULA with four elements and an inter-element spacing of 5 cm. The signal $s(t)$ in (4.5) is chosen to be the UWB pulse in

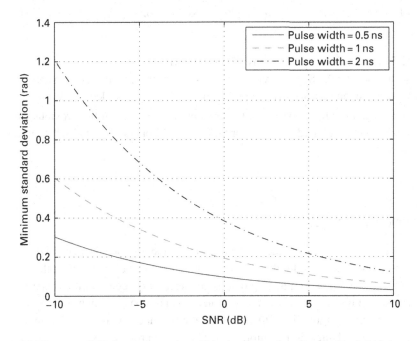

Fig. 4.7. CRLB versus SNR for various pulse widths (equivalently, effective bandwidths).

---

[7] For a signal $s(t)$ and its Fourier transform $S(f)$, the Parseval's relation can be stated as $\int_{-\infty}^{\infty} |s(t)|^2 \mathrm{d}t = \int_{-\infty}^{\infty} |S(f)|^2 \mathrm{d}f$.
[8] Angles greater than $\pi/2$ and less than π radians.
[9] Angles less than $\pi/2$ radians.

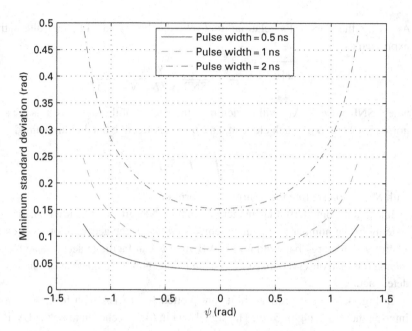

Fig. 4.8. CRLB versus ψ for various pulse widths at SNR = 5 dB.

(2.5) with various pulse widths. The signal arrives at the receiver at $\psi = \pi/4$ radians with $\alpha = 1$. As the SNR increases or the pulse width decreases (which corresponds to an increase in the bandwidth), the lower bound of the system decreases.

In Fig. 4.8, all the system parameters are the same as in the previous case, and the lower bound is plotted versus ψ for SNR = 5 dB. For smaller AOAs, better accuracy can be obtained.

4.1.3 Time of arrival (TOA)

TOA measurements provide information about the distance between two nodes by estimating the time of flight of a signal that travels from one node to the other. Therefore, a TOA measurement at a node provides an uncertainty region in the shape of a circle as shown in Fig. 4.2 for an RSS measurement (or a region around the circle, as in Fig. 4.3). To prevent ambiguity in TOA estimates, the two nodes must have a common clock, or they must exchange timing information via certain protocols, such as a two-way ranging protocol, which are studied in Chapter 6.

The conventional TOA estimation technique is performed by means of matched filtering or correlation operations [142]. Let the received signal at a node be expressed as

$$r(t) = \alpha s(t - \tau) + n(t), \qquad (4.12)$$

where τ represents the TOA, α is the channel coefficient, and $n(t)$ is white Gaussian noise with zero mean and a spectral density of $\mathcal{N}_0/2$. Then, a conventional correlator-based scheme searches for the peak[10] of the correlation of $r(t)$ with a shifted version of the template signal, $s(t-\hat{\tau})$, for various delays $\hat{\tau}$. Similarly, a matched filter scheme, in which the filter is matched to the signal, estimates the instant at which the filter output attains its largest value. These schemes are optimal for single-path AWGN channels.

Note that UWB channels are commonly more complicated than the model assumed in (4.12), as studied in Chapter 3. The TOA estimation schemes for more realistic scenarios are studied in detail in Chapter 5. Our aim here is to observe the main relations between the bandwidth and the theoretical limits for TOA estimation.

For the signal model in (4.12), the CRLB can be expressed as [140, 143]:

$$\sqrt{\text{Var}(\hat{\tau})} \geq \frac{1}{2\sqrt{2\pi}\sqrt{\text{SNR}}\,\beta}, \qquad (4.13)$$

where $\hat{\tau}$ represents an unbiased TOA estimate, $\text{SNR} = \alpha^2 E/\mathcal{N}_0$ is the signal-to-noise ratio, with E denoting the signal energy, and β is the effective signal bandwidth defined by (4.11).

Note from (4.13) that unlike RSS measurements, the accuracy of a TOA measurement can be improved by increasing the SNR and/or the effective signal bandwidth. Since a UWB signal has a very large bandwidth, this property allows highly accurate distance estimation using TOA measurements via UWB radios. For example, Fig. 4.9

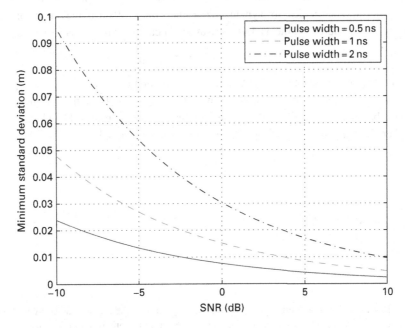

Fig. 4.9. The minimum standard deviation versus SNR for various pulse widths.

---

[10] Peak of the absolute value of the correlation in general.

illustrates the minimum standard deviations of distance estimates (which are obtained by multiplying the TOA estimate by the speed of light) according to the CRLB bound in (4.13) for various pulse widths, where the transmitted signals are as given by (2.5). It is observed that the theoretical limits are of the order of a few centimeters for reasonable SNR values, which indicates the high precision potential of UWB positioning based on TOA measurements. Furthermore, higher bandwidth, equivalently a shorter duration pulse, results in better distance estimation, as expected.

4.1.4 Time difference of arrival (TDOA)

Conventionally, TOA-based range measurements require synchronization among the target and the reference nodes.[11] However, TDOA measurements can be obtained even in the absence of synchronization between the target node and the reference nodes, if there is synchronization among the reference nodes [136]. In this case, the difference between the arrival times of two signals traveling between the target node and the two reference nodes is estimated. This locates the target node on a hyperbola, with foci at the two reference nodes, as shown in Fig. 4.10.

One way to obtain a TDOA measurement is to estimate TOA at each reference node and then to obtain the difference between the two estimates. Specifically, if the received signals are given by $r_1(t)$ and $r_2(t)$ as in (4.5), τ_1 is estimated from $r_1(t)$ and τ_2 is estimated from $r_2(t)$. Since the target node and the reference nodes are not synchronized, the TOA estimates at the reference nodes include a timing offset in addition to the time of flight. As the reference nodes are synchronized, the timing offset is the same for each TOA estimation. Therefore, the TDOA measurement can be obtained as

$$\hat{\tau}_{\text{TDOA}} = \hat{\tau}_1 - \hat{\tau}_2, \quad (4.14)$$

where $\hat{\tau}_1$ and $\hat{\tau}_2$ denote the TOA estimates at the first and second nodes, respectively.

Since it is shown in Section 4.1.3 that the accuracy of TOA measurements increases with bandwidth and SNR, the same conclusions hold true for TDOA measurements, when they are estimated from TOA measurements as in (4.14).

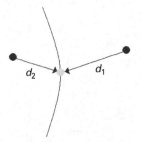

Fig. 4.10. A TDOA measurement defines a hyperbola passing through the target node (gray node) with foci at the reference nodes (black nodes) (with kind permission from Springer Science and Business Media) [135].

---

[11] It is also possible to obtain TOA-based range measurements by means of timing information exchanges via certain protocols, such as a two-way ranging protocol, as will be studied in Chapter 6.

Another way to obtain a TDOA measurement is to perform cross-correlations of the received signals $r_1(t)$ and $r_2(t)$, and to calculate the delay corresponding to the largest cross-correlation value. The cross-correlation function can be expressed as [12]

$$\phi_{1,2}(\tau) = \frac{1}{T} \int_0^T r_1(t) r_2(t+\tau) \mathrm{d}t, \qquad (4.15)$$

where T is the observation interval, and the TDOA estimate is given by

$$\hat{\tau}_{\text{TDOA}} = \arg \max_{\tau} |\phi_{1,2}(\tau)|. \qquad (4.16)$$

Although the cross-correlation-based TDOA estimation in (4.16) works well for single-path channels and white noise models as in (4.5), its performance can degrade significantly over multipath channels and/or colored noise. In order to improve the performance of the cross-correlation scheme, generalized cross-correlation (GCC) techniques are proposed [144–146]. In GCC-based TDOA estimation, filtered versions of the received signals are cross-correlated, which corresponds to shaping the cross-power spectral density (cross-PSD) of the transmitted signals. Various shaping functions can be considered for improved robustness against colored noise [147].

4.1.5 Other measurement types

Instead of performing a single measurement such as RSS or TOA, a node can estimate a combination of position-related parameters. Such *hybrid* schemes can provide more accurate information about the position of the target node than the schemes that estimate a single position parameter. Various combinations of measurements, such as TOA/AOA, TOA/RSS and TDOA/AOA, are possible depending on accuracy requirements and complexity constraints. For example, a hybrid TOA/AOA scheme estimates both distance and AOA, which can provide a unique position estimate for the target node (the intersection of a line and a circle), as shown in Fig. 4.11.

Another type of measurement involves obtaining multipath power delay profile (PDP) or channel impulse response (CIR) related to a received signal [148–153]. Compared to the distance and angle-related parameters studied in the previous sections, a PDP or a CIR measurement can contain significantly more information about the position of the target node. However, in order to extract relevant information about the position from such

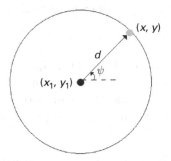

Fig. 4.11. Hybrid TOA/AOA measurements (with kind permission from Springer Science and Business Media) [135].

measurements, a regression/mapping function should be available, which is commonly obtained from a database consisting of previous PDP (or CIR) measurements. Therefore, algorithms utilizing PDP or CIR measurements usually implement training phases to obtain a mapping function from the database, before the actual position estimation process begins.

Similar to the PDP approach, multipath angular power profile measurements can be employed at nodes with antenna arrays. Note that both the PDP (CIR) and the angular power profile measurements increase the complexity of the position related parameter estimation phase significantly compared to the conventional RSS, AOA and T(D)OA schemes, since a large number of parameters need to be estimated. However, such measurements can also facilitate accurate position estimation in challenging environments [153].

4.2 Position estimation

After estimating a set of parameters from the received signals by one of the techniques described in the previous section, the second step is to estimate the position from the obtained parameters (the second block in Fig. 4.1(b)).

Position estimation techniques can be divided into two groups depending on the presence of a database that contains signal measurements at known positions. A technique that makes use of such a database, which is usually obtained by a training phase (off-line phase) before the real-time positioning starts, is called a *mapping* (*fingerprinting*) technique. Other techniques that do not utilize such a database commonly employ *geometric* or *statistical* techniques to estimate the position using only the parameter estimates from the first step in Fig. 4.1(b).

4.2.1 Mapping techniques

Mapping techniques use the available database in the system as training data and estimate the position of a target node by pattern-matching algorithms, such as k-nearest-neighbor (k-NN), support vector regression (SVR) and neural networks[12] [153–159].

A mapping technique can be considered as a regression scheme that maps an input vector to an output vector by using a training set. Let the training set be represented by

$$\mathcal{T} = \{(\mathbf{m}_1, \mathbf{p}_1), (\mathbf{m}_2, \mathbf{p}_2), \ldots, (\mathbf{m}_{N_t}, \mathbf{p}_{N_t})\}, \qquad (4.17)$$

where $\mathbf{m}_i$ represents the measurement vector for the ith position $\mathbf{p}_i$ ($\mathbf{p}_i = [x_i \ y_i]^T$ for two-dimensional positioning), and N_t is the total number of elements in the training set (the "size" of the database). For example, the measurement vector $\mathbf{m}_i$ can consist of RSS measurements at a number of reference nodes when the target is at $\mathbf{p}_i$. The problem is

---

[12] It is also possible to form a database in real-time by using measurements among reference nodes and those between target and reference nodes, and then to employ mapping techniques for position estimation [154].

4.2 Position estimation

to estimate the position **p** of a target node, by utilizing the training set in (4.17), when the measurements **m** are collected related to that target node.

One of the simplest regression techniques is to estimate the position of the target node as the position vector in the training set $\mathcal{T}$ corresponding to the measurement vector that has the shortest distance to the measurement vector **m**. In other words, the position is estimated as $\mathbf{p}_j$, with

$$j = \arg \min_{i \in \{1,\ldots,N_t\}} \|\mathbf{m} - \mathbf{m}_i\|, \quad (4.18)$$

where $\|\mathbf{m} - \mathbf{m}_i\|$ represents the Euclidean distance between **m** and $\mathbf{m}_i$.

k-NN estimation is a generalization of the approach in (4.18), which estimates the position of the target node according to the k closest measurements in set $\mathcal{T}$ to the measurement vector **m**. The position estimate $\hat{\mathbf{p}}$ is obtained by a weighted sum of positions corresponding to the k closest measurements as

$$\hat{\mathbf{p}} = \sum_{i=1}^{k} w_i(\mathbf{m}) \mathbf{p}^{(i)}, \quad (4.19)$$

where $\mathbf{p}^{(1)}, \ldots, \mathbf{p}^{(k)}$ are the locations corresponding to the k nearest measurement vectors, $\mathbf{m}^{(1)}, \ldots, \mathbf{m}^{(k)}$, to **m**, and $w_1(\mathbf{m}), \ldots, w_k(\mathbf{m})$ are the weighting factors for each position. In general, the weights are determined as a function of the measurement vector **m** and the training measurements $\mathbf{m}^{(1)}, \ldots, \mathbf{m}^{(k)}$. Various weighting functions are studied in [155]. For the uniform weighting scheme, each position is weighted equally, in which case $\hat{\mathbf{p}}$ becomes the sample mean of $\mathbf{p}^{(1)}, \ldots, \mathbf{p}^{(k)}$, that is,

$$\hat{\mathbf{p}} = \frac{1}{k} \sum_{i=1}^{k} \mathbf{p}^{(i)}. \quad (4.20)$$

The SVR [156] and neural network [153] approaches for position estimation can also be considered as different versions of the k-NN estimation in (4.19). The main idea is to perform accurate regression for a given set of training data. For example, the SVR approach first maps the measurements into a higher dimensional feature space and performs linear regression in that space. This corresponds to non-linear regression in the original space [159]. The weights for the linear regression in the feature space are determined by minimizing a combination of empirical error and regressor complexity.[13]

The main advantage of mapping techniques is that they can provide accurate position estimation in challenging environments with significant multipath and NLOS propagation. However, the database should be large enough and representative of the current environment. In other words, the database should be updated before the channel characteristics change significantly, which can be very costly in dynamic environments. Therefore, mapping techniques are not commonly employed for outdoor positioning applications.

---

[13] Very complex regressors fit the training data very closely and therefore may not fit to new measurements well enough if the size of the training set is not sufficiently large. This is called a "generalization" problem. An SVR can provide better generalization, since it imposes a constraint on the complexity of the regressor.

4.2.2 Geometric and statistical techniques

In the absence of a database consisting of previously taken measurements at known positions, the position of the target node should be estimated directly from the available measurements obtained by the first step in Fig. 4.1(b). Such positioning techniques can be considered in two groups, geometric and statistical techniques.

Geometric techniques

A geometric positioning technique solves for the position of the target node as the intersection of position lines obtained from a set of measurements at a number of reference nodes. For example, as studied in Section 4.1, a range measurement (obtained from a TOA or an RSS measurement) determines a position line for the target's location as a circle around the reference node (Fig. 4.2). Then, the intersection of three position lines, obtained from three TOA or RSS measurements, can be used to solve for the position of the target as shown in Fig. 4.12. Determination of the target position from a set of range measurements is called *trilateration*.

Let d_1, d_2 and d_3 represent the range measurements obtained from three TOA or RSS measurements. Then, the following three equations must be solved jointly in order to estimate the position of the target via trilateration:

$$d_i = \sqrt{(x_i - x)^2 + (y_i - y)^2}, \quad i = 1, 2, 3, \qquad (4.21)$$

where (x_i, y_i) is the known position of the ith reference node, and (x, y) is the position of the target node. The position (x, y) can be solved from (4.21) as

$$x = \frac{(y_2 - y_1)\gamma_1 + (y_2 - y_3)\gamma_2}{2[(x_2 - x_3)(y_2 - y_1) + (x_1 - x_2)(y_2 - y_3)]}, \qquad (4.22)$$

$$y = \frac{(x_2 - x_1)\gamma_1 + (x_2 - x_3)\gamma_2}{2[(x_2 - x_1)(y_2 - y_3) + (x_2 - x_3)(y_1 - y_2)]}, \qquad (4.23)$$

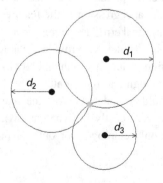

Fig. 4.12. The reference (black) nodes measure their distances (via RSS or TOA estimation) from the target node (gray node), which results in three circles passing through the target node. The intersection of the three circles can be calculated to obtain the position of the target node, which is called trilateration (with kind permission from Springer Science and Business Media) [135].

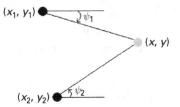

Fig. 4.13. The angles measured by the reference (black) nodes determine two lines, the intersection of which yields the target position. This technique is called *triangulation*.

where

$$\gamma_1 = x_2^2 - x_3^2 + y_2^2 - y_3^2 + d_3^2 - d_2^2, \quad (4.24)$$
$$\gamma_2 = x_1^2 - x_2^2 + y_1^2 - y_2^2 + d_2^2 - d_1^2. \quad (4.25)$$

For AOA measurements, two reference nodes are sufficient to determine the position of the target node by intersecting two lines, which is called *triangulation* (Fig. 4.13). Let ψ_1 and ψ_2 denote the angles measured by reference node 1 and 2, respectively. Then, the following two equations are solved for the position of the target:

$$\tan \psi_1 = \frac{y - y_1}{x - x_1} \quad \text{and} \quad \tan \psi_2 = \frac{y - y_2}{x - x_2}, \quad (4.26)$$

which yields

$$x = \frac{x_2 \tan \psi_2 - x_1 \tan \psi_1 + y_1 - y_2}{\tan \psi_2 - \tan \psi_1}, \quad (4.27)$$

$$y = \frac{(x_2 - x_1) \tan \psi_1 \tan \psi_2 + y_1 \tan \psi_2 - y_2 \tan \psi_1}{\tan \psi_2 - \tan \psi_1}. \quad (4.28)$$

In the case of TDOA-based positioning, each TDOA measurement determines a hyperbola for the position of the target node. For three reference nodes, two range differences (obtained from TDOA measurements) can be expressed as follows:

$$d_{i1} \triangleq d_i - d_1 = \sqrt{(x - x_i)^2 + (y - y_i)^2} - \sqrt{(x - x_1)^2 + (y - y_1)^2}, \quad (4.29)$$

for $i = 2, 3$, which define two hyperbolas as shown in Fig. 4.14.

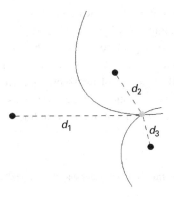

Fig. 4.14. Positioning via TDOA measurements (with kind permission from Springer Science and Business Media) [135].

The position of the target node can be obtained from the two equations in (4.29) and from the relation $d_1 = \sqrt{(x - x_1)^2 + (y - y_1)^2}$. Note that there are three unknowns in this case, x, y, and d_1. The equations in (4.29) can be expressed as two linear relations in terms of these three unknowns [160]:

$$\begin{bmatrix} x_1 - x_2 & y_1 - y_2 \\ x_1 - x_3 & y_1 - y_3 \end{bmatrix} \begin{bmatrix} x - x_1 \\ y - y_1 \end{bmatrix} = d_1 \begin{bmatrix} d_{21} \\ d_{31} \end{bmatrix} + \frac{1}{2} \begin{bmatrix} d_{21}^2 - \tilde{d}_{21}^2 \\ d_{31}^2 - \tilde{d}_{31}^2 \end{bmatrix}, \quad (4.30)$$

where $\tilde{d}_{i1}^2 = (x_i - x_1)^2 + (y_i - y_1)^2$ for $i = 2, 3$.

Combination of (4.30) and $d_1 = \sqrt{(x - x_1)^2 + (y - y_1)^2}$ yields a quadratic equation for d_1. Under geometric regularity conditions, a unique value of d_1 can be determined and the position of the target node, (x, y), can be obtained [160].

The geometric techniques can also be applied to hybrid systems, in which multiple types of measurements, such as TDOA/AOA [161] or TOA/TDOA [162], are employed in position determination. As a simple example, consider the hybrid AOA/TOA system in Fig. 4.11, where the reference node can measure both AOA and TOA of the signal from the target node. In this case, the position of the target node can be calculated as

$$x = x_1 + d \cos \psi, \quad (4.31)$$
$$y = y_1 + d \sin \psi, \quad (4.32)$$

where ψ is the AOA, and d is the range obtained via TOA estimation.

As another example of a hybrid positioning system, consider a system with three reference nodes, as in Fig. 4.12, in which two reference nodes are performing TOA estimation and the remaining one is performing RSS estimation. In such a case, the geometric solution is no different from the trilateration approach based on ranges from all TOA or all RSS measurements. However, in practice, the accuracy of TOA and RSS measurements can differ (since the measurements include noise in reality); hence, a different positioning strategy from the conventional trilateration can perform better in practice, as will be studied in the next section.

The previous case is in fact an example of the limitations of the geometric approach. Note that for the geometric solutions in this section, it is assumed that all the measurements are error free. Therefore, there always exists a single point at which all the position lines, obtained from a number of measurements, intersect. However, the measurements include noise (random errors) in practice, and the position lines may intersect at multiple points without intersecting altogether at a single point.[14] In such a case, the geometric approach does not provide any insight as to which point to choose as the position of the target node. In addition, if there are more measurements than needed (for unambiguous position estimation in the error-free case), the number of intersections increases even further. For example, Fig. 4.15 illustrates three erroneous AOA measurements from three reference nodes, which results in multiple intersections of the position lines, without all three lines intersecting at a single point.

---

[14] One exception is the intersection of two lines obtained from AOA measurements from two reference nodes, which *always* intersect at a single point.

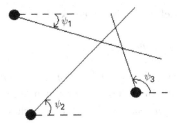

Fig. 4.15. Erroneous AOA measurements and position ambiguity (with kind permission from Springer Science and Business Media) [135].

Finally, note that since no error is assumed in the measurements, three reference nodes for TOA, RSS and TDOA-based positioning (two for AOA-based positioning) are used in the geometric solutions as more measurements would not change the estimated position. However, in a practical case of noisy measurements, more accurate position estimation can be performed by using a larger number of reference nodes. The geometric techniques do not provide a theoretical basis for such data fusion approaches. Therefore, statistical approaches should be considered, which provide a theoretical framework for position estimation using any number of measurements.

Statistical techniques

Due to the limitations of the geometric approaches, statistical positioning techniques are employed in most practical cases. In order to formulate a generic framework for statistical approaches, consider the following measurement model

$$z_i = f_i(x, y) + \eta_i, \quad i = 1, \ldots, N_m, \tag{4.33}$$

where N_m is the number of measurements, η_i is the noise at the ith measurement, and $f_i(x, y)$ is the true value of the ith signal parameter, which is a function of the position of the target, (x, y). Note that N_m is equal to the number of reference nodes for RSS, AOA and TOA-based positioning, and it is one less than the number of reference nodes for TDOA-based positioning, since each TDOA measurement is obtained with respect to a reference node.

For various positioning systems, $f_i(x, y)$ in (4.33) can be expressed as[15]

$$f_i(x, y) = \begin{cases} \sqrt{(x - x_i)^2 + (y - y_i)^2}, & \text{TOA/RSS} \\ \tan^{-1}((y - y_i)/(x - x_i)), & \text{AOA} \\ \sqrt{(x - x_i)^2 + (y - y_i)^2} - \sqrt{(x - x_0)^2 + (y - y_0)^2}, & \text{TDOA} \end{cases}, \tag{4.34}$$

where (x_i, y_i) is the position of the ith reference node and (x_0, y_0) is the position of the reference node relative to which the TDOA measurements are obtained.

---

[15] Time measurements are converted to distance measurements by scaling by the speed of light.

In vector notations, the measurement model in (4.33) can be expressed as

$$\mathbf{z} = \mathbf{f}(x, y) + \boldsymbol{\eta}, \tag{4.35}$$

where $\mathbf{z} = [z_1 \cdots z_{N_m}]^T$, $\mathbf{f}(x, y) = [f_1(x, y) \cdots f_{N_m}(x, y)]^T$ and $\boldsymbol{\eta} = [\eta_1 \cdots \eta_{N_m}]^T$.

Assume that the probability density function of the noise η is known except for a set of parameters, denoted by λ. In such a case, *parametric* approaches, such as Bayesian and maximum likelihood (ML) estimation, can be employed.[16]

Let the vector of unknown parameters be denoted by $\boldsymbol{\theta}$, which consists of the position of the target node, as well as the unknown parameters of the noise distribution;[17] i.e. $\boldsymbol{\theta} = [x \ y \ \lambda^T]^T$. Depending on the availability of prior information about $\boldsymbol{\theta}$, Bayesian or ML estimation techniques can be applied [165].

In the presence of prior information about $\boldsymbol{\theta}$, which is represented by a prior probability distribution $\pi(\boldsymbol{\theta})$, the Bayesian approach can be taken to obtain an estimate of $\boldsymbol{\theta}$ that minimizes a certain cost function [140]. Two common Bayesian estimators are the minimum mean-square error (MMSE) and the maximum *a posteriori* (MAP) estimators,[18] which estimate $\boldsymbol{\theta}$ as follows:

$$\hat{\boldsymbol{\theta}}_{\text{MMSE}} = \text{E}\{\boldsymbol{\theta}|\mathbf{z}\}, \tag{4.36}$$

$$\hat{\boldsymbol{\theta}}_{\text{MAP}} = \arg \max_{\boldsymbol{\theta}} p(\mathbf{z}|\boldsymbol{\theta}) \pi(\boldsymbol{\theta}), \tag{4.37}$$

where $\text{E}\{\boldsymbol{\theta}|\mathbf{z}\}$ is the conditional expectation of $\boldsymbol{\theta}$ given $\mathbf{z}$, and $p(\mathbf{z}|\boldsymbol{\theta})$ represents the probability density function of $\mathbf{z}$ conditioned on $\boldsymbol{\theta}$.

In some cases, there is no prior information about $\boldsymbol{\theta}$. In such cases, the ML estimation is commonly used, which finds the value of $\boldsymbol{\theta}$ that maximizes the likelihood function; i.e.

$$\hat{\boldsymbol{\theta}}_{\text{ML}} = \arg \max_{\boldsymbol{\theta}} p(\mathbf{z}|\boldsymbol{\theta}). \tag{4.38}$$

Note from (4.38) and (4.37) that the ML estimator assumes a uniform distribution for the unknown parameter $\boldsymbol{\theta}$ (sort of a worst-case prior) since there is no specific information about which parameter values are more likely than the others.

Since $\mathbf{f}(x, y)$ is a deterministic function, the likelihood function can be expressed as

$$p(\mathbf{z}|\boldsymbol{\theta}) = p_{\eta}(\mathbf{z} - \mathbf{f}(x, y)|\boldsymbol{\theta}), \tag{4.39}$$

---

[16] In the absence of information about the form of the probability density function of η, *nonparametric* techniques need to be employed. The k-NN, SVR and neural networks approaches studied in Section 4.2.1 are examples of non-parametric estimators. Although the form of the density function is unknown in the nonparametric case, there can still be generic information about certain parameters of the unknown distribution [163], such as its variance and symmetry properties, which can be used to design non-parametric estimation rules, such as the least median of squares technique in [164], the residual weighting algorithm in [16] and the variance weighted least squares technique in [4].

[17] In general, the noise components can also depend on the position of the mobile, in which case $\boldsymbol{\theta}$ includes the union of the elements x, y, and λ.

[18] In fact, the MAP estimation is not properly a Bayesian approach; but it still fits within the Bayesian framework [140].

where $p_\eta(\cdot\,|\,\boldsymbol{\theta})$ represents the conditional probability density function of the noise vector for the given parameters in $\boldsymbol{\theta}$. In the following, the ML estimators are investigated for various scenarios.

Case 1: Independent noise components: If the noise components are independent, the likelihood function in (4.39) can be expressed as

$$p(\mathbf{z}|\boldsymbol{\theta}) = \prod_{i=1}^{N_m} p_{\eta_i}(z_i - f_i(x,y)\,|\,\boldsymbol{\theta}), \qquad (4.40)$$

where $p_{\eta_i}(\cdot\,|\,\boldsymbol{\theta})$ represents the conditional probability density function for the ith noise component given the parameter vector $\boldsymbol{\theta}$.

The independent noise assumption is usually valid for TOA, RSS and AOA measurements. However, for TDOA measurements the noise components are correlated since each TDOA is computed with respect to the same reference node. Therefore, a TDOA-based system can be studied through the generic expression in (4.39), or by using a correlated Gaussian model (Case-2, below) under certain conditions.

For TOA, RSS and AOA-based systems in LOS conditions, the parameters of the noise can be assumed to be known, since the measurement noise is mainly due to thermal noise in an LOS case. In that case, the unknown parameter vector reduces to $\boldsymbol{\theta} = [x\ y]^T$. Also, it is possible to (approximately) model each noise component by a zero mean Gaussian random variable in LOS scenarios [139]; i.e.

$$p_{\eta_i}(n) = \frac{1}{\sqrt{2\pi}\,\sigma_i} \exp\left(-\frac{n^2}{2\sigma_i^2}\right). \qquad (4.41)$$

Then, the likelihood function in (4.40) can be expressed as

$$p(\mathbf{z}|\boldsymbol{\theta}) = \frac{1}{(2\pi)^{N_m/2} \prod_{i=1}^{N_m} \sigma_i} \exp\left(-\sum_{i=1}^{N_m} \frac{(z_i - f_i(x,y))^2}{2\sigma_i^2}\right). \qquad (4.42)$$

From (4.42), the ML estimator in (4.38) can be obtained as

$$\hat{\boldsymbol{\theta}}_{\mathrm{ML}} = \arg\min_{[x\ y]^T} \sum_{i=1}^{N_m} \frac{(z_i - f_i(x,y))^2}{\sigma_i^2}, \qquad (4.43)$$

which is the non-linear[19] least-squares (NLS) estimator commonly used for position estimation in the literature [136]. Note that the weights are inversely proportional to the variance of the measurements since a larger variance means a less reliable measurement. Common techniques for solving (4.43) include gradient descent algorithms and linearization techniques using Taylor series expansion [18, 136].

In the absence of LOS propagation between the target node and some reference nodes, the noise model can be significantly different for the measurements related to those

---

[19] Since $f_i(x,y)$ is a non-linear function of (x,y).

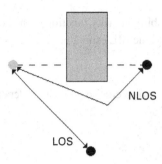

Fig. 4.16.   LOS and NLOS scenarios. The direct signal path is blocked in the NLOS case.

reference nodes compared to the measurements at the LOS nodes. If the positions of the reference nodes are sufficiently separated, the conditional independence assumption in (4.40) can still be employed. Therefore, the ML position estimation can be obtained from (4.38) and (4.40) by using appropriate noise distributions for LOS and NLOS reference nodes.[20]

In the NLOS case, the noise distribution can be considerably different from the Gaussian model in (4.41) in many cases.[21] As an example, for TOA estimation, the range measurements in the absence of LOS propagation contain an NLOS error, in addition to the errors observed in LOS measurements which are mainly due to background noise. The NLOS error is always positive since the first arriving signal component travels an extra distance as shown in Fig. 4.16. Therefore, the measurement noise can be modeled as the sum of two noise terms, one related to the background noise, and the other related to the NLOS error. Common models for the NLOS error include Gamma distribution [139] and distributions based on certain scattering models [169]. In many cases, measurement errors due to NLOS propagation dominate measurement errors due to background noise. Therefore, the measurement error due to background noise can be omitted for NLOS measurements.

Example 4.1 *Consider a TOA-based positioning scenario with N_m reference nodes, N_L of which are in LOS to the target node while the remaining ones are in NLOS. Let z_i, for $i = 1, \ldots, N_L$, denote the range measurement related to the ith LOS reference node, which includes a zero mean Gaussian measurement noise with variance σ_i^2. Similarly, z_i, for $i = N_L + 1, \ldots, N_m$, denote the range measurement related to the $(i - N_L)$th NLOS reference node, which includes an exponentially distributed measurement noise[22] with a rate parameter λ_{i-N_L}.*

---

[20] It is assumed that information about which nodes are LOS and which are NLOS is available. Such information can be obtained via NLOS detection algorithms [119], [166–68].

[21] The errors due to NLOS propagation can also be considered as fixed unknown bias terms. However, it is shown in [139] that in the absence of statistical information about NLOS errors, the measurements from NLOS nodes do not contribute to the positioning accuracy; hence they can be discarded.

[22] Exponential distribution is a special case of Gamma distribution.

Assuming independent noise at different measurements, the ML position estimate can be obtained as

$$\hat{\theta}_{\text{ML}} = \arg\max_{\theta} p(\mathbf{z}|\theta) \tag{4.44}$$

$$= \arg\max_{\theta} \left\{ \prod_{i=1}^{N_L} \frac{1}{\sqrt{2\pi}\sigma_i} \exp\left\{-\frac{1}{2\sigma_i^2}(z_i - f_i(x,y))^2\right\} \right.$$

$$\left. \times \prod_{i=N_L+1}^{N_m} \lambda_{i-N_L} \exp\left\{-\lambda_{i-N_L}(z_i - f_i(x,y))\right\} \right\}, \tag{4.45}$$

where $f_i(x,y) = \sqrt{(x-x_i)^2 + (y-y_i)^2}$.

For known noise parameters, the ML estimate can be expressed as

$$\hat{\theta}_{\text{ML}} = \arg\min_{[x\ y]^T} \left\{ \sum_{i=1}^{N_L} \frac{1}{2\sigma_i^2}(z_i - f_i(x,y))^2 + \sum_{i=N_L+1}^{N_m} \lambda_{i-N_L}(z_i - f_i(x,y)) \right\}. \tag{4.46}$$

Case 2: Correlated Gaussian noise components: if the noise vector is modeled as a multivariate Gaussian random variable with mean μ and covariance matrix Σ, the likelihood function is expressed as

$$p(\mathbf{z}|\theta) = \frac{1}{(2\pi)^{N_m/2}|\Sigma|^{1/2}} \exp\left\{-\frac{1}{2}(\mathbf{z} - \mathbf{f}(x,y) - \mu)^T \Sigma^{-1}(\mathbf{z} - \mathbf{f}(x,y) - \mu)\right\}. \tag{4.47}$$

Then, the ML position estimate can be calculated as

$$\hat{\theta}_{\text{ML}} = \arg\min_{\theta} \left\{ (\mathbf{z} - \mathbf{f}(x,y) - \mu)^T \Sigma^{-1}(\mathbf{z} - \mathbf{f}(x,y) - \mu) + \log|\Sigma| \right\}, \tag{4.48}$$

where θ includes the position of the target node and the unknown parameters related to μ and Σ.

For a noise distribution with zero mean and a known covariance matrix, (4.48) can be simplified to

$$\hat{\theta}_{\text{ML}} = \arg\min_{[x\ y]^T} (\mathbf{z} - \mathbf{f}(x,y))^T \Sigma^{-1}(\mathbf{z} - \mathbf{f}(x,y)), \tag{4.49}$$

which is called the weighted LS (WLS) solution [136].

Although the independent noise model in Case 1 is not well-suited for TDOA-based systems, the correlated Gaussian noise model in (4.47) can represent such systems quite accurately for sufficiently large SNRs.

Example 4.2 *Consider a set of range difference (equivalently, TDOA) measurements modeled as*

$$z_i = d_i - d_0 + n_i - n_0, \quad i = 1, \ldots, N_m, \quad (4.50)$$

where $d_j = \sqrt{(x - x_j)^2 + (y - y_j)^2}$ *for* $j = 0, 1, \ldots, N_m$, *and* $n_0, n_1, \ldots, n_{N_m}$ *are zero mean independent Gaussian random variables with variances* $\sigma_0^2, \sigma_1^2, \ldots, \sigma_{N_m}^2$, *respectively. Then, the measurement model can be expressed in vector notation as*

$$\mathbf{z} = \mathbf{f}(x, y) + \boldsymbol{\eta}, \quad (4.51)$$

where $f_i(x, y) = d_i - d_0$ *for* $i = 1, \ldots, N_m$, *and* $\boldsymbol{\eta} \sim \mathcal{N}(\mathbf{0}, \boldsymbol{\Sigma})$ *with*

$$\boldsymbol{\Sigma} = \begin{bmatrix} \sigma_1^2 + \sigma_0^2 & \sigma_0^2 & \cdots & \sigma_0^2 \\ \sigma_0^2 & \sigma_2^2 + \sigma_0^2 & \ddots & \vdots \\ \vdots & \ddots & \ddots & \sigma_0^2 \\ \sigma_0^2 & \cdots & \sigma_0^2 & \sigma_{N_m}^2 + \sigma_0^2 \end{bmatrix}. \quad (4.52)$$

Note that the measurement model in (4.50) expresses each TDOA as the difference of two TOA measurements. Therefore, there is a correlation between different TDOA measurements, and the covariance matrix in (4.52) is not diagonal.

In the case of NLOS propagation between the target node and a number of reference nodes, the Gaussian model in (4.47) may be inaccurate. Therefore, the generic ML estimation should be performed by means of (4.38) and (4.39) for TDOA measurements in NLOS scenarios.

Recall from Section 4.1.4 that the TOA-based approach to TDOA estimation involves measurement of individual TOAs at two reference nodes and then the estimation of TDOA as the difference between the two TOA measurements, as modeled by (4.50) in Example 4.2. Note that this is different from the cross-correlation-based TDOA estimation, which determines the TDOA directly by locating the peak of the cross-correlation between the received signals at two reference nodes. For the former case, an independent measurement model and a related ML solution can be formulated as shown in the following example.

Example 4.3 *For the TOA-based TDOA scheme, the range measurements at each reference node can be expressed as*

$$z_i = \sqrt{(x - x_i)^2 + (y - y_i)^2} + d_{\text{offset}} + \eta_i, \quad (4.53)$$

for $i = 1, \ldots, N_m$, *where* $d_{\text{offset}} = c \tau_{\text{offset}}$ *is the range bias due to the timing offset between the target node and the reference nodes (which are synchronized among themselves), and* η_i *is the noise for the ith measurement. Note that the noise components can be considered independent, since each* z_i *is obtained as in the TOA case. If position* estimation is directly obtained from the N_m measurements in (4.53) instead of calculating time differences with respect to one of the measurements, the ML solution can be obtained as

$$\hat{\theta}_{ML} = \arg \max_{[x \ y \ d_{offset}]^T} \prod_{i=1}^{N_m} p_{\eta_i}\left(z_i - \sqrt{(x-x_i)^2 + (y-y_i)^2} - d_{offset}\right), \quad (4.54)$$

where d_{offset} can be considered as an unknown parameter of the measurement noise.

For the Gaussian noise model with zero mean and variance σ_i^2 for the ith measurement, the ML solution can be expressed as

$$\hat{\theta}_{ML} = \arg \min_{[x \ y \ d_{offset}]^T} \sum_{i=1}^{N_m} \frac{1}{\sigma_i^2}\left(z_i - \sqrt{(x-x_i)^2 + (y-y_i)^2} - d_{offset}\right)^2. \quad (4.55)$$

Note that given a set of N_m TOA measurements with a common timing offset, as shown in (4.53), the optimal ML solution is given by (4.54), which jointly estimates the timing offset and the position of the target node. On the other hand, the TDOA-based positioning first obtains $N_m - 1$ time difference measurements with respect to one of the measurements, and then estimates the position based on those TDOAs via the ML solution in (4.48). In other words, the TDOA-based scheme has the benefit of not estimating the timing offset. It is shown in [139] that the TDOA-based scheme can achieve the same accuracy as the TOA-based solution in (4.54); hence, there is no need to perform joint estimation of timing offset and the position of the target node.

4.2.3 Evaluation of positioning accuracy

In order to evaluate the performance of the positioning algorithms studied in the previous sections, and to obtain theoretical lower bounds on positioning accuracy for a given system, certain measures for accuracy are defined.

CRLB

The CRLB sets a limit on the covariance matrix of any unbiased estimator [140]. It is calculated as the inverse of the Fisher information matrix (FIM) related to the conditional probability density function of the measurements given the unknown parameters, $p(\mathbf{z}|\theta)$. In other words,[23]

$$E\left\{(\hat{\theta} - \theta)(\hat{\theta} - \theta)^T\right\} \geq \mathbf{I}_\theta^{-1}, \quad (4.56)$$

where $\hat{\theta}$ is an unbiased estimate of θ, and $\mathbf{I}_\theta$ is the FIM given by

$$\mathbf{I}_\theta = E\left\{\frac{\partial \log p(\mathbf{z}|\theta)}{\partial \theta}\left(\frac{\partial \log p(\mathbf{z}|\theta)}{\partial \theta}\right)^T\right\}. \quad (4.57)$$

---

[23] For symmetric matrices $\mathbf{I}_1$ and $\mathbf{I}_2$ of the same size, $\mathbf{I}_1 \geq \mathbf{I}_2$ means that $\mathbf{I}_1 - \mathbf{I}_2$ is positive semi-definite.

Commonly, mean-square error (MSE) is used as a metric to compare performance of various positioning algorithms, which is lower bounded by the CRLB in (4.56) as follows:

$$\text{MSE} \triangleq \text{E}\{||\hat{\boldsymbol{\theta}} - \boldsymbol{\theta}||^2\} = \text{trace}\left[\text{E}\left\{(\hat{\boldsymbol{\theta}} - \boldsymbol{\theta})(\hat{\boldsymbol{\theta}} - \boldsymbol{\theta})^T\right\}\right] \qquad (4.58)$$

$$\geq \text{trace}\left[\mathbf{I}_{\boldsymbol{\theta}}^{-1}\right] \triangleq \text{MMSE}. \qquad (4.59)$$

Consider a system with N_m reference nodes, N_L of which are in LOS with the target node, and the remaining ones are in NLOS. Without loss of generality, the LOS and the NLOS nodes are indexed by $i=1,\ldots,N_L$ and $i=N_L+1,\ldots,N_m$, respectively. It is assumed that signals propagate through a single LOS or NLOS path,[24] and there is no prior statistical information related to NLOS induced errors. Let $\psi_i = \tan^{-1}((y - y_i)/(x - x_i))$, for $i = 1,\ldots,N_m$, denote the angle between the ith reference node and the target node, where (x, y) is the position of the target node and (x_i, y_i) is the position of the ith reference node.

It is shown in [138] that the MMSE for TOA-based positioning can be expressed as

$$\text{MMSE}_{\text{TOA}} = \frac{c^2 \sum_{i=1}^{N_L} \sigma_i^{-2}}{\sum_{i=1}^{N_L} \sum_{j=1}^{i-1} \sigma_i^{-2}\sigma_j^{-2} \sin^2(\psi_i - \psi_j)}, \qquad (4.60)$$

where c is the speed of light and σ_i^2, for $i = 1,\ldots,N_L$, is the variance of the zero mean Gaussian measurement noise in the LOS case.

The theoretical lower bound in (4.60) is independent of the NLOS measurements in the absence of statistical information about NLOS errors. In addition, the geometric configuration of the LOS nodes affects the theoretical limit through the sine functions of the angle differences related to the LOS nodes.

For sufficiently large SNR and/or effective bandwidth β, σ_i^{-2} can be approximated by $8\pi^2\beta^2\text{SNR}_i$, where SNR_i is the SNR at the ith node [139]. In that case, (4.60) can be expressed as

$$\text{MMSE}_{\text{TOA}} = \frac{c^2}{8\pi^2\beta^2} \frac{\sum_{i=1}^{N_L} \text{SNR}_i}{\sum_{i=1}^{N_L} \sum_{j=1}^{i-1} \text{SNR}_i \text{SNR}_j \sin^2(\psi_i - \psi_j)}, \qquad (4.61)$$

which explicitly indicates the impacts of the effective bandwidth on the MMSE.

It can be shown that an ML estimator based on LOS delay measurements obtained from matched filter (or correlation) receivers can attain the MMSE for sufficiently large SNR and/or effective bandwidth [138]. This asymptotically optimum structure is shown in Fig. 4.17.

---

[24] Refer to [170] for theoretical limits of positioning in multipath channels.

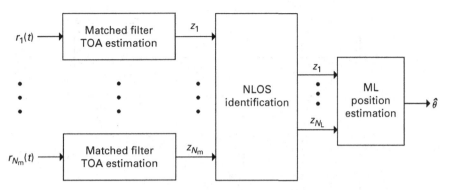

Fig. 4.17. Asymptotically optimal position estimation for the TOA-based scheme in the absence of statistical NLOS information (with kind permission from Springer Science and Business Media) [135].

Example 4.4 *Consider the positioning scenario in Fig. 4.18, where the target node is surrounded by six reference nodes that are uniformly located on a circle. Nodes use UWB pulses in (2.5) with 2 ns pulse width. It is assumed that the SNR is the same for all reference nodes. Hence, (4.61) can be simplified as*

$$\text{MMSE}_{\text{TOA}} = \frac{c^2 N_{\text{L}}}{8\pi^2 \beta^2 \text{SNR} \sum_{i=1}^{N_{\text{L}}} \sum_{j=1}^{i-1} \sin^2(\psi_i - \psi_j)}, \quad (4.62)$$

where SNR represents the common SNR for all reference nodes. After calculating the effective bandwidth of the UWB pulse from (4.11) (cf. Exercise 2), MMSE$_{TOA}$ can be evaluated for various scenarios. In Fig. 4.19, the square root of the MMSE expression in (4.62) is plotted for various numbers of NLOS nodes. Namely, for $N_L = 3$, nodes 4, 5 and 6; for $N_L = 4$, nodes 5 and 6; and for $N_L = 5$, node 6 are the NLOS nodes.

It is observed from the figure that as the number of LOS nodes or the SNR increases, the accuracy of the positioning system increases.

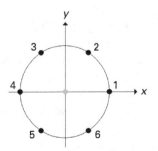

Fig. 4.18. A positioning scenario, in which six reference nodes are estimating the position of the target node in the middle via TOA measurements (with kind permission from Springer Science and Business Media) [135].

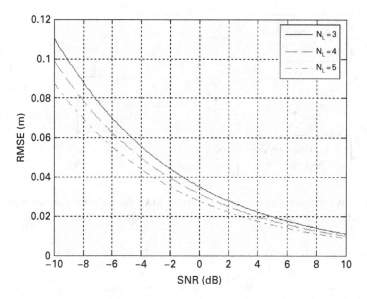

Fig. 4.19. Square root of the MMSE expression in (4.61), called RMSE, versus SNR for various numbers of NLOS nodes. For $N_L = 3$, nodes 4, 5 and 6; for $N_L = 4$, nodes 5 and 6; and for $N_L = 5$, node 6 are the NLOS nodes.

The MMSE for TDOA-based positioning can be expressed as [138]

$$\text{MMSE}_{\text{TDOA}} = c^2 \sum_{i=1}^{N_L} \sigma_i^{-2}$$

$$\times \frac{\left(\sum_{i=1}^{N_L} \sigma_i^{-2}\right)^2 - \sum_{i=1}^{N_L} \sigma_i^{-4} - \sum_{i=1}^{N_L} \sum_{j=1}^{i-1} \sigma_i^{-2} \sigma_j^{-2} \cos(\psi_i - \psi_j)}{\sum_{i=1}^{N_L} \sum_{j=1}^{i-1} \sigma_i^{-2} \sigma_j^{-2} K_{i,j}}, \quad (4.63)$$

where $K_{i,j}$ is given by

$$K_{i,j} = \sin(\psi_i - \psi_j) \sum_{k=1}^{N_L} \sigma_k^{-2} + \sum_{k=1}^{N_L} \sigma_k^{-2} \sin(\psi_i - \psi_k) + \sum_{k=1}^{N_L} \sigma_k^{-2} \sin(\psi_j - \psi_k). \quad (4.64)$$

Note that the MMSE is independent of the measurements related to NLOS nodes, as for the TOA-based positioning. Also, it is shown in [171] that the MMSE for TDOA-based positioning is always larger than or equal to that for TOA-based positioning. This is expected since there is an additional unknown parameter, timing offset, in TDOA-based positioning.

For RSS-based positioning systems, the MMSE is expressed as [138]

$$\text{MMSE}_{\text{RSS}} = \left(\frac{\ln 10}{10n}\right)^2 \frac{\sum_{i=1}^{N_m} \sigma_{\text{sh},i}^{-2} d_i^{-2}}{\sum_{i=1}^{N_m} \sum_{j=1}^{i-1} \sigma_{\text{sh},i}^{-2} \sigma_{\text{sh},j}^{-2} d_i^{-2} d_j^{-2} \sin^2(\psi_i - \psi_j)}, \quad (4.65)$$

where n is the PL exponent, $\sigma_{sh,i}^2$ is the variance of the log-normal shadowing for the ith measurement, and $d_i = \sqrt{(x - x_i)^2 + (y - y_i)^2}$ is the distance between the target node and the ith reference node. Note that the accuracy of RSS-based positioning depends heavily on the channel parameters, namely the PL exponent and the shadowing variances. Also, the accuracy depends on estimates at all nodes, LOS and NLOS, since the effects of NLOS propagation are implicitly included in the RSS signal model, as studied in Section 4.1.1.

Finally, for AOA-based positioning using a ULA, the MMSE can be expressed as [138]

$$\text{MMSE}_{AOA} = \frac{3}{4\pi^2 l^2 N_a (N_a + 1)(2N_a + 1)} \times \frac{\sum_{i=1}^{N_L} \frac{\text{SNR}_i}{d_i^2} \sin^2 \psi_i}{\sum_{i=1}^{N_L} \sum_{j=1}^{i-1} \frac{\text{SNR}_i \text{SNR}_j}{d_i^2 d_j^2} \sin^2(\psi_i - \psi_j) \sin^2 \psi_i \sin^2 \psi_j}, \quad (4.66)$$

where N_a is the number of antenna elements and l is the inter-element spacing. Similar to the time-based systems, the AOA-based positioning utilizes the estimates from LOS nodes only.

Similar expressions can be derived for hybrid positioning systems, which can be considered in two categories depending on whether estimation errors for various measurement types are correlated or not. Some lower bound expressions for hybrid systems can be found in [138] and [172].

For the positioning accuracy analysis above, it is assumed that there is no prior statistical information related to NLOS errors. In the presence of such prior information, the accuracy is evaluated by means of the generalized CRLB (G-CRLB) as investigated in [138]. In this case, an asymptotically optimal positioning receiver as shown in Fig. 4.20 can be implemented. Note that measurements from both LOS and NLOS reference nodes are utilized in the presence of NLOS error statistics.

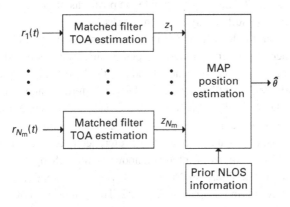

Fig. 4.20. Asymptotically optimal position estimation for the TOA-based scheme in the presence of statistical NLOS information (with kind permission from Springer Science and Business Media) [135].

Geometric dilution of precision

Another performance measure for positioning systems is geometric dilution of precision (GDOP), which is a unitless metric related to the geometric configuration of the reference nodes [173, 174]. Namely, the GDOP is defined as the ratio of the root-mean-square error (RMSE) of the position estimate to the RMSE of the ranging error; i.e.

$$\text{GDOP} = \frac{\text{RMSE}_{\text{pos}}}{\text{RMSE}_{\text{range}}} = \frac{\sqrt{\text{E}\left\{(\hat{\theta} - \hat{\mu})^{\text{T}}(\hat{\theta} - \hat{\mu})\right\}}}{\sigma_{\text{range}}}, \quad (4.67)$$

where $\hat{\theta}$ is the position estimator and $\hat{\mu}$ is the mean of the estimator. σ_{range} represents the fundamental ranging error for RSS, TOA and TDOA-based positioning systems, whereas it is the square root of the average variance of the ranges between the reference nodes and a reference point close to the true position of the target node for AOA-based positioning [136].

For an unbiased estimator, $\hat{\mu}$ is equal to the true parameter θ. Then, (4.67) can be expressed as

$$\text{GDOP} = \frac{\sqrt{\text{trace}\left[\text{Cov}\{\hat{\theta}\}\right]}}{\sigma_{\text{range}}}, \quad (4.68)$$

which reduces to

$$\text{GDOP} = \frac{\sqrt{\sigma_x^2 + \sigma_y^2}}{\sigma_{\text{range}}} \quad (4.69)$$

for $\theta = [x \ y]^{\text{T}}$, where σ_x^2 and σ_y^2 are the mean square errors for the x-axis and y-axis estimates, respectively.

From the definition of the GDOP in (4.67), it is observed that ranging errors do not cause a large positioning error if the GDOP is sufficiently low. However, for large GDOPs, even small ranging errors can cause a large positioning error. Therefore, the GDOP can be useful in network planning by helping determine the positions of the reference nodes so that the GDOP can be kept sufficiently low in most cases.

In [175], it is shown that the lowest GDOP in two-dimensions is given by $2/\sqrt{N_{\text{m}}}$, where N_{m} is the number of reference nodes, for positioning systems based on range measurements (assuming zero mean independent and identically distributed (i.i.d.) measurement errors). That lowest GDOP is achieved when the reference nodes are located at the vertices of a regular N_{m}-sided polygon, with the target node being at the center of that polygon. As an example, consider the node configuration in Fig. 4.21, where the target node is located at the origin and there are three reference nodes each at a distance r from the target node. The positions of reference nodes 1 and 2 change depending on the value of ψ as shown in the figure, and the GDOP values are obtained for various node configurations (i.e. for various ψs) as in Fig. 4.22. The minimum GDOP, $2/\sqrt{3}$, is obtained for $\psi = \pi/6$, when the reference nodes form an equilateral triangle, with the target node being at the center. As the reference nodes come closer, the GDOP increases dramatically.

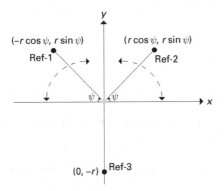

Fig. 4.21. Node configuration for a positioning scenario with the target node located at the origin, and three reference nodes at distance r from the target node.

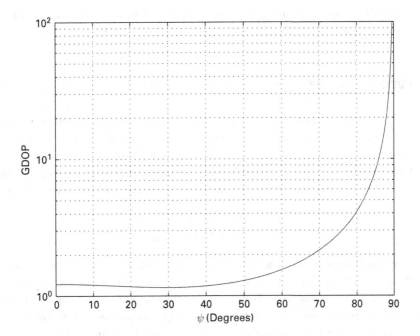

Fig. 4.22. The GDOP for the scenario in Fig. 4.21 for various ψs.

Circular error probability

The accuracy of a position estimator can also be quantified by the radius of a circle centered at the mean of the estimator, inside which half of the position estimates reside [174, 176]. In other words, if a position estimator has a CEP of r meters, each position estimate is inside a circle of radius r meters, with its center at the mean of the estimator, with a probability of 0.5.

Note that for an unbiased position estimator, the CEP is defined with respect to a circle around the true position of the target node, as shown in Fig. 4.23.

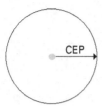

Fig. 4.23. CEP for an unbiased position estimator. With 50% probability, each position estimate is inside a circle of radius CEP around the position of the target node.

Due to the complexity of the exact expressions for CEP, the following approximation is commonly employed [136, 174]:

$$\text{CEP} \approx 0.75 \sqrt{\mathrm{E}\left\{(\hat{\boldsymbol{\theta}} - \hat{\boldsymbol{\mu}})^T (\hat{\boldsymbol{\theta}} - \hat{\boldsymbol{\mu}})\right\}}, \quad (4.70)$$

where $\hat{\boldsymbol{\theta}}$ is the position estimator and $\hat{\boldsymbol{\mu}}$ is its mean. Note from (4.67) and (4.70) that CEP is approximately related to GDOP by $\text{CEP} \approx 0.75 \sigma_{\text{range}} \text{GDOP}$.

4.3 Position tracking

In the previous section, the position of a target node is estimated based on a single observation of signals at a given time. In the presence of multiple observations over a period of time, more accurate position estimation can be performed by making use of a motion model for the target node. The problem of estimating an object's position at consecutive time instants, the *tracking* problem, can be formulated using a state-space approach [177]:

$$\mathbf{x}_k = \mathbf{g}_k(\mathbf{x}_{k-1}, \boldsymbol{v}_{k-1}), \quad (4.71)$$

$$\mathbf{z}_k = \mathbf{f}_k(\mathbf{x}_k) + \boldsymbol{\eta}_k, \quad (4.72)$$

where $\mathbf{x}_k$ is the state vector at time k, $\{\boldsymbol{v}_{k-1}\}_{k=0}^{\infty}$ is an i.i.d. process noise sequence, $\mathbf{z}_k$ is the measurement vector at time k, and $\{\boldsymbol{\eta}_k\}_{k=0}^{\infty}$ is an i.i.d. measurement noise sequence.

In the state equation in (4.71), the state vector $\mathbf{x}_k$ can, for example, be a 4×1 vector consisting of the position and the velocity of the target node at time k for a two-dimensional problem. The measurement equation in (4.72) is similar to the model in (4.35) with the generalization that the state vector can include more elements in addition to the position coordinates.

Note from (4.71) that the state equation describes a Markov process of order one, which facilitates tractable computations [178].

In the Bayesian framework, the tracking problem can be considered as calculating the conditional probability of state $\mathbf{x}_k$ given all the measurements up to time k; i.e. $p(\mathbf{x}_k | \mathbf{z}_1, \ldots, \mathbf{z}_k)$. It is assumed that the prior probability, $p(\mathbf{x}_0)$, is known.[25] Then, the state estimation can be performed recursively for each time instant k in two stages [177]:

---

[25] At time $k = 0$, no measurements are taken yet; hence, $p(\mathbf{x}_0 | \mathbf{z}_0) = p(\mathbf{x}_0)$.

$$p(\mathbf{x}_k|\mathbf{z}_1,\ldots,\mathbf{z}_{k-1}) = \int p(\mathbf{x}_k|\mathbf{x}_{k-1})p(\mathbf{x}_{k-1}|\mathbf{z}_1,\ldots,\mathbf{z}_{k-1})d\mathbf{x}_{k-1}, \tag{4.73}$$

$$p(\mathbf{x}_k|\mathbf{z}_1,\ldots,\mathbf{z}_k) = \frac{p(\mathbf{z}_k|\mathbf{x}_k)p(\mathbf{x}_k|\mathbf{z}_1,\ldots,\mathbf{z}_{k-1})}{\int p(\mathbf{z}_k|\mathbf{x}_k)p(\mathbf{x}_k|\mathbf{z}_1,\ldots,\mathbf{z}_{k-1})d\mathbf{x}_k}, \tag{4.74}$$

where the first equation describes the *prediction stage* and the second equation describes the *update (correction) stage*. In (4.73), $p(\mathbf{x}_k|\mathbf{x}_{k-1})$ is specified by the state equation (4.71) and the known statistics of v_{k-1}. Similarly, $p(\mathbf{z}_k|\mathbf{x}_k)$ in (4.74) is specified by the measurement equation (4.72) and the known statistics of η_k.

In what follows, various optimal and suboptimal approaches to the recursive tracking algorithm, described by (4.73) and (4.74), are considered.

4.3.1 Kalman filters

The Kalman filter [179] assumes that the posterior probability $p(\mathbf{x}_k|\mathbf{z}_1,\ldots,\mathbf{z}_k)$ can be expressed as a Gaussian random variable at each time instant k. This Gaussian approach yields the optimal solution to the tracking problem for linear state and measurement equations, and Gaussian noise vectors v_{k-1} and η_k [177, 180]. Under these assumptions, the state-space equations in (4.71) and (4.72) can be expressed as

$$\mathbf{x}_k = \mathbf{F}_k \mathbf{x}_{k-1} + v_{k-1}, \tag{4.75}$$

$$\mathbf{z}_k = \mathbf{G}_k \mathbf{x}_k + \eta_k, \tag{4.76}$$

where $\mathbf{F}_k$ and $\mathbf{G}_k$ are the known matrices that specify the linear relations.

For zero mean and statistically independent noise vectors v_{k-1} and η_k with respective covariances of $\mathbf{Q}_{k-1}$ and $\mathbf{R}_k$, the Kalman filter equations can be expressed as [177]:

$$\begin{aligned}
\mathbf{x}_{k-1}|\mathbf{z}_1,\ldots,\mathbf{z}_{k-1} &\sim \mathcal{N}\left(\boldsymbol{\mu}_{k-1|k-1},\boldsymbol{\Sigma}_{k-1|k-1}\right), \\
\mathbf{x}_k|\mathbf{z}_1,\ldots,\mathbf{z}_{k-1} &\sim \mathcal{N}\left(\boldsymbol{\mu}_{k|k-1},\boldsymbol{\Sigma}_{k|k-1}\right), \\
\mathbf{x}_k|\mathbf{z}_1,\ldots,\mathbf{z}_k &\sim \mathcal{N}\left(\boldsymbol{\mu}_{k|k},\boldsymbol{\Sigma}_{k|k}\right),
\end{aligned} \tag{4.77}$$

where

$$\boldsymbol{\mu}_{k|k-1} = \mathbf{F}_k \boldsymbol{\mu}_{k-1|k-1}, \tag{4.78}$$

$$\boldsymbol{\Sigma}_{k|k-1} = \mathbf{Q}_{k-1} + \mathbf{F}_k \boldsymbol{\Sigma}_{k-1|k-1} \mathbf{F}_k^T, \tag{4.79}$$

$$\boldsymbol{\mu}_{k|k} = \boldsymbol{\mu}_{k|k-1} + \mathbf{K}_k(\mathbf{z}_k - \mathbf{G}_k \boldsymbol{\mu}_{k|k-1}), \tag{4.80}$$

$$\boldsymbol{\Sigma}_{k|k} = \boldsymbol{\Sigma}_{k|k-1} - \mathbf{K}_k \mathbf{G}_k \boldsymbol{\Sigma}_{k|k-1}, \tag{4.81}$$

with

$$\mathbf{K}_k = \boldsymbol{\Sigma}_{k|k-1} \mathbf{G}_k^T \left(\mathbf{G}_k \boldsymbol{\Sigma}_{k|k-1} \mathbf{G}_k^T + \mathbf{R}_k\right)^{-1} \tag{4.82}$$

denoting the Kalman gain matrix.

Note from (4.77) that the Kalman filter estimates a Gaussian distribution, specified by its mean and covariance, at each time instant, which can be used for optimal estimation of the state. For example, the conditional mean in (4.80) is the optimal estimate for the state according to the MMSE, MAP and minimum mean absolute error (MMAE) criteria [140].

One of the main advantages of the Kalman filter is its computational efficiency in obtaining the optimal solution. However, the assumptions under which the Kalman filter provides the optimal solution are not commonly satisfied in practical situations. For example, the measurement equation in (4.72) is not linear as in (4.76) because of the non-linear relations between the position of the target node and the measurements, as studied in Section 4.2.2.

In order to apply Kalman filtering to non-linear tracking problems, local linearization approaches can be applied [177]. The extended Kalman filter (EKF) and its variants based on the unscented transform (the unscented Kalman filters) [181, 182] employ various linearization techniques for solving non-linear state-space equations.

Although non-linearities can be linearly approximated in certain cases, the Gaussianity assumption is still needed for accurate estimates. The multihypothesis tracking (MHT) approach generalizes the Kalman filtering technique to the cases in which the posterior probability can be represented by a mixture of Gaussian random variables at each time instant [183]. Therefore, the MHT technique is more widely applicable than the Kalman filter. However, it has higher computational complexity.

Example 4.5 *Consider a tracking scenario in which three reference nodes are tracking a target node moving on the black solid line (from left to right) in Fig. 4.24 at a constant velocity of 1 m/s. Each reference node is taking four range measurements per second ($\Delta t = 0.25$ s), and each measurement is modeled by a Gaussian random variable around the true range with a variance of 0.5 m^2.*

At each 0.25 s interval, a position estimate is obtained by the LS algorithm as in (4.43). These estimates are shown in Fig. 4.24 as cross signs. Note that these estimates are quite scattered around the true target path. Let the position estimate from the LS estimator at the kth instant be denoted as

$$\mathbf{z}_k = \begin{bmatrix} z_k^{(1)} & z_k^{(2)} \end{bmatrix}^T. \tag{4.83}$$

These position estimates are used by a Kalman filter as in (4.75) and (4.76), where $\mathbf{z}_k$ forms the measurement vector, and the state vector consists of the position and the velocity of the target in two dimensions; i.e.

$$\mathbf{x}_k = \begin{bmatrix} x_k^{(1)} & x_k^{(2)} & x_k^{(3)} & x_k^{(4)} \end{bmatrix}^T, \tag{4.84}$$

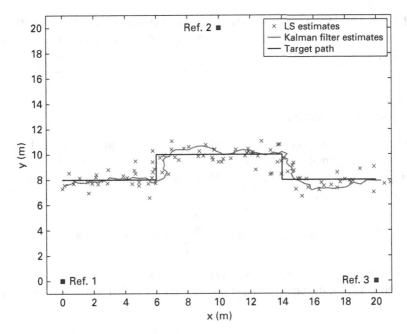

Fig. 4.24. A target tracking scenario, in which three reference nodes are tracking the target node. The cross signs represent the position estimates from the LS algorithm and the heavy solid line is the smoothed position estimate using the Kalman filter.

with $\left(x_k^{(1)}, x_k^{(2)}\right)$ representing the position and $\left(x_k^{(3)}, x_k^{(4)}\right)$ representing the velocity. The matrices in (4.75) and (4.76) are specified as follows:

$$\mathbf{F}_k = \begin{bmatrix} 1 & 0 & \Delta t & 0 \\ 0 & 1 & 0 & \Delta t \\ 0 & 0 & 1 & 0 \\ 0 & 0 & 0 & 1 \end{bmatrix}, \quad (4.85)$$

$$\mathbf{G}_k = \begin{bmatrix} 1 & 0 & 0 & 0 \\ 0 & 1 & 0 & 0 \end{bmatrix}, \quad (4.86)$$

and the covariance A matrix of $\boldsymbol{\eta}_k$ in (4.76) is given by diag$\{5, 5\}$. Similarly, the covariance of $\boldsymbol{v}_{k-1}$ in (4.75) is expressed as diag$\{0, 0, (\Delta t)^2, (\Delta t)^2\}$, which models an additive random acceleration component for the velocity state variables [156].

The conditional mean in (4.80) is used as the position estimate of the Kalman filter. The initial value for the mean estimate is obtained from the first position estimate $[\hat{x}_k^{(1)} \ \hat{x}_k^{(2)}]$ of the LS algorithm as $\boldsymbol{\mu}_{0|0} = [\hat{x}_k^{(1)} \ \hat{x}_k^{(2)} \ 0 \ 0]^T$, where the initial velocity is assumed to be zero. Similarly, the initial covariance estimate, $\boldsymbol{\Sigma}_{0|0}$ in (4.79), is assumed to be diag$\{1, 1, 1, 1\}$.

In Fig. 4.24, the position estimates obtained from the Kalman filter are plotted by the heavy solid line,[26] which indicates smoother position estimates compared to the

---

[26] The discrete estimates are connected by straight lines to show a path estimate.

LS algorithm. Averaged over 100 independent simulations, the average RMSE of the LS algorithm reduces from 0.86 to 0.55 m after applying Kalman filtering on the LS position estimates.

4.3.2 Grid-based approaches

For discrete state spaces that consist of a finite number of states, the grid-based methods provide the optimal solution to the tracking problem. Let $x_{k-1}^i, i = 1, \ldots, N_g$, represent the number of states at time $k-1$. Then, the posterior probability at that time instant can be expressed as

$$p(\mathbf{x}_k | \mathbf{z}_1, \ldots, \mathbf{z}_{k-1}) = \sum_{i=1}^{N_g} w_{k-1|k-1}^i \delta(\mathbf{x}_{k-1} - \mathbf{x}_{k-1}^i), \quad (4.87)$$

where $w_{k-1|k-1}^i = \Pr(\mathbf{x}_{k-1} = \mathbf{x}_{k-1}^i | \mathbf{z}_1, \ldots, \mathbf{z}_{k-1})$.

From (4.87), the prediction and update stages in (4.73) and (4.74) can be expressed as [177]

$$p(\mathbf{x}_k | \mathbf{z}_1, \ldots, \mathbf{z}_{k-1}) = \sum_{i=1}^{N_g} w_{k|k-1}^i \delta(\mathbf{x}_k - \mathbf{x}_k^i), \quad (4.88)$$

$$p(\mathbf{x}_k | \mathbf{z}_1, \ldots, \mathbf{z}_k) = \sum_{i=1}^{N_g} w_{k|k}^i \delta(\mathbf{x}_k - \mathbf{x}_k^i), \quad (4.89)$$

where

$$w_{k|k-1}^i = \sum_{j=1}^{N_g} w_{k-1|k-1}^j p(\mathbf{x}_k^i | \mathbf{x}_{k-1}^j), \quad (4.90)$$

$$w_{k|k}^i = \frac{w_{k|k-1}^i p(\mathbf{z}_k | \mathbf{x}_k^i)}{\sum_{j=1}^{N_g} w_{k|k-1}^j p(\mathbf{z}_k | \mathbf{x}_k^j)}. \quad (4.91)$$

The optimality condition for the grid-based approach is not usually satisfied for position tracking problems, since the position of the target node can take on a continuum of values. In such a case, approximate grid-based techniques can be employed [177], which decompose the continuous state space into a number of sub-regions (called "cells") and apply the grid-based approach above to approximate the posterior probabilities. Hidden Markov model (HMM) filters are also based on this approximate grid-based principle [184–186].

4.3.3 Particle filters

Particle filters, also known as sequential Monte Carlo (SMC) methods, represent the posterior probability of the state at a given time by a set of samples, called *particles*, and related weights, called *importance factors* [187].

A particle filter approximates the posterior density at each time instant by a discrete random variable which takes on the value of the ith sample with a probability that is equal to the ith weight for $i = 1, \ldots, N_g$, where N_g is the number of samples. Since the weights define a probability mass function, they add up to one at each time instant. The weights are determined according to the *importance sampling* principle [187].

Particle filters have been applied successfully in many tracking problems [177, 188–193]. The main advantage of the particle filtering is its ability to represent any probability distribution. In other words, particle filters can converge to the optimal Bayesian solution in non-Gaussian and non-linear environments. Also, they are more efficient than the grid-based approaches since more samples can be collected from the regions in state space with high probability [178]. However, the main disadvantage of particle filtering is related to its computational complexity, which is exponential in the number of elements in the state vector.

Particle filters have been employed in UWB tracking scenarios in order to mitigate the effects of NLOS propagation [194, 195]. Specifically, they can be used to estimate the state of the target node and the measurement biases due to NLOS so that accurate position tracking can be performed.

4.4 Problems

(1) For a reference node performing RSS measurements, calculate the expected value of the lower bound in (4.4) in terms of the PL exponent and the standard deviation of the log-normal shadowing,
   (a) if the position of the target node is uniformly distributed inside a circle of radius d around the reference node;
   (b) if the position of the target node is distributed according to the following probability density function

$$p(x, y) = \frac{1}{2\pi\sigma^2} \exp\left\{-\frac{1}{2\sigma^2}\left[(x - x_0)^2 + (y - y_0)^2\right]\right\}, \quad (4.92)$$

where (x_0, y_0) is the position of the reference node.

(2) A reference node receives the signal $r(t) = s(t - \tau) + n(t)$, where $s(t)$ is the signal transmitted from the target node and $n(t)$ denotes white Gaussian noise with zero mean. Assume that $s(t)$ is given by

$$s(t) = A\left(1 - \frac{4\pi t^2}{\zeta^2}\right) e^{-2\pi t^2/\zeta^2}, \quad (4.93)$$

with $\zeta = 0.1$ ns and A being a positive constant.
   (a) Calculate the effective bandwidth of $s(t)$.
   (b) Calculate the CRLBs on the standard deviations of unbiased TOA estimators for SNR= 0, 5, 10 dB.

(3) Consider an AOA estimation scenario in which the reference node employs a ULA as in Fig. 4.6 to determine the angle of the signal transmitted from the single antenna of the target node. Assume that the spectral densities of noise and the channel coefficients are the same for all antenna elements, and the AOA is confined to the interval $(-\pi/4, \pi/4)$. For the UWB pulse in Problem 2, find the minimum number of antenna elements at the reference node such that the CRLB for AOA estimation is never more than 0.05 radians for an inter-element spacing of 6 cm and SNR $=10$ dB.

(4) Consider a positioning scenario in which three reference nodes are trying to locate a target node. Reference nodes 1 and 2 are performing TOA measurements, whereas the third reference node is measuring AOA. It is assumed that there is no noise in the measurements; i.e. TOA and AOA measurements yield the true TOA and AOA values. For such a scenario:
  (a) Express the position of the target node, (x, y), in terms of the positions of the reference nodes, (x_1, y_1), (x_2, y_2) and (x_3, y_3), and the TOA (τ_1 and τ_2) and AOA (ψ) measurements.
  (b) Is it always possible to obtain a unique solution for the position of the target? If not, what are the conditions to guarantee a unique solution?

(5) Consider a positioning scenario in which four reference nodes are located as shown in Fig. 4.25. Each node is estimating its distance (range) to a target node in the environment (not shown in the figure), and those range measurements are modeled as

$$z_i = d_i + \eta_i, \qquad (4.94)$$

for $i = 1, 2, 3, 4$, where d_i is the distance between the target node and the ith reference node, and η_i is the measurement noise, which is modeled as $\mathcal{N}(0, \sigma_i^2)$. It is assumed that the measurement noise is independent for different reference nodes

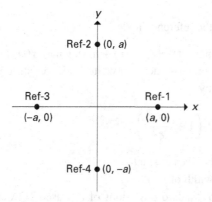

Fig. 4.25. Reference node configuration for Problem 5.

and the variances $\sigma_1^2, \sigma_2^2, \sigma_3^2$ and σ_4^2 are known. Also, the position of the target node, (x, y), is distributed according to a multivariate Gaussian distribution around the origin as

$$p(x, y) = \frac{1}{2\pi \sigma_x \sigma_y} \exp\left\{-\frac{x^2}{2\sigma_x^2} - \frac{y^2}{2\sigma_y^2}\right\}, \quad (4.95)$$

with known σ_x and σ_y values.

(a) Express the MAP estimator for the position of the target, as a minimization problem, in terms of $\{z_i\}_{i=1}^4$, a, $\{\sigma_i\}_{i=1}^4$, σ_x and σ_y.

(b) How does the result simplify for $\sigma_i = \sigma\ \forall i$ and $\sigma_x, \sigma_y \to \infty$?

(6) Obtain the MAP estimator for the position of the target node in Problem 5, if the prior distribution of the target position is expressed as follows:

$$p(x, y) = \begin{cases} 1/a^2, & -a \leq x \leq a,\ -a \leq y \leq a \\ 0, & \text{otherwise} \end{cases}. \quad (4.96)$$

(7) Consider a TOA-based positioning scenario with N_m reference nodes, N_L of which are in LOS to the target node while the remaining ones are in NLOS. Let z_i, for $i=1,\ldots,N_L$, denote the range measurement related to the ith LOS reference node, and z_i, for $i = N_L + 1, \ldots, N_m$, denote the range measurement related to the ith NLOS reference node. The measurements are modeled as

$$z_i = \begin{cases} d_i + \eta_i, & i = 1, \ldots, N_L, \\ d_i + \eta_i + \epsilon_{i-N_L}, & i = N_L + 1, \ldots, N_m, \end{cases} \quad (4.97)$$

where d_i is the distance between reference node i and the target node, $\eta_i \sim \mathcal{N}(0, \sigma_i^2)$ and ϵ_{i-N_L} represents an exponential distribution with parameter λ_{i-N_L} that is independent of η_i.

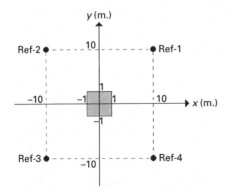

Fig. 4.26. Reference node configuration for Problem 8. Target nodes are located inside the shaded square.

Assuming independent noise at different measurements and known noise parameters, show that the ML estimator for the target position (x, y) is given by the following:

$$(x_{\text{ML}}, y_{\text{ML}}) = \arg\min_{(x,y)} \left\{ \sum_{i=1}^{N_L} \frac{(z_i - d_i)^2}{2\sigma_i^2} + \sum_{i=N_L+1}^{N_m} \lambda_{i-N_L}(z_i - d_i) \right.$$

$$\left. - \sum_{i=N_L+1}^{N_m} \log\left(Q\left(\lambda_{i-N_L} - \frac{z_i - d_i}{\sigma_i} \right) \right) \right\}, \quad (4.98)$$

where $d_i = \sqrt{(x - x_i)^2 + (y - y_i)^2}$ and $Q(r) = \frac{1}{\sqrt{2\pi}} \int_r^\infty e^{-t^2/2} dt$ is the Q-function.

Note that (4.98) reduces to (4.46) when $\sigma_i = 0$ for $i = N_L + 1, \ldots, N_m$.

(8) *(programming exercise)* Consider the TOA-based positioning scenario in Fig. 4.26, where four reference nodes are trying to locate target nodes in the shaded square around the origin. It is assumed that SNRs are the same for all target positions inside the shaded square, and the targets are always in LOS of all the reference nodes.
   (a) Generate 500 target positions inside the shaded square according to a uniform distribution that assigns equal probabilities to all the points inside the square.
   (b) Evaluate the MMSE in (4.62) for each of the 500 target positions generated in part (a) for an SNR of 5 dB and an effective bandwidth of 5 GHz. Find the positions corresponding to maximum and minimum MMSE values, and calculate the average MMSE.
   (c) Find, theoretically, the target positions that yield the minimum and maximum MMSEs. Are the theoretical results close to the ones obtained in the previous part?
   (d) Plot average MMSE versus SNR for SNR values ranging from 0 dB up to 15 dB in steps of 0.5 dB or less. For each SNR value, average MMSE should be obtained from averaging over 500 random target node positions as in part (b).
   (e) Plot the average MMSE versus the effective bandwidth β, for β ranging from 3.5 to 7 GHz in steps of 250 MHz or less.

5 Time-based ranging via UWB radios

As discussed in the previous chapter, the position of a mobile node in a wireless network can be estimated based on AOA, RSS, TOA, and/or TDOA of received signals. Due to their large bandwidths, UWB signals have very high time resolution, hence individual multipath components (MPCs) can be resolved at the receiver. While AOA, TOA, and TDOA approaches all benefit from this high resolution, AOA-based implementations have high complexity. Therefore, positioning based on TOA[1] (or TDOA) estimation is the method of choice in UWB-based positioning systems [196] as opposed to AOA or RSS (which has low ranging accuracy)-based approaches. Therefore, the emphasis of this chapter is on time-based UWB ranging techniques.

The chapter is organized as follows. In Section 5.1, the time-based positioning problem is briefly re-visited and importance of accurate ranging for precise positioning is emphasized. The error sources in time-based ranging are discussed in Section 5.2. In Section 5.3, the time-based ranging problem is formulated and models for various transceiver types are studied. Section 5.4 reviews fundamental limits on the accuracy of time-based ranging, Section 5.5 investigates maximum likelihood (ML)-based techniques, and Section 5.6 presents alternative low-complexity ranging algorithms for UWB systems.

5.1 Time-based positioning

Consider a wireless network in which there are N_m reference nodes (RNs). The ith RN is located at (x_i, y_i), and a target node (TN) at position (x, y). The measured distance $\hat{d}_i$ between the TN and the ith RN is commonly modeled as [18, 138, 197]

$$\hat{d}_i = d_i + e_i + n_i = c\, t_i, \quad i = 1, 2, \ldots, N_\mathrm{m}, \tag{5.1}$$

where t_i is the time of flight estimate (measurement) of the signal at the ith RN, c is the speed of light, d_i is the true distance between the TN and the ith RN, $n_i \sim \mathcal{N}(0, \sigma_i^2)$ is the zero mean Gaussian measurement noise with variance σ_i^2, and e_i is a non-negative distance bias introduced due to the obstructed line-of-sight (LOS),[2] given by

---

[1] In the sequel, terms "ranging" and "TOA estimation" will be used interchangeably since they imply each other.

[2] Note that in essence, a non-negative distance bias can be present even when the first path is not completely obstructed, but experiences a different propagation speed due to obstacles [198].

$$e_i = \begin{cases} 0, & \text{if } i\text{th RN is LOS}, \\ |\tilde{e}_i|, & \text{if } i\text{th RN is NLOS}. \end{cases} \quad (5.2)$$

As studied in Chapter 4, in the absence of noise and NLOS bias e_i, the actual distance d_i between the TN and the ith RN defines a circle centered at (x_i, y_i), and this circle corresponds to all possible locations of the TN. The joint solution of the N_m circles (from N_m RNs) yields the position of the TN. However, due to noisy measurements and NLOS biases, the circles may not intersect at a single point (see Fig. 5.1), resulting in the following equations

$$(x - x_i)^2 + (y - y_i)^2 = \hat{d}_i^2, \quad i = 1, 2, \ldots, N_m. \quad (5.3)$$

There are numerous ways to estimate the position of the TN. For example, a simple solution is obtained using the non-linear least squares (NLS) technique, studied in Section 4.2.2, as follows

$$\begin{aligned}(\hat{x}, \hat{y}) &= \arg\min_{(x, y)} \left\{ \sum_{i=1}^{N_m} \varsigma_i \left(\hat{d}_i - \sqrt{(x - x_i)^2 + (y - y_i)^2} \right)^2 \right\} \\ &= \arg\min_{(x, y)} \left\{ \varepsilon_{\text{res}}(x, y) \right\}, \end{aligned} \quad (5.4)$$

where $\varepsilon_{\text{res}}(x, y)$ is the residual error corresponding to the TN location (x, y), and ς_i characterizes the reliability of the measurement [4, 119].

It is apparent from (5.1), (5.3), and Fig. 5.1 that having accurate range estimates at each RN is crucial for precise positioning. Fine time resolution of UWB signals makes accurate identification of the first MPC possible. However, this may not be easy in many scenarios due to NLOS propagation and a vast number of MPCs. In the next section, a discussion on different error sources for UWB ranging is presented, followed by formulation of the time-based UWB ranging problem.

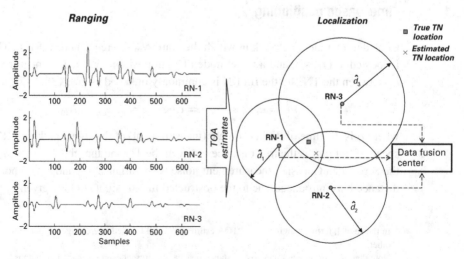

Fig. 5.1. Relation between ranging and positioning for $N_m = 3$.

5.2 Error sources in time-based ranging

There are a number of error sources that may seriously degrade the accuracy of the range estimation. Before getting into different ranging algorithms and fundamental bounds, some of those error sources will be briefly overviewed in this section.

5.2.1 Multipath propagation

Multipath propagation introduces challenges for UWB ranging due to a large number of MPCs and relatively long excess delays compared to the transmitted pulse duration. In the absence of multipath propagation, estimation of the arrival time of a signal is relatively easy: the cross-correlation of the received signal with a local template[3] is obtained, and the TOA is given by the correlation peak (see Section 5.5.1 for details).

In practice, reflections from scatterers in an environment arrive at a receiver as replicas of the transmitted signal with various attenuation levels and delays. As discussed in Chapter 3, the maximum excess delay of the received multipath signal can be on the order of a hundred nanoseconds, and the strongest MPC may arrive much later than the first path. For such a signal, the TOA is no longer given by the correlation peak, and

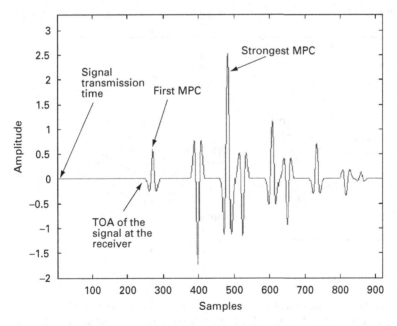

Fig. 5.2. Illustration of the TOA estimation problem in a multipath channel. The receiver might already be synchronized to the strongest MPC and tries to identify the first arriving MPC using different signal processing techniques (receiver noise is not shown).

---

[3] Ideally, the received pulse shape is used as the correlation template since it may be different from the transmitted pulse shape in practice. Since it may be difficult to estimate the received pulse shape in practical scenarios, the transmitted pulse shape is commonly used as a suboptimal template.

first path detection algorithms are required. In typical UWB channels, the first path may be considerably weaker than the strongest component (see Fig. 5.2), and it may arrive several tens of nanoseconds earlier than the strongest [100]. As illustrated in various simple scenarios in Fig. 5.3, the first path may be weaker than the later paths due to NLOS signal propagation or antenna effects.

An important problem in multipath channels is that, if the MPCs are not resolvable (i.e. in the absence of a sufficient time delay between any two consecutive MPCs), there may be interference from one received pulse to another. Hence, the first arriving path and other paths may overlap, which may shift a TOA estimate from the true TOA.[4]

Another error source related to multipath propagation is associated with the correlation characteristics of the spreading sequences employed in ranging. In a typical UWB receiver that uses a correlator (or a matched filter), the receiver first uses the spreading

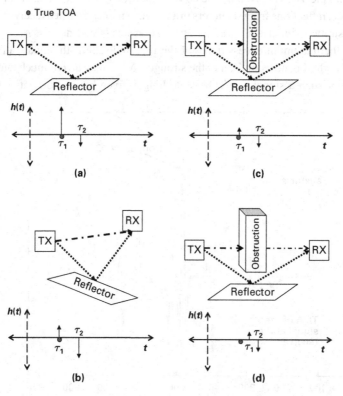

Fig. 5.3. Four different simple scenarios for channel realizations in LOS and NLOS situations. (a) The direct path is unobstructed (LOS). (b) The direct path is unobstructed, but attenuated due to antenna pattern (LOS). (c) The direct path is obstructed. The attenuation due to the obstruction makes the direct path weaker than the strongest path. The delay due to the obstruction is neglected since it is insignificant (NLOS). (d) The direct path is obstructed. The obstruction both attenuates and delays the direct path. Note that in certain cases the direct path may completely disappear (NLOS) (After [198]).

---

[4] Note that this problem is much more significant in narrowband systems due to longer pulse durations.

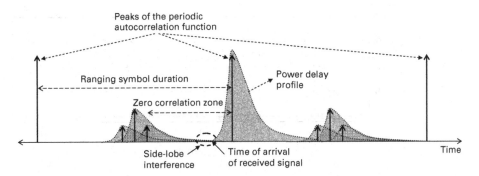

Fig. 5.4. Illustration of the side-lobe interference (SLI) due to imperfect autocorrelation characteristics of ranging preamble and the large delay spread of the multipath channel.

sequence of the desired user[5] in order to de-spread the received waveform. It then locks onto the correlation peak, and tries to identify the first arriving MPC preceding the correlation peak. If the transmitted waveform has imperfect autocorrelation characteristics, correlation side-lobes appear between the correlation peaks. In Fig. 5.4, such an imperfect periodic autocorrelation of a spreading sequence is illustrated. The autocorrelation peaks are repeated periodically with the symbol length and one can observe the autocorrelation side-lobes in between. In the presence of multipath, the channel profile is replicated at each of the autocorrelation peaks as well as the autocorrelation side-lobes. If the zero correlation zone prior to the correlation peak is not large enough, extension of the channel profile from the autocorrelation side-lobes may interfere with the leading MPCs corresponding to the autocorrelation peak. In here, such an effect is referred to as the side-lobe interference (SLI) and it may be a limiting factor on the ranging accuracy.

In order to prevent the SLI, the periodic autocorrelation characteristics of the transmitted waveforms should have a sufficiently large zero correlation zone (ZCZ) compared to the maximum excess delay of the channel. In [200], four different ranging waveforms and their correlation characteristics (as well as other design trade-offs) are discussed.[6]

Example 5.1 *Periodic correlation characteristics of two common ranging waveforms are compared in Fig. 5.5. Fig. 5.5a shows the periodic correlation output of an M-ary ternary orthogonal keying (MTOK) sequence of length 31, where the sequence is given by {1,−1,−1,0,0,0,1,−1,0,1,1,1,0,1,0,−1,0,0,0,0,1,0,0,−1,0,−1,1,0,0,−1,−1}. The sequence is processed with a bipolar template (BPT) at the receiver rather than the transmitted MTOK sequence itself [199]. The figure shows that MTOK sequences have optimal correlation characteristics when processed with a BPT. In other words, there are no correlation side lobes between any two correlation peaks. This characteristic makes MTOK sequences very suitable for ranging applications.*

---

[5] Or, a specially designed template such as the bi-polar templates for de-spreading M-ary ternary orthogonal keying (MTOK) sequences [199].
[6] Namely, M-ary ternary orthogonal keying (MTOK), time hopping (TH), direct-sequence (DS), and transmitted reference (TR).

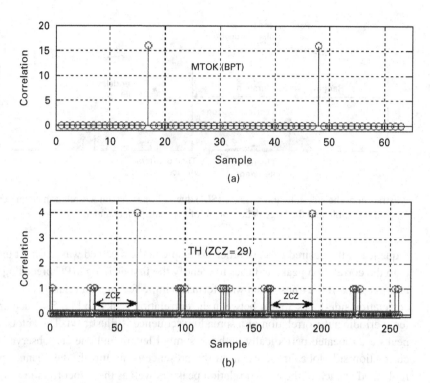

Fig. 5.5.   Periodic code correlations for MTOK-IR and TH-IR (2 periods): (a) periodic MTOK correlation using a bi-polar template (BPT), (b) periodic TH-IR autocorrelation (© 2005 IEEE) [200].

As opposed to MTOK sequences, TH sequences have a much smaller ZCZ. It is possible to find TH sequences with good correlation characteristics (e.g. using a brute-force search). For example, Fig. 5.5b shows the periodic autocorrelation output of a TH sequence {0, 3, 3, 2} with $N_h = 32$ (i.e., the number of chip positions is similar to that of an MTOK sequence of length 31).[7] *While there are correlation side lobes between two correlation peaks, there is a periodic ZCZ of length 29. Such a ZCZ duration may or may not be satisfactory for ranging systems depending on factors such as the maximum excess delay of the channel and how far the strongest path arrives relative to the first path.*

Note that for coherent receivers, polarity coding may also be used along with TH to improve the autocorrelation characteristics of ranging waveforms. Some design trade-offs regarding the ranging transmit-waveform will be discussed in more detail in Chapter 8.

---

[7] As will be discussed in Section 5.3, N_h denotes the number of chip positions per frame. A pulse may be transmitted at one of the chip positions determined by a TH sequence.

5.2.2 Multiple-access interference

Performing ranging under multiple-access interference (MAI) is another challenging issue. The accuracy of the TOA-based ranging may be significantly degraded in the presence of MAI.

Effects of MAI can be mitigated by assigning orthogonal channels to different users, either in time, frequency, code, or space domains in a network. However, even with orthogonal channel assignments, there may still be interference from other users in simultaneously operating networks (SONs). In such cases, MAI mitigation algorithms can be employed. For example in [201], a non-linear filtering technique is proposed to improve the ranging accuracy for non-coherent receivers in the presence of SONs. Mitigation of the MAI will be discussed in more detail in Chapter 7.

As discussed in Section 5.2.1, the autocorrelation characteristics of the transmitted waveforms are important in order to mitigate the effects of multipath propagation, and they should have a large ZCZ. On the other hand, the *cross*-correlation characteristics of the transmitted waveforms are also important to minimize the degradation due to MAI. In particular, the ranging waveforms employed by different users should have low cross-correlation values for lower MAI.

5.2.3 Obstructed line of sight propagation

When the direct path between a transmitter and a receiver is obstructed, the first arriving MPC can be attenuated more than some other MPCs. This makes its detection quite challenging. Since the strongest MPC is not necessarily the first MPC, intelligent first-path detection techniques are required to mitigate the effects of NLOS propagation.

In some cases, the direct signal is attenuated so much that it cannot be observed at the receiver at all. Then, the TOA estimate at the receiver includes a positive bias. This NLOS error is modeled in the literature as an exponentially [16, 202], uniformly [197, 203], or Gaussian distributed random variable [204], constant along a time interval [205], or a random variable of which the distribution is derived from empirical data [206, 207]. Typically, the model depends on the wireless propagation channel and the receiver type.

NLOS identification and mitigation techniques have been discussed extensively in the literature, especially within a cellular network framework [166, 168, 208]. For example in [208], the standard deviation of range measurements are compared against a threshold for NLOS identification, where the measurement noise variance is assumed to be known. In [166], a decision-theoretic NLOS identification framework is presented, where various hypothesis tests are developed for known and unknown probability density functions (PDFs) of the TOA measurements. A non-parametric NLOS identification approach is discussed in [168], where a suitable distance metric is used between the known measurement error distribution and the non-parametrically estimated distance measurement distribution in order to determine if a given base station (BS) is LOS or NLOS.

It is also possible to use the information embedded within the MPCs of the received signal for NLOS mitigation purposes [119, 209, 210]. The large number of received MPCs in a UWB signal enables NLOS identification techniques to exploit different statistics of the MPCs, such as the mean excess delay and root mean square (RMS) delay spread [119]. As shown in Fig. 3.7 in Chapter 3, compared to residential and indoor-office environments, the PDFs of the mean excess delay and RMS delay spread in outdoor and industrial environments are quite distinct, which implies reliability of the LOS/NLOS identification. Using prior information such as the data in Table 3.5, a likelihood ratio test can be developed easily to determine whether a channel realization is LOS or NLOS [119].

5.2.4 Other error sources

Timing imperfections among reference devices in infrastructure-based ranging and clock drifting between transmitter and receiver devices in the round-trip TOA measurements induce additional errors on range estimates. Due to their large signal bandwidth, UWB receivers are more sensitive to timing jitter and clock drifting effects [211].

It is desirable that accurate ranging is achieved with low sampling rates to facilitate low-cost and low-power designs. This is especially important for UWB signals since sampling them at or above the Nyquist rate proves to be quite expensive. However, sampling UWB signals at sub-Nyquist rates commonly factors in as an additional error source and degrades the accuracy of UWB ranging systems.

5.3 Time-based ranging

Detecting the first signal path requires intelligent signal processing techniques and the related literature dates back a few decades. One of the earlier works by Coppens develops a technique for *picking of first arrivals*, which operates well at high signal-to-noise ratios (SNRs) [212]. Vidal *et al.* develop first arriving path detection techniques using generalized likelihood ratio tests (GLRTs) and minimum variance (MV) estimators in [213–215] for cellular systems. Techniques based on minimum mean square error (MMSE) estimation [216], least squares (LS) estimation [217], and maximum-likelihood (ML) estimation [218] of the signal arrival time are also available in the literature. These techniques are mostly developed for relatively narrowband systems. However, large bandwidths of UWB signals introduce additional challenges.

Coarse timing of a received signal can be obtained by acquisition and by locking onto the strongest MPC [219]. The acquisition performance of IR-UWB systems for coherent receivers [219, 220], differential/dirty-template receivers [221–224], and noncoherent receivers [219, 225, 226] have been studied in the literature. After acquisition and coarse synchronization, refinement of the arrival time requires some processing gain to improve the SNR first, and then signal processing to detect the leading (i.e. first) path (see Fig. 5.2). As discussed in Section 5.2.1, this first path may not be the strongest component.

In order to define a ranging problem and develop a framework for a ranging system, let the received IR-UWB signal in a multipath environment be represented as follows:

$$r(t) = \sum_{l=1}^{L} \alpha_l s(t - \tau_l) + n(t), \quad (5.5)$$

where $s(t)$ represents a ranging signal to be defined below, L is the number of MPCs, α_l and τ_l are the channel coefficient and the delay of the lth MPC, respectively, and $n(t)$ is zero-mean additive white Gaussian noise (AWGN) with double-sided power spectral density $\mathcal{N}_0/2$.

The signal $s(t)$ in (5.5) is given by

$$s(t) = \sqrt{\frac{E_s}{N_f}} \sum_{j=-\infty}^{\infty} a_j \omega(t - jT_f - c_j T_c), \quad (5.6)$$

where E_s represents the energy of a ranging symbol, $a_j \in \{\pm 1\}$ is the polarity code, $c_j \in \{0, 1, \ldots, N_h - 1\}$ is the time-hopping (TH) code, $\omega(t)$ denotes the received UWB pulse with unit energy, T_f is the frame duration, N_f is the number of pulses (frames) per ranging symbol, T_c is the chip duration, and N_h is the number of chips per frame. The width of the received pulse $\omega(t)$ is represented by T_p, and it is assumed that $T_p \leq T_c$.

Note that the ranging signal in (5.6) is assumed to consist of ranging symbols, each of which is composed of N_f UWB pulses. The duration of a ranging symbol is given by $T_s = N_f T_f$, and the energy of a UWB pulse is represented as

$$E_p = \int_{-\infty}^{\infty} |\omega(t)|^2 dt = 1, \quad (5.7)$$

where a unit-energy pulse is assumed without loss of generality. In Fig. 5.6, an example ranging symbol is illustrated for $N_f = 6$, $N_h = 5$, $a_j = 1 \, \forall \, j$, and $c_j = \{0, 2, 1, 0, 4, 2\}$.

It should be noted that if a single pulse is transmitted per symbol (i.e. $N_f = 1$), and if no TH or polarity codes are employed, (5.6) becomes

$$s(t) = \sqrt{E_s} \sum_{j=-\infty}^{\infty} \omega(t - jT_s). \quad (5.8)$$

Then, the received signal in (5.5) can be expressed as

$$r(t) = r_{\text{des}}(t) + n(t), \quad (5.9)$$

with

$$r_{\text{des}}(t) = \sum_{j=-\infty}^{\infty} \sum_{l=1}^{L} \alpha_l \omega(t - \tau_l - jT_s), \quad (5.10)$$

where the $\sqrt{E_s}$ term is omitted without loss of generality, as it can be considered to be included in the channel coefficient terms.

In the rest of the chapter, unless otherwise stated, the signal model in (5.9) and (5.10) is considered for simplicity. The main difference of this signal model from the

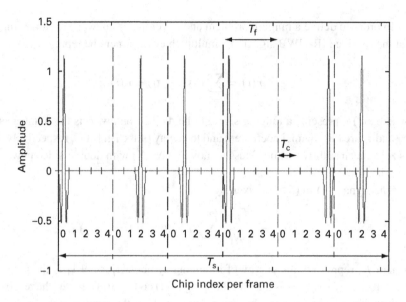

Fig. 5.6. An example transmitted signal for a time-hopping impulse radio UWB system. The frame duration is denoted by T_f, chip duration is denoted by T_c, and a time-hopping sequence $c_j = \{0, 2, 1, 0, 4, 2\}$ is used.

one obtained from (5.6) is that the TH codes in the latter can affect the ranging performance by changing the correlation properties of the ranging signal as discussed in Section 5.2.

Stacking the unknown path gains α_l and delays τ_l in (5.5) into a $2L \times 1$ unknown parameter vector yields

$$\boldsymbol{\theta} = [\boldsymbol{\alpha} \ \boldsymbol{\tau}]^T, \tag{5.11}$$

where

$$\boldsymbol{\alpha} = [\alpha_1 \cdots \alpha_L], \tag{5.12}$$

$$\boldsymbol{\tau} = [\tau_1 \cdots \tau_L]. \tag{5.13}$$

Furthermore, upon defining $\tau_{l,1} = \tau_l - \tau_1$, a $(2L - 1) \times 1$ nuisance parameter vector can be defined as

$$\tilde{\boldsymbol{\theta}} = [\alpha_1 \cdots \alpha_L \ \tau_{2,1} \cdots \tau_{L,1}]^T. \tag{5.14}$$

Note that while $\boldsymbol{\theta}$ captures the absolute delays of all the MPCs, $\tilde{\boldsymbol{\theta}}$ captures the delays relative to the first MPC of the received signal.

The problem of TOA estimation is to estimate the delay of the first path, τ_1, which is equivalent to the TOA estimate $\hat{t}_i$ in (5.1) for the ith node. In practice, this is a challenging problem due to a number of error sources. Typically, given some geographical constraints, the range of τ_1 can be assumed to be bounded. For the rest of the chapter, it is assumed that the TOA is uniformly distributed in $[0, T_a]$, and that the received signal observed within an interval $[0, T_{\text{obs}}]$ where $T_{\text{obs}} \geq T_a$. Also, the observation interval consists of N_r ranging symbols; that is, $T_{\text{obs}} = N_r T_s$.

In the rest of this section, different ways to obtain the decision variables for TOA estimation, starting from the continuous-time signal in (5.9), will be discussed. The ranging algorithms to be introduced later in the chapter will employ one of these techniques or a modified version of them.

5.3.1 Direct sampling receiver

In order to estimate the TOA of the received signal, the samples directly obtained from the received signal in (5.9) may be employed. Assume that the observation interval consists of a single ranging symbol[8]; i.e. $N_r = 1$.

Let the received pulse $\omega(t)$ and received signal $r(t)$ in (5.9) be sampled at a rate of $1/T_{smp}$. This yields $N_\omega = T_p/T_{smp}$ samples of $\omega(t)$ and $N = T_{obs}/T_{smp}$ samples of $r(t)$, assuming T_p and T_{obs} are integer multiples of T_{smp}. Then, the samples of $r(t)$ can be collected in an $N \times 1$ vector $\mathbf{r}$ as follows [227]

$$\mathbf{r} = \mathbf{\Omega}(\tau)\boldsymbol{\alpha}^T + \mathbf{n}, \tag{5.15}$$

where $\boldsymbol{\alpha} = [\alpha_1 \cdots \alpha_L]$, $\mathbf{n}$ is an $N \times 1$ noise vector consisting of the samples of $n(t)$, the ith sample in $\mathbf{r}$ is given by

$$r[i] = r(iT_{smp}) = \sum_{l=1}^{L} \alpha_l \omega(iT_{smp} - \tau_l) + n(iT_{smp}), \tag{5.16}$$

for $i = 1, \ldots, N$, and

$$\mathbf{\Omega}(\tau) = [\boldsymbol{\omega}^{(D_1)} \boldsymbol{\omega}^{(D_2)} \cdots \boldsymbol{\omega}^{(D_L)}], \tag{5.17}$$

with

$$\boldsymbol{\omega}^{(D_l)} = [\mathbf{0}_{D_l} \, \boldsymbol{\omega} \, \mathbf{0}_{N-N_\omega-D_l}]^T, \tag{5.18}$$

for $l = 1, \ldots, L$. Note that $\mathbf{0}_{D_l}$ is a $1 \times D_l$ vector of zeros, $D_l = \lfloor \tau_l/T_{smp} \rfloor$, and the elements of $\boldsymbol{\omega}$ are given by $\omega[i] = \omega(iT_{smp})$ for $i = 1, \ldots, N_\omega$.

Note again that, as discussed in Section 5.2, in order to accurately reconstruct the received signal from its samples, the received signal should be sampled at or above the Nyquist rate. For UWB systems, this implies sampling rates on the order of several GHz, which may considerably increase the cost and complexity of the receiver.

While ranging can be achieved directly from the samples of the received signal in (5.5), different analog front-end processing techniques can also be employed before sampling the signal. Three of the common receiver architectures are matched filter, energy detection, and delay-and-correlate receivers, which are illustrated in Fig. 5.7. For all receiver types, it is assumed that the signal is first passed through a band-pass filter of bandwidth B to remove out-of-band noise, and through a low-noise amplifier (LNA) to improve the signal quality. The samples may further be averaged over a number of ranging symbols to improve the ranging accuracy.

---

[8] In the case of multiple symbols in the observation interval, independent observations of the ranging symbol are obtained, which provides increased processing gain as will be considered in the next section.

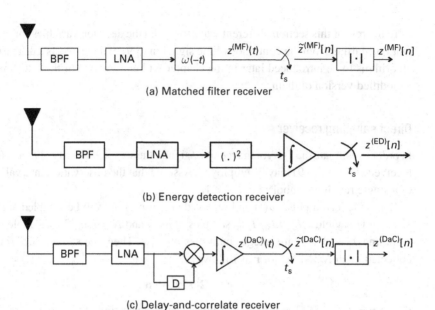

Fig. 5.7. Block diagrams of (a) matched filter (MF) receiver, (b) energy detection (ED) receiver, and (c) delay-and-correlate (DnC) receiver. Functionalities of the bandpass filter (BPF) and the low noise amplifier (LNA) are discussed in more detail in Chapter 8 (After [229]).

5.3.2 Matched filter receiver

If the received pulse shape is known at the receiver, a matched filter (MF) can be used to obtain decision variables for TOA estimation. Passing the received signal in (5.9) through an MF yields [219, 228, 229]

$$z^{(\text{MF})}(t) = \sum_{j=0}^{N_r-1} \sum_{l=1}^{L} \alpha_l R_\omega(t - \tau_l - jT_s) + n_\omega(t), \quad (5.19)$$

where N_r is the number of ranging symbols in the observation interval. The impulse response of the filter is given by $\omega(-t)$, $R_\omega(\tau)$ is the autocorrelation function of $\omega(t)$,

$$R_\omega(\tau) = \int_{-\infty}^{\infty} \omega(\tau + t)\omega(t)dt, \quad (5.20)$$

and $n_\omega(t)$ is colored Gaussian noise with its autocorrelation function given by $\mathcal{N}_0 R_\omega(\tau)/2$. Note that since $R_\omega(\tau) = 0$ for $|\tau| \geq T_c$, noise samples taken at intervals larger than $2T_c$ will be independent [228].

If (5.19) is sampled at every T_{smp} seconds, the MF outputs can be obtained as

$$\tilde{z}^{(\text{MF})}[n] = \frac{1}{N_r} \sum_{j=1}^{N_r} z^{(\text{MF})}(t) \Big|_{t=nT_{\text{smp}}+(j-1)T_s}, \quad n = 1, 2, \ldots, N, \quad (5.21)$$

where $N = T_{\text{obs}}/T_{\text{smp}}$ (assume that T_{obs} is an integer multiple of T_{smp}). Note that averaging the MF outputs over N_r symbols provides processing gain and improves the SNR.

In many TOA estimation algorithms, the absolute values of the MF outputs are compared to a threshold. Let the absolute value of (5.21) be represented by

$$z^{(\text{MF})}[n] = \left| \tilde{z}^{(\text{MF})}[n] \right|. \tag{5.22}$$

Commonly, it is convenient to define the complementary cumulative distribution function (CDF) of (5.22) for error analysis of threshold-based ranging algorithms. For noise-only samples, the complementary CDF of $z^{(\text{MF})}[n]$ can be obtained as

$$P\left(z^{(\text{MF})}[n] > \xi\right) = 2Q\left(\frac{\xi}{\sigma}\right), \tag{5.23}$$

where ξ denotes the threshold, $\sigma^2 = \mathcal{N}_0/(2N_r)$, and $Q(x)$ denotes the Q-function given by

$$Q(x) = \frac{1}{\sqrt{2\pi}} \int_x^\infty e^{-t^2/2} dt. \tag{5.24}$$

5.3.3 Energy detection receiver

The MF receiver requires the knowledge of the received pulse shape which may not be available in practice. Another key disadvantage of the MF solution is that, for accurate ranging, it may require Nyquist rate sampling, hence complex analog-to-digital converters (ADCs). A low complexity alternative to an MF receiver is the energy detection (ED) receiver, which does not assume the knowledge of the received pulse shape.

The integrator output samples for an ED receiver can be expressed as follows [219, 228–231]

$$z^{(\text{ED})}[n] = \frac{1}{N_r} \sum_{j=1}^{N_r} \int_{(j-1)T_s + (n-1)T_{\text{smp}}}^{(j-1)T_s + nT_{\text{smp}}} |r(t)|^2 dt, \quad n = 1, 2, \ldots, N, \tag{5.25}$$

where n denotes the sample index, and the samples are averaged over N_r ranging symbols to increase the SNR as in (5.21) for an MF receiver.

The output $z^{(\text{ED})}[n]$ in (5.25) has a centralized chi-squared distribution for noise-only samples, and its complementary CDF is given by [231–233]

$$P(z^{(\text{ED})}[n] > \xi) = \exp\left(-\frac{\xi N_r}{\mathcal{N}_0}\right) \sum_{x=0}^{M/2-1} \frac{1}{x!} \left(\frac{\xi N_r}{\mathcal{N}_0}\right)^x, \tag{5.26}$$

where the degrees of freedom (DOF) is approximately given by [234]

$$M = N_r(2BT_{\text{smp}} + 1), \tag{5.27}$$

with B denoting the bandwidth of the BPF in Fig. 5.7, and the DOF M is assumed to be an even number. On the other hand, $z^{(\text{ED})}[n]$ has a non-centralized chi-squared distribution when signal and noise are both present, in which case the complementary CDF becomes [231–233][9]

$$P(z^{(ED)}[n] > \xi) = Q_{M/2}\left(\frac{E_n}{\sigma}, \frac{\sqrt{\xi}}{\sigma}\right), \quad (5.28)$$

where $\sigma^2 = \mathcal{N}_0/(2N_r)$, E_n represents the (noise-free) ranging symbol energy that falls within the nth sample (which may be zero if no signal is present within the sample interval), and $Q_M(y_1, y_2)$ denotes the Marcum-Q function with parameter M, given by

$$Q_M(y_1, y_2) = \frac{1}{y_1^{M-1}} \int_{y_2}^{\infty} x^M \exp\left(-\frac{x^2 + y_1^2}{2}\right) I_{M-1}(y_1 x) \, dx, \quad (5.29)$$

with $I_M(x)$ representing the modified Bessel function of the first kind [233].

As shown in (5.25), the samples can be averaged over N_r symbols[10] in order to increase the SNR. For sufficiently large N_r, from the central limit theorem (CLT), the distribution of $z^{(ED)}[n]$ can be approximated as

$$z^{(ED)}[n] \sim \mathcal{N}(\mu_n, \sigma_n^2), \quad (5.30)$$

with

$$\mu_n = \tilde{M}\mathcal{N}_0/2 + E_n, \quad (5.31)$$

$$\sigma_n^2 = \frac{\tilde{M}\mathcal{N}_0^2/2 + 2E_n\mathcal{N}_0}{N_r}, \quad (5.32)$$

where

$$\tilde{M} = 2BT_{\text{smp}} + 1 \quad (5.33)$$

is the DOF for each term in the summation in (5.25). Note that the approximation in (5.30)–(5.32) is valid for both the noise-only ($E_n = 0$) and the signal-plus-noise cases.

A major drawback of an ED receiver is due to the noise-squared and signal-cross-noise terms that appear at the output of the square-law device. These terms make the decision variable more noisy than that of an MF receiver. Therefore, an ED receiver typically requires a large N_r value for accurate ranging compared to an MF receiver. On the other hand, at low sampling rates, ED receivers can have better energy capture compared to the MF receiver [229] (see Example 5.2).

5.3.4 Delay-and-correlate receiver

Another receiver that does not require the knowledge of the received pulse shape to construct a local template is the delay-and-correlate (DaC) receiver. In one possible implementation of the DaC receiver, pairs of pulses with a known time delay between each pair are transmitted. The first arriving pulse is delayed and then used as a reference

---

[9] Note that this is valid only when M is an even number. If M is not an even number, the complementary CDF is given by an integral that does not have a closed form solution [233].
[10] It is assumed that only one pulse is used per symbol for simplicity.

template to correlate the later arriving pulse to obtain the decision variable, which is commonly referred as the transmitted-reference (TR) receiver in the literature [229]. Alternatively, symbol-long portions of the signal can be used as a template as in the dirty-template scheme [224], which will be discussed in more detail in Section 5.6.3.

The samples after correlating the received signal with the delayed version of itself can be written as

$$\tilde{z}^{(\text{DaC})}[n] = \frac{1}{N_r} \sum_{j=1}^{N_r} \int_{(j-1)T_s+(n-1)T_{\text{smp}}}^{(j-1)T_s+nT_{\text{smp}}} r(t)r(t-D)dt, \quad n = 1, 2, \ldots, N, \quad (5.34)$$

where D represents the delay between the pulse pairs. Note that for a dirty-template receiver, $D = T_s$, and $r(t)$ and $r_{\text{des}}(t)$ are as given in (5.9) and (5.10), respectively. On the other hand, for a TR receiver, a pair of pulses are transmitted together, with a pre-determined delay D in between them. In other words, $r_{\text{des}}(t)$ in a TR receiver becomes

$$r_{\text{des}}(t) = \sum_{j=-\infty}^{\infty} \sum_{l=1}^{L} \frac{\alpha_l}{\sqrt{2}} \Big[\omega(t - \tau_l - jT_s) + \omega(t - \tau_l - jT_s - D) \Big], \quad (5.35)$$

which can be plugged into (5.9) to obtain the received signal $r(t)$.

As in the ED receiver, for sufficiently large N_r, the decision statistics in a DaC receiver can be approximately modeled by a Gaussian random variable,

$$\tilde{z}^{(\text{DaC})}[n] \sim \mathcal{N}\left(E_n, \frac{\tilde{M}\mathcal{N}_0/4 + E_n\mathcal{N}_0}{N_r}\right), \quad (5.36)$$

where E_n denotes the cross-correlation value (in the absence of noise) for the nth sample and $\tilde{M}$ is given in (5.33). Similar to the MF receiver, the absolute values of the samples can be used for ranging purposes in order to detect the first arriving path,

$$z^{(\text{DaC})}[n] = \left|\tilde{z}^{(\text{DaC})}[n]\right|. \quad (5.37)$$

For noise-only samples, the complementary CDF of $z^{(\text{DaC})}[n]$ can be expressed as

$$P\left(z^{(\text{DaC})}[n] > \xi\right) = 2Q\left(\frac{4N_r\xi}{\tilde{M}\mathcal{N}_0}\right). \quad (5.38)$$

As in the ED receiver, a major disadvantage of the DaC receiver is the enhanced noise terms. In particular, noise-cross-noise terms and signal-cross-noise terms can make the decision variable quite noisy. Similar to the ED receiver, the DaC receiver can have better energy capture than the MF receiver at low sampling rates [229].

Example 5.2 *Consider a root-raised cosine (RRC) pulse shape with $T_p = 1$ ns, which is generated using a roll-off factor of $v = 0.5$. The RRC pulse is given by*

$$\omega_{\text{RRC}}(t) = \begin{cases} \frac{1}{\sqrt{T_p}} \frac{\sin\left((1-v)\pi t/T_p\right) + 4v(t/T_p)\cos\left((1+v)\pi t/T_p\right)}{(\pi t/T_p)\left(1-(4vt/T_p)^2\right)}, & t \neq 0, t \neq \pm\frac{T_p}{4v}, \\ \frac{1}{\sqrt{T_p}}\left[1 - v + \frac{4v}{\pi}\right], & t = 0, \\ \frac{v}{\sqrt{2T_p}}\left[\left(1 + \frac{2}{\pi}\right)\sin\left(\frac{\pi}{4v}\right) + \left(1 - \frac{2}{\pi}\right)\cos\left(\frac{\pi}{4v}\right)\right], & t = \pm\frac{T_p}{4v}. \end{cases}$$

(5.39)

The pulse is illustrated in Fig. 5.8 when sampled at 0.125 ns intervals.

In practice, a receiver might not have a capability to sample the signal at such high sampling rates. Instead, it may be sampled at lower sampling rates after analog front end processing. Figure 5.8 also illustrates sample outputs after processing the received signal with an MF receiver, an ED receiver, and a TR receiver. Both the ED and the TR receivers collect the energy within 1 ns windows, while the MF receiver uses the RRC pulse in (5.39) as a template. For the TR receiver, it is assumed that half of the energy is

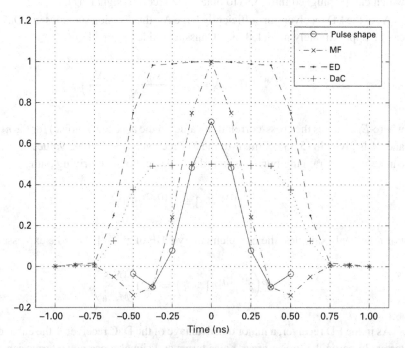

Fig. 5.8. Received *normalized* pulse shape, and sampled outputs corresponding to MF, ED, and DaC receivers for different timing offsets (1 ns pulse is sampled at 8 GHz, and energy is collected within 1 ns windows and different time offsets) (After [229]).

spared for the reference pulse. The sample outputs at time shift increments of $0.125\,ns$ are plotted in Fig. 5.8 in the absence of noise. It is observed from the figure that the MF receiver requires sampling rates on the order of the Nyquist rate in order to accurately capture the peak energy. On the other hand, the ED and the TR receivers can still capture a sufficient amount of energy at lower sampling rates.

For example, for a sampling period of $0.5\,ns$ and assuming that the receiver starts collecting the samples at time instant $-0.75\,ns$, the MF receiver samples are approximately given by $(0.01, 0.25, 0.25, 0.01)$ at time instants $(-0.75, -0.25, 0.25, 0.75)\,ns$, respectively. On the other hand, the samples after processing with the ED receiver are approximately given by $(0.01, 0.99, 0.99, 0.01)$, while the samples after processing with the TR receiver are approximately given by $(0.005, 0.485, 0.485, 0.005)$, at the same time instants. Hence, at lower than Nyquist sampling rates, the ED and the TR receivers can capture a larger amount of energy at timings closer to the true TOA of the signal.

Note again that in the ED and the TR receivers, the enhanced noise terms at the low/medium SNR regions become problematic. Hence, despite better energy capture at low sampling rates, the MF receiver may still outperform these receivers below certain SNR values. In other words, which receiver performs better than the other depends on both the SNR and the sampling period.

5.4 Fundamental limits for time-based ranging

Fundamental lower bounds such as the Cramer–Rao Lower Bound (CRLB) are commonly used for setting a lower bound on an estimator's MSE. Use of the CRLB is justified by the asymptotic optimality properties of the ML estimator. In other words, the ML estimator[11] asymptotically achieves the CRLB (and is asymptotically unbiased) under certain regularity conditions [140, 236]. For the problem of time-delay estimation, the CRLB yields a tight bound commonly at large SNRs and/or for sufficiently long observation intervals [235].

For moderate and low SNRs, or when the observation interval is not sufficiently long, CRLBs may not be very tight. This is because the cross-correlation outputs for a received signal are quasi-periodic with respect to f_c, where f_c is the center frequency of the received signal. In other words, while there is a unique correlation peak at the true TOA, there are also weaker correlation peaks (which may be comparable to the global peak for narrowband signals) that oscillate with a period of $1/f_c$. Therefore, the receiver may erroneously lock onto one of those weaker peaks at low/medium SNRs. In order to get close to the CRLB, the receiver must unambiguously distinguish between adjacent

---

[11] The ML estimator for the problem of time-delay estimation simply forms the cross-correlation of the received signal with a correlation template [235]. Ideally, the cross-correlation output is maximized at the true TOA of the received signal.

peaks of the correlation function. Otherwise, the achievable MSE may be drastically inferior to that predicted by the CRLB [236].

Example 5.3 *In order to better visualize the ambiguity problem, consider a signal whose autocorrelation function is modeled as follows [237]*

$$R(\tau) = f_H \operatorname{sinc}(f_H \tau)/\pi - f_L \operatorname{sinc}(f_L \tau)/\pi \tag{5.40}$$
$$= B\cos(f_c \tau)\operatorname{sinc}(B\tau/2)/\pi, \tag{5.41}$$

where $f_L = f_c - B/2$ and $f_H = f_c + B/2$ are, respectively, the lower and upper cut-off frequencies, f_c is the center frequency, and B is the signal bandwidth. Note that (5.41) has a flat power spectral density, and B can be changed in order to have sample realizations of narrowband, wideband, and UWB signals.

The autocorrelation function in (5.41) is plotted for a narrowband, wideband, and UWB signal in Fig. 5.9. For a narrowband signal, two types of ambiguities are observed: (1) local ambiguity around the global correlation peak, and (2) ambiguities due to

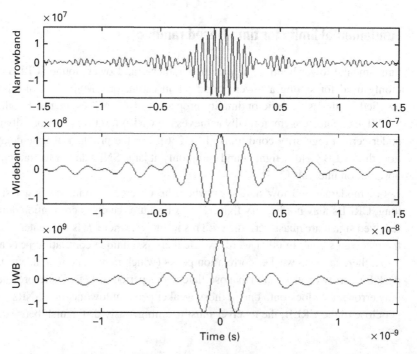

Fig. 5.9. The autocorrelation functions for narrowband, wideband, and UWB signals. The system parameters are $f_c = 0.4$ GHz and $B = 60$ MHz for the narrowband signal, $f_c = 0.9$ GHz and $B = 400$ MHz for the wideband signal, and $f_c = 7$ GHz and $B = 6$ GHz for the UWB signal.

multiple correlation peaks at the envelope of the autocorrelation function. For wideband and UWB signals, the ambiguity problem is lessened compared to a narrowband system.

Bounds other than the CRLB have also been investigated extensively in the literature for the time-delay estimation problem [235, 238–242]. In [236], the lower bounds for time-delay estimation are presented for baseband and bandpass signals. These bounds are summarized in Fig. 5.10. There are typically three distinct regimes of operation [241].

(i) At high SNRs, the noise variance at the output of the cross-correlator is smaller than the difference between the adjacent correlation peaks. Hence, the estimation error is due to small shifts from the correlation peak, which is bounded by the CRLB.
(ii) At intermediate SNRs, the noise variance may be higher than the difference between the adjacent correlation peaks; hence, the receiver may not distinguish the difference between the local and global maxima, and the achievable lower bound becomes much larger than that predicted by the CRLB.
(iii) At very low SNRs, the noise variance is much larger than the signal energy, and the estimator is useless.

Figure 5.10 shows that at very low SNRs, the MSE is the variance of a random variable uniformly distributed in $(0, T_a]$ and is given by $T_a^2/12$. For baseband systems, the lower bound has a threshold region at intermediate SNR values (due to the ambiguity problem), and converges to the CRLB at high SNRs. For bandpass systems, there is an intermediate region which is modeled by the Baranakin bound [236], and the MSE depends on the center frequency f_c and the signal bandwidth B. Within this region, the MSE bound is $12(f_c/B)^2$ dB larger than that predicted by the CRLB. For a more detailed discussion and exact evaluation of these bounds, the reader is referred to [235, 236, 240, 241].

When multiple replicas of the transmitted signal arrive at the receiver, the simple cross-correlator is no longer the optimal receiver. However, the three distinct regions

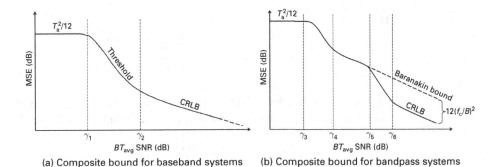

(a) Composite bound for baseband systems   (b) Composite bound for bandpass systems

Fig. 5.10. Composite bounds for time delay estimation in baseband and bandpass systems (γ_1 to γ_6 denote the SNR thresholds for different regions, B is signal bandwidth, f_c is the center frequency of the signal, and TOA is uniformly distributed in $(0, T_a]$). It is assumed that the received signal is averaged over an interval T_{avg} before obtaining the desired delay from the peak of the cross-correlation function (© 1984 IEEE) [236].

of operation are still expected [241]. As for the UWB signals, the receiver can observe on the order of one hundred MPCs; hence, new interpretations of the prior results on improved lower bounds remains an active research area.

In the remainder of this section, the CRLBs for time delay estimation in UWB systems are studied for both single-path and multipath channels, which is followed by a discussion on the improved lower bounds for UWB ranging.

5.4.1 Cramer–Rao lower bounds for single-path channels

As discussed in Chapter 2, the CRLB for a single-path AWGN channels is given by [243]

$$\sqrt{\text{Var}(\hat{d})} \geq \frac{c}{2\sqrt{2\pi}\sqrt{\text{SNR}}\,\beta}, \qquad (5.42)$$

where SNR is the signal-to-noise ratio, c is the speed of light, and β is the effective signal bandwidth defined as

$$\beta = \left[\frac{\int_{-\infty}^{\infty} f^2 |S(f)|^2 \mathrm{d}f}{\int_{-\infty}^{\infty} |S(f)|^2 \mathrm{d}f}\right]^{\frac{1}{2}}, \qquad (5.43)$$

with $S(f)$ denoting the Fourier transform of the transmitted signal. The CRLB for time-based ranging decreases with the square-root of the SNR and effective signal bandwidth. For example, with a received pulse bandwidth of 1.5 GHz and an SNR of 0 dB, less than an inch of accuracy can be obtained using UWB signals [243].

As implied by (5.43), the CRLB depends on the Fourier transform of the transmitted signal. A UWB pulse can be conveniently generated from the nth derivative of a Gaussian pulse,

$$\omega_0(t) = \exp\left(-2\pi \frac{t^2}{\kappa_p^2}\right), \qquad (5.44)$$

as [244]

$$\omega_n(t) = \omega_0^{(n)}(t) \sqrt{\frac{(n-1)!}{(2n-1)! \pi^n \kappa_p^{(1-2n)}}}, \qquad (5.45)$$

where $\omega_0^{(n)}(t)$ is the nth-order derivative of the Gaussian pulse $\omega_0(t)$, and κ_p is a parameter that adjusts the pulse duration. For the nth-order pulse, (5.43) simplifies to [244]

$$\beta_{(n)}^2 = \frac{2\pi(2n+1)}{\kappa_p^2}. \qquad (5.46)$$

Then, for the same pulse order n and with different parameters $\kappa_{p,1}$ and $\kappa_{p,2}$,

$$\frac{\text{CRLB}_{\kappa_{p,1}}}{\text{CRLB}_{\kappa_{p,2}}} = \frac{\kappa_{p,1}^2}{\kappa_{p,2}^2}, \qquad (5.47)$$

which implies that the CRLB is proportional to the square of the pulse width. On the other hand, for the same κ_p (i.e. the same pulse width), the CRLB for different pulse orders is given by [244]

$$\frac{\text{CRLB}_{n+1}}{\text{CRLB}_n} = \frac{2n+1}{2n+3}, \qquad (5.48)$$

which implies that the CRLB is lower for higher order Gaussian pulses. Note that (5.48) converges to 1 for large n.

5.4.2 Cramer–Rao lower bounds for multipath channels

When the received signal in (5.5) is observed over an interval $t \in [0, T_{\text{obs}}]$, with $T_{\text{obs}} = N_r T_s$, the log-likelihood function of θ in (5.11) is given by [245]

$$\Lambda(\theta) = \tilde{k} - \frac{1}{\mathcal{N}_0} \int_0^{N_r T_s} \left| r(t) - \sum_{l=1}^{L} \alpha_l s(t - \tau_l) \right|^2 dt, \qquad (5.49)$$

where $\tilde{k}$ represents a term that is independent of θ. Hence, the ML estimation of the unknown parameter vector θ is achieved by maximizing (5.49). The Fisher information matrix (FIM) is obtained from the second-order derivatives of (5.49), which results in (with the assumption of no inter-pulse interference) [243]

$$\mathbf{I}_\theta = \begin{bmatrix} \mathbf{I}_{\alpha\alpha} & \mathbf{I}_{\alpha\tau} \\ \mathbf{I}_{\alpha\tau}^T & \mathbf{I}_{\tau\tau} \end{bmatrix}, \qquad (5.50)$$

$$\mathbf{I}_{\alpha\alpha} = \frac{2N_r E_s E_p}{\mathcal{N}_0} \operatorname{diag}\{1, \ldots, 1\}, \qquad (5.51)$$

$$\mathbf{I}_{\alpha\tau} = -\frac{2N_r E_s E_p'}{\mathcal{N}_0} \operatorname{diag}\{\alpha_1, \ldots, \alpha_L\}, \qquad (5.52)$$

$$\mathbf{I}_{\tau\tau} = \frac{2N_r E_s E_p''}{\mathcal{N}_0} \operatorname{diag}\{\alpha_1^2, \ldots, \alpha_L^2\}, \qquad (5.53)$$

where

$$E_p' = \int_0^{T_p} \omega(t)\omega'(t)dt, \qquad (5.54)$$

$$E_p'' = \int_0^{T_p} |\omega'(t)|^2 dt, \qquad (5.55)$$

with E_p being given by (5.7), and $\omega'(t)$ denoting the first derivative of the pulse shape $\omega(t)$. From (5.50), the CRLB for each of the delay values τ_l for $l = 1, 2, \ldots, L$ is obtained as [243]

$$\text{CRLB}(\tau_l) = \left[\left(\mathbf{I}_{\tau\tau} - \mathbf{I}_{\alpha\tau}^T \mathbf{I}_{\alpha\alpha}^{-1} \mathbf{I}_{\alpha\tau} \right)^{-1} \right]_{l,l}$$

$$= \frac{\mathcal{N}_0}{2N_r E_s (E_p'' - E_p'^2/E_p)\alpha_l^2}. \qquad (5.56)$$

As implied by (5.56), the CRLB in multipath channels depends on the pulse shape, the path gains, and the SNR.

For the special case of a single-path channel,[12] $L=1$ and $E'_p=0$, and the CRLB becomes [246]

$$\text{CRLB}(\tau_1) = \frac{\mathcal{N}_0}{2 N_r E_s E''_p \alpha_1^2}. \tag{5.57}$$

Since

$$E''_p = \frac{\int_{-\infty}^{\infty} (2\pi f)^2 |\Phi(f)|^2 \mathrm{d}f}{\int_{-\infty}^{\infty} |\Phi(f)|^2 \mathrm{d}f} = 4\pi^2 \beta^2, \tag{5.58}$$

where $\Phi(f)$ denotes the Fourier transform of $\omega(t)$, (5.57) is equivalent to the CRLB expression in (5.42).

Alternative derivations of the CRLB for UWB systems in multipath environments are presented in [247–249]. In [247], it is shown that for a received signal with a certain number of MPCs, the CRLB will be inferior to that for a signal with a smaller number of MPCs. Another parameter affecting the CRLB is the autocorrelation characteristics of the spreading sequences (e.g. TH sequences); if they have ideal autocorrelation characteristics, the CRLB for a multipath channel converges to the CRLB for single-path channels. In [248], the CRLB for time delay estimation has been presented for log-normal distributed path gains and Poisson distributed path delays. In addition to the ordinary CRLB in (5.56) that is obtained from the ML criterion, three different *a-posteriori* CRLBs that depend on three different *a-priori* knowledge scenarios are presented. In [249], frequency-dependent features and phase of the MPCs are included in the CRLB derivation, which is solved through the Whittle's formula.

Even though the CRLB is useful for serving as a benchmark for other practical estimators, it may not always be achievable. In particular, for UWB signals, sampling rates above the Nyquist rate are required in order to achieve the CRLB. This implies sampling rates on the order of tens of GHz, which may not be possible in practical receivers.

5.4.3 Ziv–Zakai lower bounds for single-path channels

A disadvantage of the CRLB is that it is usually tight only at high SNRs. Contrary to the CRLB, the Ziv–Zakai lower bound (ZZLB) is tight for a wide range of SNRs [236, 238]. The ZZLB can be derived from the following identity for the MSE of an estimator [244]

$$\mathrm{E}\{\epsilon^2\} = \frac{1}{2} \int_0^{\infty} \zeta P\left(|\epsilon| \geq \zeta/2\right) \mathrm{d}\zeta, \tag{5.59}$$

where ϵ represents the error term. The key issue in the evaluation of (5.59) is that a lower bound on $P(|\epsilon| \geq \zeta/2)$ has to be found [238]. The $P(|\epsilon| \geq \zeta/2)$ expression is identical to

---

[12] Note that ideally, (5.57) is also valid in multipath channels as long as the MPCs are resolvable. However, in practical algorithms, estimation of the first arriving path may be dependent on estimation of the other paths (such as the strongest path) even in resolvable channels.

the error probability of a binary hypothesis testing (BHT) problem with a sub-optimum decision rule, given by [238]

$$\mathcal{H}_0 : r(t) \sim P(r(t)|\tau)$$
$$\mathcal{H}_1 : r(t) \sim P(r(t)|\tau + \zeta), \qquad (5.60)$$

where $\mathcal{H}_0$ and $\mathcal{H}_1$ are known to be equally probable, and $P(r(t)|\tau)$ is the PDF of $r(t)$ conditioned on τ. Then,

$$P\left(|\epsilon| \geq \zeta/2\right) \geq \int_{-\infty}^{\infty} [p_\tau(\tilde{\tau}) + p_\tau(\tilde{\tau} + \zeta)] P_{\text{opt}}(\tilde{\tau}, \tilde{\tau} + \zeta) d\tilde{\tau}, \qquad (5.61)$$

where $p_\tau(\tilde{\tau})$ is the PDF of τ, which is considered to be uniformly distributed in $(0, T_a]$, and $P_{\text{opt}}(\tilde{\tau}, \tilde{\tau} + \zeta)$ is the error probability obtained from the optimum decision rule (i.e. a likelihood ratio test for the two hypotheses) [244].[13] For example, in a single-path AWGN channel, it is well known that the probability of error for a binary modulation with an optimum receiver is given by[14]

$$P_{\text{opt}}(\zeta) = Q\left(\sqrt{\frac{E_p}{N_0}(1 - R_\omega(\zeta))}\right), \qquad (5.62)$$

where the filter matched to the received signal is given by $\omega_\tau(t - \tau) - \omega_\tau(t - \tau - \zeta)$, and $R_\omega(\zeta)$ denotes the auto-correlation function of the received pulse $\omega(t)$ as given in (5.20).

From (5.59) and (5.61), the ZZLB can be derived as [244]

$$\text{ZZLB} = \frac{1}{T_a} \int_0^{T_a} \zeta(T_a - \zeta) P_{\text{opt}}(\zeta) d\zeta, \qquad (5.63)$$

where $P_{\text{opt}}(\zeta)$ is as given in (5.62).

Example 5.4 *In Fig. (5.11), the CRLB and the ZZLB of the ranging error are analyzed when the UWB pulse in (2.5) is employed with various pulse durations. A single-path AWGN channel is considered and T_a is set to 100 ns.*

While the ZZLBs and the CRLBs overlap in the high SNR region, the ZZLB is much tighter than the CRLB at low SNRs. This is because at low SNRs, the received signal is unreliable; hence the estimator may not always lock on the correct signal peak but may rather lock onto the side-lobes. Note that the overall accuracy improves as shorter pulse durations are used.

---

[13] Since $P_{\text{opt}}(\tilde{\tau}, \tilde{\tau} + \zeta)$ does not usually depend on $\tilde{\tau}$, $P_{\text{opt}}(\zeta)$ will be used in the sequel.
[14] Note that (5.62) assumes single pulse transmission per symbol. For multiple pulses per symbol, it is sufficient to replace E_p with E_s.

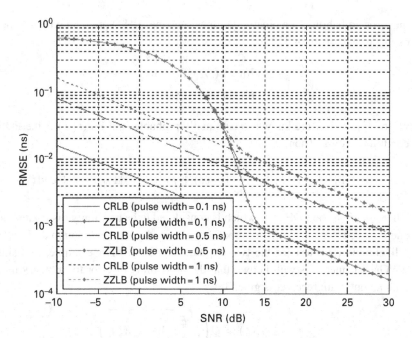

Fig. 5.11. ZZLBs and CRLBs in AWGN channels for different pulse widths.

5.4.4 Ziv–Zakai lower bounds for multipath channels

In multipath channels, the ZZLB depends on the receiver's *a-priori* knowledge of the nuisance parameter vector $\tilde{\theta}$ in (5.14). First, assume that the receiver has perfect knowledge of $\tilde{\theta}$. While this is not practical in realistic scenarios, it yields a perfect measurement bound (PMB) as discussed in [250] and sets a lower-bound on any TOA estimator. The error probability for PMB is given by [244]

$$P_{\text{opt}}^{(\text{PMB})}(\zeta|\tilde{\theta}) = Q\left(\sqrt{\frac{E_{\text{p}}}{\mathcal{N}_0}\left(\tilde{R}(0|\tilde{\theta}) - \tilde{R}(\zeta|\tilde{\theta})\right)}\right), \quad (5.64)$$

where the auto-correlation function for the multipath signal is given by

$$\tilde{R}(\zeta|\tilde{\theta}) = \sum_{l=1}^{L}\sum_{k=1}^{L} \alpha_l \alpha_k R_\omega(\tau_l - \tau_k - \zeta). \quad (5.65)$$

Then, (5.64) can be plugged into (5.63) in order to obtain the ZZLB for a given channel realization. To obtain the average ZZLB (AZZLB) for a particular environment, the ZZLBs can be averaged over a large number of channel realizations representative of that environment.

On the other hand, if $\tilde{\boldsymbol{\theta}}$ is a set of random variables with known distributions, the conditional PDF $P(r(t)|\tau, \tilde{\boldsymbol{\theta}})$ shall be averaged over $\tilde{\boldsymbol{\theta}}$ as follows [244]

$$P(r(t)|\tau) = \int_{\mathcal{R}^{2L-1}} p_{\tilde{\boldsymbol{\theta}}}(\check{\boldsymbol{\theta}}) P(r(t)|\tau, \check{\boldsymbol{\theta}}) d\check{\boldsymbol{\theta}}, \qquad (5.66)$$

where $p_{\tilde{\boldsymbol{\theta}}}(\check{\boldsymbol{\theta}})$ is the joint PDF of $\tilde{\boldsymbol{\theta}}$. Then, (5.66) can be used to obtain hypothesis $\mathcal{H}_0$ and $\mathcal{H}_1$ in order to evaluate the ZZLB. Since (5.66) may not be analytically tractable, the ZZLB can be evaluated using empirical CIRs or Monte Carlo simulations.

Alternative analysis of the ZZLBs for UWB ranging systems can be found in [237, 251]. A simplified bound that has a closed-form expression is derived in [237], which is exact at high SNRs, and close to the actual ZZLB at medium and low SNRs. In [251], second-order statistics of received signals are used to evaluate the ZZLB for UWB ranging systems.

5.5 Maximum likelihood-based ranging techniques

Fundamental bounds such as the CRLB and the ZZLB can be used to find a lower bound on the ranging accuracy of UWB signals. However, these bounds may be very difficult to achieve with practical estimators. Both the CRLB and the ZZLB discussed in the previous section assume perfect knowledge of $\tilde{\boldsymbol{\theta}}$ or its probability distribution; such information may not be available in practice.

Maximum likelihood (ML)-based estimators may achieve accuracies that are close to fundamental lower bounds provided that certain *a-priori* information is available. In this section, different ML-based ranging techniques are described with varying *a-priori* information requirements.

5.5.1 ML estimation with full *a-priori* information

The TOA can be estimated optimally by using an MF that is perfectly matched to the received multipath signal, and choosing the time delay of the template that maximizes the correlation output (see Fig. 5.12) [142, 243]. The optimal template can be defined as

$$s_{\text{tmp}}(t) = \sum_{l=1}^{L} \alpha_l s(t - \tau_l). \qquad (5.67)$$

However, this optimal receiver is not possible to implement in practice since the received waveform has unknown parameters to be estimated. In particular, the nuisance parameter vector $\tilde{\boldsymbol{\theta}}$ and the pulse shape at each MPC (which may be different from the transmitted pulse shape in practice) have to be available. Therefore, the optimal correlation template cannot be obtained in practice.

Alternatively, the transmitted waveform itself can be used as a correlation template at the receiver. However, since the multipath received waveform is different from the correlation template, such a receiver is obviously suboptimal. In fact, even in the absence

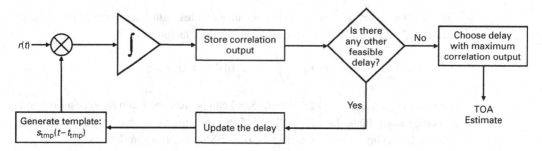

Fig. 5.12. Block diagram for a correlation-based TOA estimator. The correlator template is denoted by $s_{\text{tmp}}(t)$ and its delay is denoted by t_{tmp}.

of noise, the correlation peak does not necessarily correspond to the TOA of the received signal,[15] because the first arriving MPC is not always the strongest one [100]. In order to identify the first arriving path, first path detection algorithms need to be employed, which will be discussed later in this chapter.

5.5.2 ML estimation with no prior information

While ranging requires estimation of the TOA corresponding to the first arriving signal path, τ_1, the received signal itself depends on the unknown parameter vector θ in (5.11). Hence, the TOA estimation problem is closely related to the channel estimation problem where all parameters captured by θ are estimated [227, 252].

In order to formulate the ML solution, consider that the samples of the received signal are as given in (5.15). In the presence of Gaussian noise, the ML solution is equivalent to a minimum mean-squared error (MMSE) solution, which is given by [227]

$$\hat{\theta} = \arg\min_{\theta} \left\{ \frac{1}{N} \sum_{i=1}^{N} |r[i] - \hat{r}[i]|^2 \right\}, \tag{5.68}$$

where $\hat{r}[i]$ are the samples of the reconstructed received signal, given by

$$\hat{r}[i] = \sum_{l=1}^{L} \hat{\alpha}_l \omega(i T_{\text{smp}} - \hat{\tau}_l). \tag{5.69}$$

Then, the ML estimates for the elements of the unknown parameter vector $\theta = [\tau \; \alpha]$ are given by [252]

$$\hat{\tau} = \arg\max_{\tau} \left\{ \left[\mathbf{\Omega}^{\mathrm{T}}(\tau)\mathbf{r} \right]^{\mathrm{T}} \mathbf{R}_{\Omega}^{-1}(\tau) \left[\mathbf{\Omega}^{\mathrm{T}}(\tau)\mathbf{r} \right] \right\}, \tag{5.70}$$

$$\hat{\alpha} = \mathbf{R}_{\Omega}^{-1}(\hat{\tau})\mathbf{\Omega}^{\mathrm{T}}(\tau)\mathbf{r}, \tag{5.71}$$

---

[15] Note that rather than the transmitted signal, if (5.67) is adopted as the reference signal, the maximum peak (in the absence of noise) always gives the correct TOA even when the first MPC is not the strongest one.

where

$$\mathbf{R}_\Omega(\boldsymbol{\tau}) = \mathbf{\Omega}^T(\boldsymbol{\tau})\mathbf{\Omega}(\boldsymbol{\tau}) \qquad (5.72)$$

$$= \begin{bmatrix} [\omega^{(D_1)}]^T\omega^{(D_1)} & [\omega^{(D_1)}]^T\omega^{(D_2)} & \cdots & [\omega^{(D_1)}]^T\omega^{(D_L)} \\ [\omega^{(D_2)}]^T\omega^{(D_1)} & [\omega^{(D_2)}]^T\omega^{(D_2)} & \cdots & [\omega^{(D_2)}]^T\omega^{(D_L)} \\ \vdots & \vdots & \ddots & \vdots \\ [\omega^{(D_L)}]^T\omega^{(D_1)} & [\omega^{(D_L)}]^T\omega^{(D_2)} & \cdots & [\omega^{(D_L)}]^T\omega^{(D_L)} \end{bmatrix}, \qquad (5.73)$$

and $\mathbf{\Omega}^T(\boldsymbol{\tau})$ and $\omega^{(D_l)}$ are as in (5.17) and (5.18), respectively. Note that the ML estimation of the delays and the channel coefficients requires high computational complexity since it requires a search over different values of $\boldsymbol{\tau}$.

Dependence of the estimation of τ_1 on the other channel parameters is subject to the resolvability of the channel. If $|\tau_i - \tau_j| \geq T_p \; \forall i \neq j$, then the channel is called resolvable. For a resolvable channel, the unknown parameter vector can be estimated as [227]

$$\hat{\boldsymbol{\tau}} = \arg\max_{\boldsymbol{\tau}} \left\{ \sum_{l=1}^{L} \frac{([\omega^{(D_l)}]^T\mathbf{r})^2}{R_\Omega(0)} \right\}, \qquad (5.74)$$

$$\hat{\boldsymbol{\alpha}} = \frac{\mathbf{\Omega}^T(\hat{\boldsymbol{\tau}})\mathbf{r}}{R_\Omega(0)}, \qquad (5.75)$$

where $R_\Omega(0) = E_p = 1$. Note that for a resolvable channel, $\hat{\boldsymbol{\tau}}$ in (5.74) can be estimated by maximizing each term in the summation independently. In other words, estimation of the TOA τ_1 might be decoupled from the estimation of $\tilde{\boldsymbol{\theta}}$. This result is similar to what happens when determining the CRLB in (5.57) in multipath channels (i.e. the CRLB for τ_1 can be ideally decoupled from the CRLB of other parameters in resolvable channels). In fact, the ML estimator in (5.74) and (5.75) is an efficient estimator; i.e. it achieves the CRLB asymptotically.

On the other hand, if for some $i \neq j$ there are some τ_i and τ_j such that $|\tau_i - \tau_j| < T_p$, then the channel is non-resolvable, and estimation of τ_1 may depend on estimation of other parameters.

5.5.3 Ranging with generalized maximum likelihood ratio test

The ML estimation is primarily designed for channel parameter estimation purposes and inherently yields an estimate of τ_1. However, it may be impractical for UWB channels with a vast number of MPCs because the estimation of all the channel parameters may be computation intensive. With some prior assumptions such as perfect synchronization of the receiver to the strongest MPC, it is possible to design alternative ML techniques that may decrease the parameter search space. For example, a receiver that uses a generalized maximum likelihood (GML) technique is presented in [253]. It searches only the paths prior to the strongest MPC.

The signal model for the GML receiver can be obtained as follows. The received signal in (5.5) can be re-written as a sum of the first path, the remaining paths, and the noise as [253]

$$r(t) = \alpha_1 \omega(t - \tau_1) + \sum_{l=2}^{L} \alpha_l \omega(t - \tau_l) + n(t), \qquad (5.76)$$

where $\tau_1 < \tau_2 < \cdots < \tau_L$. Also assume that the delay τ_{peak} and the channel coefficient α_{peak} of the strongest MPC are already estimated. Then, the normalized signal prior to and including the strongest MPC can be defined as

$$\tilde{r}(t) = \frac{r(t + \tau_{\text{peak}})}{|\alpha_{\text{peak}}|}$$

$$= \tilde{\alpha}_1 \omega(t + \tilde{\tau}_1) + \sum_{l=2}^{L_{\text{max}}} \tilde{\alpha}_l \omega(t + \tilde{\tau}_l) + \tilde{n}(t), \qquad (5.77)$$

where $\tilde{\tau}_l = \tau_{\text{peak}} - \tau_l$, $\tilde{\alpha}_l = \alpha_l / |\alpha_{\text{peak}}|$, L_{max} is the number of MPCs prior to and including the strongest MPC, and $\tilde{n}(t)$ is the white Gaussian noise. After wideband filtering and sampling the signal in (5.77) at above the Nyquist rate, $\tilde{r}(t)$ can be written in vector form as

$$\tilde{\mathbf{r}} = \tilde{\alpha}_1 \boldsymbol{\omega}_{\tilde{\tau}_1} + \sum_{l=2}^{L_{\text{max}}} \tilde{\alpha}_l \boldsymbol{\omega}_{\tilde{\tau}_l} + \tilde{\mathbf{n}}, \qquad (5.78)$$

with $\boldsymbol{\omega}_{\tilde{\tau}_l}$ denoting the samples of $\omega(t + \tilde{\tau}_l)$. Then, given that $\tilde{\mathbf{n}}$ is a white Gaussian noise vector, the GML estimate of $\tilde{\tau}_1$ is given by [253]

$$\hat{\tilde{\tau}}_1 = \arg\max_{\tilde{\tau}_1} \left[\min_{\tilde{\alpha}_1, L_{\text{max}}, \tilde{\boldsymbol{\alpha}}, \tilde{\boldsymbol{\tau}}} \left\| \tilde{\mathbf{r}} - \tilde{\alpha}_1 \boldsymbol{\omega}_{\tilde{\tau}_1} - \sum_{l=2}^{L_{\text{max}}} \tilde{\alpha}_l \boldsymbol{\omega}_{\tilde{\tau}_l} \right\|^2 \right], \qquad (5.79)$$

where $\tilde{\boldsymbol{\alpha}} = [\tilde{\alpha}_2, \ldots, \tilde{\alpha}_{L_{\text{max}}}]$ and $\tilde{\boldsymbol{\tau}} = [\tilde{\tau}_2, \ldots, \tilde{\tau}_{L_{\text{max}}}]$. Note that (5.79) has very high computational complexity since a search over the unknown parameter set $\{\tilde{\tau}_1, \tilde{\alpha}_1, L_{\text{max}}, \tilde{\boldsymbol{\alpha}}, \tilde{\boldsymbol{\tau}}\}$ is required.

In order to have a lower complexity implementation, the GML algorithm is modified in [253] as outlined in Table 5.1. The $\Delta_{\tilde{\tau}}$ and $\Delta_{\tilde{\alpha}}$ are two critical thresholds that are obtained from channel statistics, and are used to define the search space and the stopping rule, respectively. In summary, the algorithm searches $\tilde{\tau}_1$ in the region where $t \geq -\Delta_{\tilde{\tau}}$, and stops when $|\tilde{\alpha}_1| < \Delta_{\tilde{\alpha}}$. The GML algorithm is a recursive algorithm and yields very high ranging accuracies as reported in [253]. However, it requires very high sampling rates which may be a major drawback in practical applications.

5.5.4 Sub-Nyquist sampling ML estimation with different levels of *a-priori* information

Computational complexity, sampling rate requirements, and *a-priori* knowledge requirements of the ML techniques discussed in Sections 5.5.2 and 5.5.3 may be prohibitive for their practical implementation. In this section, ML estimators that can operate at low sampling rates and with different levels of *a-priori* information are described. While

Table 5.1. Steps of the simplified GML algorithm in [253].

(i) Set $n = 1$, $\delta_1 = 0$, and $\mu_{11} = 1$.
(ii) Increment n by 1.
(iii) Find δ_n that satisfies

$$\delta_n = \arg \max_{\delta_{n-1} < \delta < \Delta_{\tilde{\tau}}} \left[\tilde{\mathbf{r}} - \sum_{i=1}^{n-1} \mu_{(n-1)i}\, \boldsymbol{\omega}_{\delta_i} \right]^{\mathrm{T}} \boldsymbol{\omega}_\delta . \tag{5.80}$$

(iv) Find $[\mu_{n1}, \ldots, \mu_{nn}]$ that satisfies

$$[\mu_{n1}, \ldots, \mu_{nn}] = \arg \min_{\mu'_1, \ldots, \mu'_n} \left\| \tilde{\mathbf{r}} - \sum_{i=1}^{n} \mu'_i \boldsymbol{\omega}_{\delta_i} \right\|^2 . \tag{5.81}$$

(v) If $\mu_{nn} \geq \Delta_{\tilde{a}}$, go to step 2. Else, proceed to step 6.
(vi) The estimate of $\tilde{\tau}_1$ is given by $\hat{\tilde{\tau}}_1 = \delta_{n-1}$.

---

any of the receiver architectures illustrated in Fig. 5.7 can be used to obtain the decision variables, in the system below, an ED receiver is considered. First, the TOA estimation problem is interpreted as a multiple hypothesis testing problem and a decision theoretical framework is provided. Then, different ML-based TOA estimators with various levels of complexity are presented.

Multiple hypothesis testing system model

Let an ED receiver collect N_b samples from the received signal at a sampling rate of $1/T_{\mathrm{smp}}$ and $\mathbf{z}$ represent a $1 \times N_b$ vector that consists of those samples. Furthermore, let $n_{\mathrm{le}} = \lceil \tau_1/T_{\mathrm{smp}} \rceil \sim \mathcal{U}\{1, 2, \ldots, T_a/T_{\mathrm{smp}}\}$ denote the TOA of the signal in samples, $N_e = \lceil T_{\mathrm{med}}/T_{\mathrm{smp}} \rceil$ represent the maximum excess delay of the channel in samples, T_{med} is the maximum excess delay of the channel in seconds, and $\mathcal{H}_k$ denotes the hypothesis that the signal arrives at the kth sample. Then, for $k = 1, 2, \ldots, T_a/T_{\mathrm{smp}}$ (assume that T_a is an integer multiple of T_{smp}), different hypotheses can be written as follows

$$\mathcal{H}_k : \begin{aligned} z[n] &= \int_{(n-1)T_{\mathrm{smp}}}^{nT_{\mathrm{smp}}} \eta^2(t)dt, & n &= 1, \ldots, k-1 \\ z[n] &= \int_{(n-1)T_{\mathrm{smp}}}^{nT_{\mathrm{smp}}} [r_{\mathrm{des}}(t) + \eta(t)]^2 dt, & n &= k, \ldots, k+N_e-1, \\ z[n] &= \int_{(n-1)T_{\mathrm{smp}}}^{nT_{\mathrm{smp}}} \eta^2(t)dt, & n &= k+N_e, \ldots, N_b \end{aligned} \tag{5.82}$$

where $r_{\mathrm{des}}(t)$ is the desired signal in (5.10), $z[n]$ is the nth element of $\mathbf{z}$, $\eta(t)$ is the noise after the BPF,[16] and $\mathcal{H}_{n_{\mathrm{le}}}$ is the true hypothesis.

For notational convenience in the rest of the section, define $\mathcal{E}_{n-n_{\mathrm{le}}+1} = E_n$ for $n \in \{n_{\mathrm{le}}, \ldots, n_{\mathrm{le}}+N_e-1\}$ (where E_n represents the desired signal energy at the nth sample as discussed in Section 5.3.3, which is zero for noise-only samples), which removes the TOA

---

[16] Signal part is assumed to be undistorted by the BPF.

offset from the desired signal's sample index. In addition, define $\mathcal{E}_{N_e} = [\mathcal{E}_1, \mathcal{E}_2, \ldots, \mathcal{E}_{N_e}]$ to be a channel energy vector that captures the desired signal samples. Furthermore, let $\mathbf{z}_k^{(\text{no})}$ and $\mathbf{z}_k^{(\text{sn})}$ denote the hypothesized noise-only energy vector and signal-plus-noise energy vector of sizes $1 \times (N_b - N_e)$ and $1 \times N_e$, respectively, for the kth hypothesis. The vectors on the two sides of signal-plus-noise vector $\mathbf{z}_k^{(\text{sn})}$ are concatenated to yield $\mathbf{z}_k^{(\text{no})}$. Note that as discussed in Section 5.3.3, under certain conditions, it may be possible to model $z[n]$ with a Gaussian random variable as in (5.30).

Maximum energy selection

Typically, N_e in (5.82) is much larger than 1 for T_{smp} values of the order of a pulse duration, and the signal energy is spread over many samples. The simplest way to obtain a TOA estimate from these samples is the maximum energy selection (MES) from the sample vector $\mathbf{z}$ by neglecting the information in the neighboring samples, which yields

$$\hat{n}_{\text{toa}}^{(\text{mes})} = \arg \max_{k \in \{1, \ldots, N_b\}} \{z[k]\}. \tag{5.83}$$

However, the MES is susceptible to noise since the energy in only a single sample is used; hence, it may not provide high timing resolution as there may be a large delay between the first path and the strongest path.

Maximum energy sum selection

In order to exploit the energy in the neighboring MPCs, energy samples within a window can be aggregated. With a window duration of $N_w \leq N_e$ samples, the leading edge estimate using maximum energy sum selection (MESS) is given by

$$\hat{n}_{\text{toa}}^{(\text{mess})} = \arg \max_{k \in \{1, \ldots, N_b\}} \left\{ \mathbf{z}_k^{(\text{sn}, N_w)} \mathbf{1}_{N_w} \right\}, \tag{5.84}$$

where $\mathbf{1}_{N_w}$ is an $N_w \times 1$ column vector of ones, and MESS reduces to MES for $N_w = 1$. The vector $\mathbf{z}_k^{(\text{sn}, N_w)}$ is composed of the first N_w elements of $\mathbf{z}_k^{(\text{sn})}$. Since a very large window length captures a large amount of noise and small window length may not capture sufficient energy, there exists an optimum window length that depends on the channel realization and the $E_s/\mathcal{N}_0$.

In essence, the MESS is nothing but a sliding window of length $N_w T_{\text{smp}}$, where the window is shifted with increments of T_{smp}. The shift that captures the largest energy is used to determine the TOA of the received signal. In [210], it is shown that such a sliding window based solution that uses an energy detector is actually similar to a GML-based estimator.

Example 5.5 *Consider a UWB system that transmits a root-raised cosine transmitted pulse of duration $T_p = 4$ ns. Let the wireless channel that signal traverses before it arrives at the receiver be modeled by the first channel model (CM-1) of the IEEE 802.15.4a [100]. The signal arriving at the receiver is first filtered with a BPF of bandwidth $B = 0.5$ GHz, and passed through an ED receiver with $T_{\text{smp}} = 4$ ns (hence, DOF is 2×4 ns $\times 0.5$ GHz $+ 1 = 5$). The samples are further averaged over $N_r = 100$ ranging symbols in order to*

5.5 Maximum likelihood-based ranging techniques

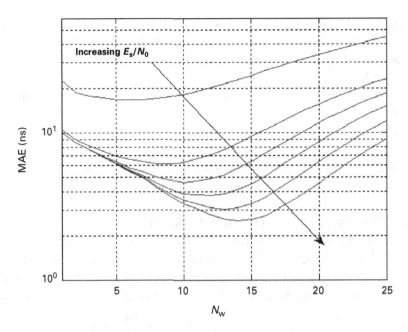

Fig. 5.13. Simulated MAEs corresponding to different lengths of sliding windows and at different SNRs ($E_s/\mathcal{N}_0 \in \{7, 10, 13, 16, 19, 22\}$ dB). The CM-1 channel model of IEEE 802.15.4a is used.

increase the SNR.[17] Also let $\tau_1 \sim \mathcal{U}(0, 512)$ ns, and the receiver searches the interval between $(0, 1024)$ ns with a sliding window and using MESS. The results are averaged over 100 different noise realizations and 1000 different realizations of CM-1 channels.

In Fig. 5.13, the simulated ranging accuracy of the MESS algorithm is investigated for various window sizes N_w using the mean absolute error (MAE) of range estimates as an error metric.[18] The optimum sliding window size increases as the SNR increases. At very high SNRs, it is of the order of maximum excess delay of the channel. For example, for $E_s/\mathcal{N}_0 = 22$ dB, the optimum window size is approximately $16T_{smp} = 64$ ns.

Weighted maximum energy sum selection

If *a-priori* knowledge of the channel power delay profile is available, it may be used to weigh the hypothesized energy vector, which yields

$$\hat{n}_{\text{toa}}^{(\text{wmess})} = \arg \max_{k \in \{1,\ldots,N_b\}} \left\{ \mathbf{z}_k^{(\text{sn}, N_e)} \boldsymbol{\rho}_{N_e} \right\}, \tag{5.85}$$

---

[17] Hence, Gaussian approximation can be used to model the statistics of the energy detector.
[18] Note that different metrics can be used to characterize the ranging accuracy. Typically root mean square error (RMSE) and MAE are preferable metrics since they are in the same units as the original data. The MAE is usually easier to understand and interpret compared to RMSE for the mathematically challenged [254]. On the other hand, the RMSE gives a relatively larger weight to large errors, and hence is most useful when large errors are particularly undesirable. The MAE will always be smaller than or equal to the RMSE.

where $\boldsymbol{\rho}_{N_e}$ is a column vector of $1 \times N_e$ mean energy values for a particular channel model and block duration. Note that the weighted-MESS (W-MESS) in (5.85) is actually equivalent to correlating the received energy vector with the power delay profile before peak selection.

Alternatively, if a perfect knowledge of the channel energies is available, the TOA estimate can be obtained as

$$\hat{n}_{\text{toa}}^{(\text{wmess}')} = \arg \max_{k \in \{1,\ldots,N_b\}} \left\{ \mathbf{z}_k^{(\text{sn},N_e)} \boldsymbol{\mathcal{E}}_{N_e} \right\}. \tag{5.86}$$

While it may be impractical to obtain the perfect knowledge of the channel vector $\boldsymbol{\mathcal{E}}_{N_e}$, it is possible to utilize its noisy estimate.

Double-weighted maximum energy sum selection

Careful study of (5.82) supports that for correct $\mathcal{H}_k$, the mean $\hat{\mu}_k^{(\text{no})}$ and variance $\hat{\sigma}_k^{(\text{no})}$ of $\mathbf{z}_k^{(\text{no})}$ are minimized. Therefore, weighing the energy sum in (5.85) with the inverse of $\hat{\mu}_k^{(\text{no})} \hat{\sigma}_k^{(\text{no})}$ will increase the likelihood of the correct hypothesis. This estimator is referred to as double-weighted MESS (DW-MESS) and it yields the following TOA estimate

$$\hat{n}_{\text{toa}}^{(\text{dw-mess})} = \arg \max_{k \in \{1,\ldots,N_b\}} \left\{ \frac{\mathbf{z}_k^{(\text{sn},N_e)} \boldsymbol{\rho}_{N_e}}{\hat{\mu}_k^{(\text{no})} \hat{\sigma}_k^{(\text{no})}} \right\}. \tag{5.87}$$

Bayesian estimation

If the distribution of $\mathcal{E}_m$ is known *a-priori* for each energy block m, and the noise variance σ^2 is known accurately, the TOA estimate can be obtained using a Bayesian approach.[19] Then, the leading energy block can be estimated as follows

$$\hat{n}_{\text{toa}}^{(\text{Bys})} = \arg \max_{k \in \{1,\ldots,N_b\}} \left\{ \int_{\mathcal{E}_1} \int_{\mathcal{E}_2} \cdots \int_{\mathcal{E}_{N_e}} p(\mathbf{z}|k, \sigma^2, \boldsymbol{\mathcal{E}}_{N_e}) p(\mathcal{E}_1) \ldots p(\mathcal{E}_{N_e}) d\mathcal{E}_{N_e} \ldots d\mathcal{E}_1 \right\}, \tag{5.88}$$

where the probability distribution function is expressed as

$$p(\mathbf{z}|k, \sigma^2, \boldsymbol{\mathcal{E}}_{N_e}) = \prod_{n=1}^{k-1} \frac{1}{\sqrt{2\pi \tilde{\sigma}^2}} \exp\left(-\frac{(z[n] - \tilde{\mu})^2}{2\tilde{\sigma}^2}\right)$$

$$\times \prod_{n=k}^{k+N_e-1} \frac{1}{\sqrt{2\pi \tilde{\sigma}_m^2}} \exp\left(-\frac{(z[n] - \tilde{\mu}_m)^2}{2\tilde{\sigma}_m^2}\right)$$

$$\times \prod_{n=k+N_e}^{N_{\text{toa}}} \frac{1}{\sqrt{2\pi \tilde{\sigma}^2}} \exp\left(-\frac{(z[n] - \tilde{\mu})^2}{2\tilde{\sigma}^2}\right), \tag{5.89}$$

with $m = n - k + 1$. The noise-only parameters are denoted by $(\tilde{\mu}, \tilde{\sigma})$, and signal-plus-noise parameters at the mth energy sample are denoted by $(\tilde{\mu}_m, \tilde{\sigma}_m)$, which are calculated using $\boldsymbol{\mathcal{E}}_{N_e}$ and σ. For example, the histograms of elements of $\boldsymbol{\mathcal{E}}_{N_e}$ are presented

---

[19] The reader is referred to [255] for a detailed discussion on Bayesian estimators.

in [256] for CM-1 channel model of the IEEE 802.15.4a channels. Similar semi-analytic techniques can be used to evaluate (5.89) for a given channel model.

Note that in order to keep the problem analytically tractable, (5.88) assumes that the energies $\mathcal{E}_m$ are uncorrelated. The ideal Bayesian estimator should consider the joint PDFs of the energies. Since it is usually very hard to know the prior PDFs of the parameters, and it requires multidimensional integration over the PDF of each parameter yielding a very complex implementation, Bayesian analysis is usually of theoretical interest and serves as a benchmark for other sub-optimal estimators rather than for practical consideration.

5.6 Low-complexity UWB ranging techniques

While the ML techniques discussed in the previous section can yield good ranging accuracies, they may not be very practical due to the *a-priori* knowledge requirements and/or implementation complexities. In this section, various low-complexity ranging algorithms specifically introduced for UWB systems are reviewed.

5.6.1 Ranging with largest-$\tilde{N}$ peak-detection techniques

The reason that the peak selection discussed above may not yield an accurate range estimate is the possible existence of other MPCs prior to the strongest one. One way to improve the performance of the peak detector is to consider the largest $\tilde{N}$ correlation peaks prior to making a decision for a range estimate. Three algorithms based on this principle are (1) single search, (2) search and subtract, and (3) search, subtract, and readjust [227].

Single search

The single search algorithm first calculates the absolute values of the MF outputs as given in (5.22). Then, $\tilde{N}$ strongest correlation peaks are calculated, which correspond to $\tilde{N}$ strongest MPCs. Let the time indices corresponding to these MPCs be represented by $\tilde{k}_1, \tilde{k}_2, \ldots, \tilde{k}_{\tilde{N}}$, where $\tilde{k}_i$ represents the time index for the ith strongest component. Then, the TOA of the received signal is estimated as [227]

$$\hat{\tau}_1 = T_{\text{smp}} \min\{\tilde{k}_1, \tilde{k}_2, \ldots, \tilde{k}_{\tilde{N}}\} \tag{5.90}$$

where T_{smp} denotes the sampling period of the receiver.

Search and subtract

For resolvable channels, the single search algorithm discussed above may successfully lock onto the correct correlation peaks. However, for non-resolvable channels, estimation of individual MPCs and their TOA may be cumbersome.

In order to improve the TOA estimation performance in non-resolvable channels, the single search algorithm can be modified as follows. After estimating the TOA corresponding to the strongest MPC (i.e. $\tilde{k}_1 T_{\text{smp}}$), this MPC is regenerated using the received pulse shape $\omega(t)$, and subtracted from the received signal. The channel coefficient to be used for regenerating the strongest path before subtracting it from the received signal is calculated as [227]

$$\hat{\alpha}_{\tilde{k}_1} = \left([\boldsymbol{\omega}^{(k_1)}]^T \boldsymbol{\omega}^{(k_1)}\right)^{-1} [\boldsymbol{\omega}^{(k_1)}]^T \mathbf{r}, \qquad (5.91)$$

where $\boldsymbol{\omega}^{(k_i)}$ and $\mathbf{r}$ are as defined in Section 5.3.1.

In the next step, the TOA of the second strongest MPC (i.e. $\tilde{k}_2 T_{\text{smp}}$) is estimated using this updated received signal (i.e. after the strongest MPC is subtracted from the received signal). Again, this MPC is reconstructed, and subtracted from the signal. After the same procedure iterates $\tilde{N}$ times, the TOA of the received signal is given by the minimum of the TOA values as in (5.90). Note that this is similar to the successive interference cancelation technique in code division multiple-access (CDMA) systems, where at each step, the strongest interferer is detected, reconstructed, and subtracted from the received signal [257].

Search, subtract, and readjust

The search and subtract algorithm determines the channel coefficient of the MPCs independently from each other. However, it is possible to improve the performance of the search and subtract algorithm by joint estimation of the channel coefficients at each iteration of the algorithm. For example, when calculating the channel coefficient for the first MPC, there is no information about the channel coefficients of the other MPCs; and hence, (5.91) is used. On the other hand, at the second step, the channel coefficient for the second strongest MPC is calculated as follows [227]

$$\begin{bmatrix} \hat{\alpha}_{k_1} \\ \hat{\alpha}_{k_2} \end{bmatrix} = \left([\boldsymbol{\omega}^{(k_1)} \ \boldsymbol{\omega}^{(k_2)}]^T [\boldsymbol{\omega}^{(k_1)} \ \boldsymbol{\omega}^{(k_2)}]\right)^{-1} [\boldsymbol{\omega}^{(k_1)} \ \boldsymbol{\omega}^{(k_2)}]^T \mathbf{r}, \qquad (5.92)$$

where the channel coefficients of the strongest two MPCs are jointly estimated. In a similar way, the algorithm calculates the channel coefficients of all the $\tilde{N}$ strongest MPCs, and the TOA is calculated from (5.90). Observe that when $\tilde{N} = L$, (5.92) becomes the ML estimator in Section 5.5.2.

A critical parameter that affects the accuracy of the above algorithms is the selection of $\tilde{N}$. While very small values of $\tilde{N}$ may not yield accurate range estimates, choosing a large $\tilde{N}$ will increase the computational complexity. Hence, its value should be optimized according to trade-off between accuracy and complexity. Furthermore, accurate knowledge of the received pulse shape is needed, which may vary at different MPCs in practice.

In terms of computational complexity, the single search algorithm has the lowest complexity, but yields the worst accuracy compared to the other two techniques. While the latter two algorithms can perform better in non-resolvable channels, they require matrix inversion operations, and their implementation may be computationally intensive, especially for large values of $\tilde{N}$.

5.6.2 Ranging with two-step TOA estimators

One of the most challenging issues in UWB ranging systems is to perform accurate range estimation without employing high sampling rates. In order to have a low-power and low-complexity implementation, typically, symbol-rate or frame-rate samples can be considered [243], which can however increase the time it takes to perform TOA estimation significantly.

Two-step TOA estimators can be used in an intelligent way to relax the sampling rate requirements without having to compromise from the ranging accuracy. As illustrated in Fig. 5.14, at the first step, a rough timing estimate is obtained using low sampling rates. Then, a second step refines the TOA estimate using higher sampling rates.

First step

In the first step, a low-complexity receiver with a low sampling rate is employed in order to obtain a rough estimate of the TOA. For example, as discussed in the subsection within Section 5.5.4 entitled "Maximum energy selection," an ED receiver that employs the RSS measurements in (5.25) can be used to provide a rough TOA estimate and to reduce the uncertainty region for the TOA:

$$\hat{n}_b = \arg \max_{1 \leq n \leq N_b} z^{(ED)}[n], \tag{5.93}$$

where $\hat{n}_b$ represents the index of the block that has the largest energy sample and $z^{(ED)}[n]$ is as given in (5.25).

A critical parameter is the selection of the sampling interval T_{smp} for the energy detector. If T_{smp} is selected very large, then it can accurately lock to the desired signal; however, the ambiguity region remains very large. If T_{smp} is selected very small, the ambiguity region is narrowed, but it becomes more likely that the first MPC may be missed. As an alternative to an energy detector, other low-rate techniques such as the dirty-template scheme, which employs symbol-rate samples (see Section 5.6.3), can be considered to obtain a coarse TOA estimate.

Second step

After obtaining a rough estimate of the signal's TOA from its low-rate samples, the second step uses higher sampling rates and more accurate techniques in order to precisely determine the TOA. For example, in [258], a method-of-moments estimator with chip-rate samples is employed to find the TOA accurately. Alternatively, correlation-based techniques and searchback algorithms with higher sampling rates can also be considered.

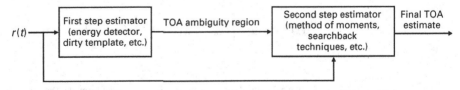

Fig. 5.14. Block diagram for a two-step TOA estimator.

Note that since the RSS in block $\hat{n}_b$ is the strongest, ideally, the TOA should be within the block $\hat{n}_b$. However, due to multipath effects, the first arriving MPC may be within one of the blocks preceding block $\hat{n}_b$. For example, the signal may arrive closer to the end of a block, which implies that the RSS of the TOA block can be weaker than the next blocks. Hence, for an improved ranging accuracy, the second step of the two-step algorithm searches a number of blocks including and prior to the $\hat{n}_b$th block.

The major advantage of a two-step technique is that it narrows down the TOA search space in its low-complexity first step, and the second step searches for the TOA only within a considerably smaller time interval.

5.6.3 Ranging with dirty templates

Another low complexity TOA estimator is the *dirty-template* receiver introduced in [224, 259, 260] which operates on *symbol-rate* samples. The basic idea behind this receiver is as follows. The optimal template signal in (5.67) is not available during TOA estimation. However, the received signal itself can be used as a correlator template, which is noisy ("dirty"). Then, cross-correlations of the symbol-length portions of the received signal are obtained in order to estimate the TOA.[20]

In order to analyze the dirty template receiver, first consider the following modifications to the received signal model in (5.5)

$$r(t) = \sqrt{E_s} \sum_{k=-\infty}^{\infty} b_k \omega_R(t - kT_s - \tau_1) + n(t) , \quad (5.94)$$

$$\omega_R(t) = \sum_{l=1}^{L} \alpha_l \sum_{j=0}^{N_f-1} \omega(y - jT_f - c_j T_c - \tau_{l,1}) , \quad (5.95)$$

where $T_s = N_f T_f$ denotes the symbol duration, b_k denotes the training data, $\tau_{l,1} = \tau_l - \tau_1$, and it is assumed that the TOA τ_1 is within one symbol duration, $[0, T_s)$, without loss of generality. Then, with the assumption that there is no ISI between consecutive symbols, the cross-correlation between symbol-length portions of the received signal can be obtained as follows [260]

$$x_k(\tau) = \int_0^{T_s} r\!\left(t + 2kT_s + \tau\right) r\!\left(t + (2k-1)T_s + \tau\right) dt , \quad (5.96)$$

where $\tau \in [0, T_s)$. After some calculations, (5.96) can be expressed as [260]

$$x_k(\tau) = b_{2k-1}\!\left[b_{2k-2} E_A(\tilde{\tau}_1) + b_{2k} E_B(\tilde{\tau}_1)\right] + \eta_k(\tau) , \quad (5.97)$$

---

[20] As discussed in Section 5.3.4, the dirty-template receiver is a special form of a DaC receiver with $D = T_s$, which is illustrated in Fig. 5.7c.

where

$$\tilde{\tau}_1 = (\tau_1 - \tau) \bmod T_s , \quad (5.98)$$

$$E_A(\tau) = E_s \int_{T_s-\tau}^{T_s} \omega_R^2(t) dt , \quad (5.99)$$

$$E_B(\tau) = E_s \int_0^{T_s-\tau} \omega_R^2(t) dt , \quad (5.100)$$

and $\eta(\tau)$ is the noise term, which is a zero-mean Gaussian variable with variance σ_η^2.

At this point, both non-data aided (blind) and data-aided approaches can be considered for the dirty-template scheme. For the non-data aided case, the mean square of (5.97) is calculated as [260]

$$\mathrm{E}\{X^2{}_k(\tau)\} = \frac{1}{2}[E_A(\tilde{\tau}_1) + E_B(\tilde{\tau}_1)]^2 + \frac{1}{2}[E_A(\tilde{\tau}_1) - E_B(\tilde{\tau}_1)]^2 + \sigma_\eta^2 , \quad (5.101)$$

where the symbols are assumed to be equiprobable. Observing that $E_A(\tilde{\tau}_1) + E_B(\tilde{\tau}_1) = E_s \int_0^{T_s} \omega_R^2(t) dt$ is a constant and $E_B(\tilde{\tau}_1) - E_A(\tilde{\tau}_1)$ is maximized at $\tau = \tau_1$, the TOA can be estimated as follows [260]

$$\hat{\tau}_1 = \arg\max_{\tau \in [0,T_s)} \left\{ \frac{1}{N_r} \sum_{k=1}^{N_r} x_k^2(\tau) \right\} , \quad (5.102)$$

where the expected value in (5.101) is obtained from sample mean estimate of N_r symbol-long pairs of received segments.

Using special training sequences improves the convergence time for a data-aided version of the dirty-template scheme. For example, letting $b_k = (-1)^{\lfloor k/2 \rfloor}$, (5.97) can be expressed as

$$x_k(\tau) = [E_A(\tilde{\tau}_1) - E_B(\tilde{\tau}_1)] + \eta_k(\tau) , \quad (5.103)$$

which has a mean square value of $\mathrm{E}\{X^2{}_k(\tau)\} = [E_A(\tilde{\tau}_1) - E_B(\tilde{\tau}_1)]^2 + \sigma_\eta^2$. Since the mean square value becomes maximum at $\tau = \tau_1$, (5.102) can be used for finding the TOA.

The dirty template approach is favorable due to its unique multipath energy collection capability; the correlation template is basically a delayed version of the received signal itself, and no multipath parameter estimation (i.e., estimation of vector $\tilde{\theta}$) is required. However, since the signal itself is noisy, additional signal-cross-noise and noise-cross-noise terms are introduced, which yields some performance degradation. A drawback of the dirty-template scheme is that the resulting TOA estimate will have an ambiguity that depends on the noise-only region between consecutive symbols. This is because the symbol-length cross-correlation outputs due to the noise-only region will be similar. In such a case, a more precise timing offset estimation can be implemented to resolve the ambiguity, and the dirty-template algorithm may be used as a first step of a two-step algorithm as in Section 5.6.2.

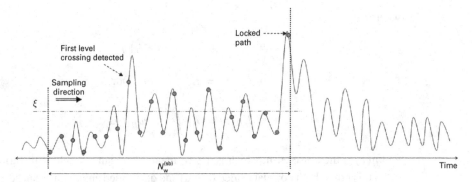

Fig. 5.15. Illustration of threshold-based first path detection based on direct samples of the received signal where ε denotes a threshold and N_w^sb denotes the length of a search-back window.

5.6.4 Threshold-based ranging

Threshold-based search algorithms compare individual signal samples with a certain threshold in order to identify the first arriving MPC and obtain the range information [200, 228, 232, 261]. One of the main advantages of threshold-based TOA estimators is that they can potentially be implemented in the analog domain (i.e. without using any ADC or sampling).

In [261], a simple threshold-based algorithm is introduced which compares absolute values of the direct samples of the received signal (or, direct samples of the MF output) with a threshold as illustrated in Fig. 5.15. The algorithm starts a search from the earliest locked MPC (which is obtained via an acquisition algorithm),[21] and the TOA of the signal is determined by the first threshold-exceeding sample. Since the signal energy and the channel impulse response are unknown, the threshold is determined based only on the noise floor.

In general, using the total probability theorem, the mean and variance of the TOA estimation error using such a threshold-based estimator can be obtained as follows [228, 231]

$$\mu_{\tau_1} = \mathrm{E}\{\hat{\tau}_1 - \tau_1\} = \sum_{n=n_\text{sb}}^{n_\text{max}} P_\text{D}(n) \mathrm{E}\left\{(\hat{\tau}_1 - \tau_1)|n\right\} + P_\text{miss} \mathrm{E}\left\{(\hat{\tau}_1 - \tau_1)|\text{miss}\right\}, \quad (5.104)$$

$$\sigma_{\tau_1}^2 = \mathrm{E}\left\{(\hat{\tau}_1 - \tau_1)^2\right\} = \sum_{n=n_\text{sb}}^{n_\text{max}} P_\text{D}(n) \mathrm{E}\left\{(\hat{\tau}_1 - \tau_1)^2|n\right\} + P_\text{miss} \mathrm{E}\left\{(\hat{\tau}_1 - \tau_1)^2|\text{miss}\right\},$$
(5.105)

where $P_\text{D}(n)$ is the detection probability of the nth sample, P_miss is the probability that no sample is larger than the threshold ξ, n_max denotes the sample index for the strongest path, and n_sb denotes the first sample that the search algorithm compares with a threshold.

---

[21] An acquisition algorithm may operate in different ways and it does not necessarily lock onto the strongest MPC. With a threshold-based acquisition scheme, the receiver may initially lock onto a path which arrives earlier than the strongest path.

One way to set the $\hat{\tau}_1$ for evaluating P_{miss} is to set it to the middle of the observation interval, i.e. $\hat{\tau}_1 = T_a/2$. Alternatively, estimation process can be repeated by decreasing the threshold until a level crossing occurs [228].

Note that the GML technique discussed in Section 5.5.3 also uses a threshold $\Delta_{\tilde{\alpha}}$ to determine a stopping rule. However, this requires the knowledge of the statistics of the relative path strength (i.e. $\alpha_{\text{peak}}/\alpha_1$). Furthermore, the GML algorithm requires sampling rates at or above the Nyquist rate, while the threshold-based algorithm may operate at lower sampling rates.

In the following subsections, two different approaches for a threshold-based ranging algorithm are presented, both of which can operate below the Nyquist rate.

Threshold-based ranging with jump back and search forward algorithm

Rather than using direct samples of the received signal, it may be possible to improve the energy capture at low sampling rates by using some analog front end processing as illustrated in Fig. 5.7. While any of the three receiver techniques may be considered, in the following text an ED receiver is considered.

The threshold-based ranging algorithm considered here assumes that the receiver is synchronized to the strongest path. First, the algorithm jumps to a sample prior to the strongest path and then searches for the leading edge in the forward direction by comparing the samples against a threshold. The leading edge sample using the jump back and search forward (JBSF) algorithm is estimated as follows

$$\hat{n}_{\text{JBSF}} = \min\{n \mid z[n] > \xi\}, \qquad (5.106)$$

where $n \in \{n_{\text{sb}}, n_{\text{sb}}+1, \ldots, n_{\text{max}}\}$, $n_{\text{sb}} = n_{\text{max}} - N_{\text{w}}^{(\text{sb})}$ with $N_{\text{w}}^{(\text{sb})}$ denoting the searchback window length in samples, n_{max} is the index for the strongest sample, and ξ is a predetermined threshold. The operation of the algorithm is illustrated in Fig. 5.16.

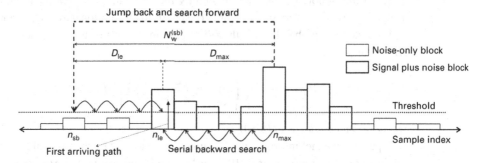

Fig. 5.16. Illustration of the JBSF algorithm in the present subsection and the SBS algorithm in the next subsection using an ED receiver. $N_{\text{w}}^{(\text{sb})}$ denotes the searchback window length in samples, n_{max} is the index of the strongest sample, n_{le} is the index of the first arriving path's sample, n_{sb} is the index of the first sample within the searchback window, D_{max} is the delay between the first arriving path's sample and the strongest sample, and D_{le} is the delay between the index of the first sample within the search window and the first arriving path's sample (After [232]).

The MAE of the TOA estimate can then be calculated, after weighting the average timing error corresponding to each block with the probability of detecting that block, as

$$\text{MAE}(n_{\text{le}}, \xi) = \sum_{i=n_{\text{sb}}}^{n_{\text{max}}} P_{\text{D}}(i|\xi)\bar{e}_i \,, \tag{5.107}$$

where n_{le} is the index for the true leading edge sample, the average timing error for choosing block i is approximately given by

$$\bar{e}_i \approx \begin{cases} |n_{\text{le}} - i|T_{\text{smp}} & \text{if } i \neq n_{\text{le}}, \\ T_{\text{smp}}/4 & \text{if } i = n_{\text{le}}, \end{cases} \tag{5.108}$$

and the set of samples between n_{sb} and n_{max} can all be estimated using the searchback algorithm as the leading edge.

The justification for (5.108) is as follows. Let $\epsilon_{\text{le}} = \tau_1 - T_{\text{smp}} n_{\text{le}}$ denote the delay of the first arriving path within the leading signal block. Then, for uniformly distributed ϵ_{le} and correct block detection, choosing the center of the block as the TOA estimate yields an average timing error given by

$$\text{E}\left\{\left|\epsilon_{\text{le}} - \frac{T_{\text{smp}}}{2}\right|\right\} = \frac{T_{\text{smp}}}{4} \,, \tag{5.109}$$

while the mean and variance of the time estimate for correct block detection are respectively given by

$$\text{E}\left\{\epsilon_{\text{le}} - \frac{T_{\text{smp}}}{2}\right\} = 0 \,, \tag{5.110}$$

$$\text{E}\left\{\left(\epsilon_{\text{le}} - \frac{T_{\text{smp}}}{2}\right)^2\right\} = \frac{T_{\text{smp}}^2}{12} \,. \tag{5.111}$$

Note that (5.108) is only an approximation since ϵ_{le} and the event of detecting $z[n_{\text{le}}]$ as the leading edge are correlated. For example, as $\epsilon_{\text{le}} \to T_{\text{smp}}$, it becomes more likely to detect the sample $n_{\text{le}} + 1$, which slightly increases the MAE. In addition, a simple analysis shows that every block error corresponds to an additional average error of T_{smp}.

The detection probability for the ith sample in (5.107) is algorithm dependent. For JBSF, it is given by

$$P_{\text{D}}(i|\xi) = \left[\prod_{j=1}^{i-1}\left(1 - Q\left(\frac{\xi - \mu_j}{\sigma_j}\right)\right)\right] Q\left(\frac{\xi - \mu_i}{\sigma_i}\right), \tag{5.112}$$

where μ_j and σ_j are as in (5.30).

If perfect channel information is assumed to be available, the optimal threshold that minimizes the ranging error in (5.107) can be found using a brute-force search as follows[22]

$$\xi_{\text{opt}} = \arg\min_{\xi}\left\{\text{MAE}(n_{\text{le}}, \xi)\right\}. \tag{5.113}$$

---

[22] Note that while perfect channel knowledge is not possible in practice, the resulting ranging error can be used as a lower bound on the JBSF algorithm that uses any other method for ranging threshold selection.

Since it is typically difficult to have *a-priori* knowledge of the channel parameters, a simple (but sub-optimal) way to practically set the threshold is to use only the standard deviation of the, noise which can be obtained from the noise-only region of the received signal. Then, for a certain P_{FA}, the threshold that is based solely on the noise level can be defined as

$$\xi = \tilde{\sigma} Q^{-1}(P_{FA}) + \tilde{\mu}, \quad P_{FA} = Q\left(\frac{\xi - \tilde{\mu}}{\tilde{\sigma}}\right), \tag{5.114}$$

where the mean and the variance of noise-only samples are given by $\tilde{\mu} = M\sigma^2$, and $\tilde{\sigma}^2 = 2M\sigma^4/N_r$, respectively, as discussed in Section 5.3.3, and P_{FA} denotes the probability of a single noise sample being larger than a threshold ξ.

It is worth emphasizing that for some channel realizations it may be preferable to detect a sample that is after the leading edge sample, since the leading edge sample(s) may be very weak. Beside NLOS effects, this might also be due to the fact that only a small portion of the first arriving path may fall within the first sampling duration. In such cases, the optimal threshold that minimizes the MAE may be larger than the leading edge energy value(s). In other words, since setting the threshold to a very small value may yield early false alerts, the MAE may be minimized by the detection of a stronger sample later than the first sample using a larger threshold (see Example 5.7).

Threshold-based ranging with serial backward search algorithm

As an alternative to the JBSF algorithm for finding the leading edge of the signal, the paths/samples can be searched one-by-one in the backward direction starting from the strongest sample n_{max}, which is referred to as the SBS algorithm (see Fig. 5.16 for an illustration). Note that there may be noise-only samples between the strongest path and the first arriving path. These noise-only regions may occur due to multi-cluster structure of UWB channels where there may be time delays between two clusters. Moreover, it may also be the case that within the same cluster, there are gaps between the MPCs, which are relatively large (i.e. with respect to the sampling duration). Hence, the SBS algorithm should handle the existence of such noise-only regions for accurate leading edge detection.

Example 5.6 *The multi-cluster structure of UWB channels is simulated and depicted in Fig. 5.17 for the CM-1 channel model of the IEEE 802.15.4a. First, 1000 channel realizations are generated and sampled with an energy detector for a sampling period of 4 ns. The histogram of the strongest path delay is depicted in Fig. 5.17(a), which shows that the strongest path may arrive as late as 60 ns with respect to the first arriving path. On the other hand, for more than half of the realizations, the delay is smaller than 10 ns.*

In Fig. 5.17(b), a histogram of the delay between any two clusters is presented. While the delay is typically smaller than 10 ns, it may be on the order of 30 ns in some rare cases.

In the following, two different cases are considered for the SBS algorithm. For Case 1, a single-cluster channel is considered, where there is no noise-only region between the

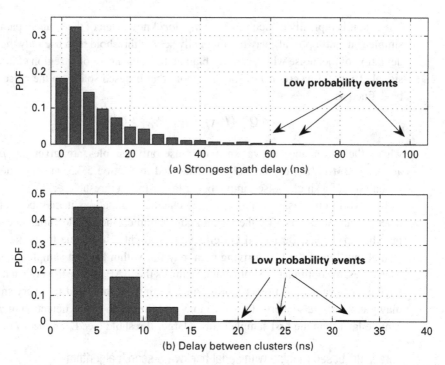

Fig. 5.17. Different statistics of IEEE 802.15.4a CM-1 channel models. (a) PDF of the delay of the strongest path, (b) PDF of delays between cluster pairs if there is at least one cluster prior to the peak ($T_c = T_{smp} = 4$ ns) (After [200]).

strongest sample and the leading edge sample. Note that this may typically occur in very dense channel environments such as the CM-8 channel of the IEEE 802.15.4a [100]. For Case 2, a multiple-cluster channel structure is considered where there may be noise-only regions between the strongest path and the first path. Differences for the two cases are illustrated in Fig. 5.18.

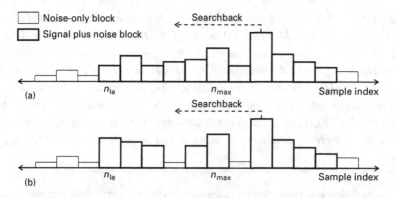

Fig. 5.18. Illustration of the searchback scheme: (a) single cluster, (b) multiple clusters (After [200]).

Case 1: dense single cluster analysis. In a single-cluster (SC) scenario, in which there are no noise-only samples between the leading edge and the strongest sample, as illustrated in Fig. 5.18(a), the leading block estimate for SBS-SC is given by

$$\hat{n}_{\text{SBS-SC}} = \max\left\{n \in \{n_{\max}, \ldots, n_{\max} - N_{\text{w}}^{(\text{sb})}\} \,\Big|\, z[n] > \xi \text{ and } z[n-1] < \xi\right\}. \tag{5.115}$$

Case 2: multiple clusters (with noise-only region) analysis. The received MPCs in typical UWB channels usually arrive at the receiver in multiple clusters (MCs), i.e. groups of MPCs that are separated by noise-only samples. The PDF of the delays between any two clusters if there is at least one cluster prior to the peak energy sample is presented in Fig. 5.17(b).[23] Since there may be delays as large as 32 ns between two sequential clusters, the algorithm discussed in the previous section may lock to a sample that arrives later than the leading edge (but is at the beginning of its own cluster).

It is therefore proposed to allow a number of consecutive occurrences of noise samples while continuing the backward search to handle the clustering problem. The false alarm probability when K multiple consecutive noise samples exist is expressed as

$$P_{\text{FA}} = 1 - \left[1 - Q\left(\frac{\xi - \tilde{\mu}}{\tilde{\sigma}}\right)\right]^K, \tag{5.116}$$

which leads to a threshold given by

$$\xi = \tilde{\sigma} Q^{-1}\left(1 - (1 - P_{\text{FA}})^{\frac{1}{K}}\right) + \tilde{\mu}, \tag{5.117}$$

where the optimum threshold is now a function of K. The leading edge estimation for SBS-MC is then modified as follows

$$\hat{n}_{\text{SBS-MC}} = \max\left\{n \in \{n_{\max}, \ldots, n_{\max} - N_{\text{w}}^{(\text{sb})}\} \,\Big|\, z[n] > \xi \text{ and}\right.$$

$$\left.\max\{z[n-1], z[n-2], \ldots, z[\max(n-K, n_{\text{sb}})]\} < \xi\right\}. \tag{5.118}$$

Similar to D_{le}, the statistics of K may be obtained from measurement campaigns in a certain environment or from certain channel models that attempt to model a particular environment (e.g. see Fig. 5.17(b) for relevant statistics of CM-1 channel model). Note that choosing K too large may increase the probability of false alarms due to the noise-only region prior to the first arriving path.

Example 5.7 *In Fig. 5.19, the MAEs of the JBSF algorithm when simulated for $P_{\text{FA}} \in \{0.001, 0.005, 0.01\}$ and the optimal thresholds in (5.113) are given. The simulation parameters in Example 5.5 are used with $N_{\text{r}} = 1000$ and a searchback window*

---

[23] These statistics can be used to determine a parameter K for setting the ranging threshold as will be discussed below.

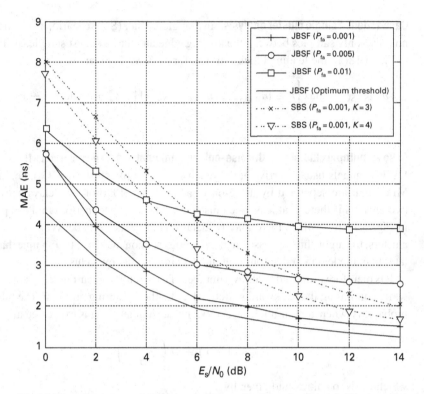

Fig. 5.19. The MAE performances of different algorithms for the optimal thresholds that minimize the MAE (simulation) (After [116]).

length of 60 ns. Simulation results for SBS-MC are also presented for two different values of K, and for $P_{FA} = 0.001$. When the noise-only-based threshold is used with $P_{FA} = 0.001$, the MAE is of the order of a fraction of a nano-second close to the optimal MAE for sufficiently high E_s/N_0. On the other hand, as E_s/N_0 (or N_r) decreases, choosing a larger P_{FA} becomes more desirable. The accuracy of the SBS-MC algorithm is observed to be inferior to that of the JBSF algorithm for $K \in \{3, 4\}$.

The MAEs obtained from (5.107) for JBSF are plotted in Fig. 5.20 for different threshold settings which are averaged over 1000 channel realizations. The results are slightly worse than the simulated performances, which is due to the approximation in (5.108). They are closer to the simulation results for larger P_{FA} values.

Also included in Fig. 5.20 is the ranging error when using a threshold that maximizes the probability of leading edge detection. It shows that the error is considerably worse than the noise-based thresholds and MAE-minimizing threshold. The reason for this is that, in certain cases, it may be easier to detect a sample that arrives later than the leading edge. If a threshold is set too low to maximize $P_D(n_{le})$, the probability of false alarm in the noise-only region of the signal may be larger compared to, for example, when using a threshold that maximizes $P_D(n_{le} + 1)$.

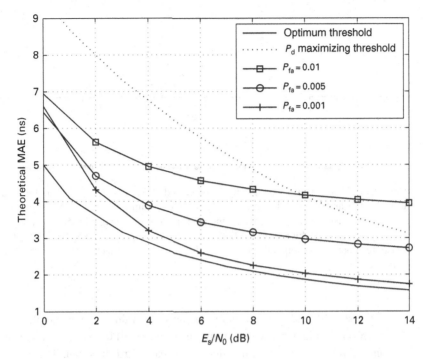

Fig. 5.20. Theoretical MAE performance of the JBSF at optimum threshold settings that minimize the MAE (After [116]).

5.7 Problems

(1) Consider a UWB ranging system that uses the second derivative of the Gaussian pulse shape given in (2.5) with a pulse duration of 2 ns. Let the TOA be uniformly distributed within [0, 200] ns.

(a) Calculate the CRLB and ZZLB on the MSE of the TOA estimate at SNRs of 2 dB, 12 dB, and 22 dB.

(b) Repeat part (a) for a pulse duration of 0.2 ns.

Solve questions 2 − 5 based on Fig. 5.21.

(2) Consider a threshold-based detector which first stores all the samples starting from the first sample at $t = 0$ ns in a vector. It then compares the absolute value of each sample with a threshold starting from the first sample, and chooses the first threshold-exceeding sample as the leading edge of the signal.

(a) Plot a realization of the received signal samples in MATLAB or another programming language, using an AWGN channel with zero mean and variance 0.01.

(b) Calculate the probability of detection of the leading edge of the received signal for ranging thresholds of $\xi \in \{0.2, 0.3, 0.4\}$.

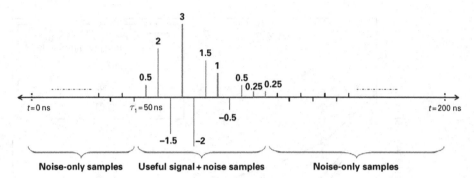

Fig. 5.21. Consider a UWB receiver which samples a received signal at a sampling period of 1 ns (e.g. using an MF, an ED, or a TR receiver). The receiver assumes that the TOA of the received signal is uniformly distributed in [0, 200] ns, and the true TOA of the signal is 50 ns. The channel coefficients are given by $\alpha = [0.5, 2, -1.5, 3, -2, 1.5, 1, -0.5, 0.5, 0.25, 0.25]$ as illustrated above. For simplicity, let each of the samples be subject to AWGN with mean 0 and variance 0.01.

(3) A threshold-based detector stores the absolute values of the samples in a vector; however, it initiates a search from the strongest MPC (which is assumed known), and searches for the leading edge in the backward direction. The channel vector α in Fig. 5.21 is subject to AWGN with mean 0 and variance 0.01. Then, calculate the detection probability for the leading edge of the received signal for a threshold of
   (a) $\xi = 0.3$,
   (b) $\xi = 0.4$.

(4) Assume that the receiver has perfect knowledge of the channel vector α. Then, write down the expression for the maximum likelihood estimator which gives the TOA of the received signal.

(5) The channel impulse response is $\alpha = [3]$. Given the receiver settings in Fig. 5.21, if the receiver adopts peak selection for synchronizing to the received signal, calculate the probability of incorrect TOA estimation.

(6) Study the TH and MTOK sequences presented in Fig. 5.5.
   (a) Design another TH sequence with $N_h = 32$, $N_f = 4$, and a ZCZ of 29.
   (b) Design another MTOK sequence of length 31 with similar ZCZ characteristics.

(7) *(programming exercise)* Consider the CM-1 model of IEEE 802.15.4a channels.
   (a) Generate 1000 different realizations of CM-1 channels and calculate the mean excess delay and RMS delay spread of each channel.
   (b) Using the parameters of the log-normal functions for the mean excess delay of the channels in Table 3.5 for CM-1(LOS) and CM-2 (NLOS) channels, simulate and report the probability of correct identification of the LOS hypothesis for these 1000 channel realizations.

5.7 Problems

(8) *(programming exercise)* Consider the fundamental lower bounds presented in Fig. 5.11 for AWGN channels. Compute and plot the same bounds for the first derivative of a Gaussian pulse shape: plot the ZZLBs and CRLBs for pulse durations of
   (a) $T_p = 1$ ns,
   (b) $T_p = 0.5$ ns,
   (c) $T_p = 0.1$ ns.

How do they compare with the results in Fig. 5.11? Why?

6 Ranging protocols

The previous chapter deals with detecting time of arrival of first signal path. Even though it is an essential and the very first step in TOA-based ranging and positioning systems, more need to be done to obtain range and position estimates. The TOA information makes sense only if the signal's time of transmission is known. Then, time of flight (TOF) of the signal can be easily computed. The TOF is directly proportional to the distance between a device that transmits the signal and the device that receives it. There are various protocols to transform any TOA information to a TOF and range estimate. Ranging protocols require actions to be taken at devices that are involved in ranging and positioning. The focus of this chapter is to study these protocols in detail.

The chapter consists of three parts. The first part provides an overview of communication protocol layers and explains functionalities of the service and management interfaces between adjacent layers. It is important to know what intra- and inter-device events take place for obtaining ranging information and how ranging-related information is passed from the physical layer to the upper layers for application's use. Interfaces play a key role in achieving this. A good example of an intra-device event is management of ranging signal parameters at the MAC sub-layer and then generation of the signal accordingly at the PHY. What constitutes an inter-device event is transmission and reception of ranging signals and ranging-related messages (e.g., time-stamps).

The second part takes a detailed look into well-known time-based ranging protocols and analyzes their advantages and drawbacks. There are numerous ranging protocols. Their only common goal at the simplest is to find separation of two devices. Some protocols also aim to mitigate clock-frequency offset-induced range errors and minimize the number of ranging messages transmitted. Some protocols require fine time synchronization among ranging devices (e.g., TDOA protocol), but others do not rely on this requirement (e.g., two-way TOA protocol, symmetric double-sided ranging protocol, differential TOA protocol). Generally, it is the application that determines the protocol. The TDOA protocol is a favorable choice for indoor tracking of assets and people if deploying an infrastructure is not considered an issue and scalability is important. On the other hand, for ad-hoc network applications, the TW-TOA protocol would be a better decision. A part of this chapter discusses pros and cons of these ranging protocols. Conventional and bi-static radars are outside the scope of this section.

The third part expounds the ranging features of the new IEEE 802.15.4a standard and studies packet structure, symbol waveform design and management of timing counters. The standard defines a PHY layer with precision-ranging capability. It adopts IR-UWB and chirp spread spectrum (CSS) as underlying signal waveforms. Only ranging via IR-UWB PHY is allowed in the standard for non-technical reasons, even though the CSS design is also capable of ranging.

6.1 Layered protocols

In any network, communication and ranging are accomplished by exchanging messages between involved parties. A ranging protocol is an agreement between two or more ranging devices on a ranging procedure. Let a device that initiates ranging be referred to as the *local device* and a responding device as the *peer*. Prior to transmitting and after receiving a ranging message, certain events are triggered and actions are taken by both local and peer devices to manage ranging-related data.

To reduce design complexity and provide flexibility, traditionally most network devices are considered and implemented as multiple layers, each layer offering its services to the higher layers [262]. This approach isolates the upper layers from the implementation details of the services it receives.

In n-layer implementation of a network, ith layers of local and peer devices communicate using layer i protocol (see Fig. 6.1); and the entities at the same layers of two devices are called *peer* entities. Between any two adjacent layers, communication is managed via function calls referred to as primitives (services). A primitive may specify an action to be taken or it may report the result of a previous action.

The PHY, as the first layer according to the widely practiced open systems interconnection (OSI) model, provides two services. The first one is the PHY data service. It operates through an interface called PHY data service access point (*PD-SAP*). The

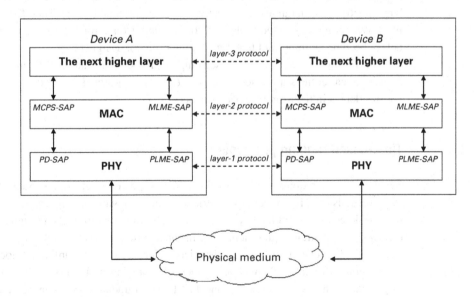

Fig. 6.1. Architecture of a typical ranging device, including layers, protocols and interfaces.

second one is the PHY management service. It deals with transmission and reception of PHY protocol data units (PPDU) through an interface called PHY layer management entity service access point (*PLME-SAP*).

Similarly, layer 2, which is the MAC sublayer, provides MAC data and MAC management services through interfaces that are referred to, in most IEEE standard specifications, as MAC common part sublayer service access point (*MCPS-SAP*) and MAC sublayer management entity service access point (*MLME-SAP*), respectively. Within the scope of this chapter are the interactions for ranging between the PHY layer, MAC sublayer and the next higher layer (NHL).

Ranging-related data are passed between the protocol layers via primitives. For instance, the NHL can tell its local MAC sublayer to initiate ranging with a peer device only using a specific signal waveform. In this case, parameters that define the desired signal waveform should be passed to the local MAC sublayer entity via a primitive. In [262] and [263], primitives are divided into four groups as *request*, *confirm*, *indication* and *response*.

The *request* primitive is generated at the MAC sublayer of a device to inform its PHY layer of a request. It can be a request to transmit a packet. The *confirm* primitive is generated at the PHY layer of a device to inform its MAC sublayer that a corresponding request is handled. A good example is a confirmation issued by the PHY after completing the transmission of a packet. The *indication* primitive is generated at the PHY layer of a device to inform its MAC sublayer that an event occurred. For instance, reception of a request from a peer entity can be an event. The *response* primitive is generated in reply to an indication primitive, in case a response is required.

The time of flight of a signal between two devices is proportional to the distance between themselves. Precise measurement of this flight time is challenging. In addition, timing imperfections even in the order of nanoseconds can easily induce an undesirably large positive bias in range estimates (30 cm per nanosecond error). Therefore, handling primitives as quickly as possible and compensating for errors due to protocol-related delays become an important task for ranging protocol designers.

In the following sections, time-based ranging protocols are analyzed. Communications between local entities are also illustrated for subject protocols to visualize impacts of various delays in their ranging performance.

6.2 Time-based ranging protocols

Synchronization requirement, channel occupancy, power efficiency, tolerance to crystal imperfections (e.g. frequency offset, drifting) and achievable maximum update rate can be considered as some of the key design criteria for ranging protocols. Design criteria are driven by application requirements. For instance, infrastructure-based tracking applications typically depend on time-synchronized reference devices, unlike ad-hoc network applications. Typically, channel occupancy of a ranging protocol is desired to be low to provide high energy efficiency and fast location updates. This can be achieved by both minimizing the number of messages required to obtain a single range measurement and using shorter frames. However, note that shorter frames may not provide sufficient processing gain to improve SNR at long distances. Low channel occupancy gives more devices an opportunity to perform ranging within a given time period and allows more frequent ranging updates. There are also ranging protocols and techniques to mitigate range errors induced by crystal imperfections. All in all there is always a trade-off between performance and complexity. In the following subsections, time-based ranging protocols are evaluated in various aspects.

6.2.1 Two-way time-of-arrival-based ranging

Let RDEV represent a ranging capable device and RFRAME a frame that is tailored and used for ranging. In a so-called two-way time-of-arrival (TW-TOA) ranging protocol, the range between two RDEVs is determined via exchanging RFRAMEs and tracking their arrival times. The major advantage of this protocol is that it does not require time synchronization between ranging devices. Assume that $RDEV\ A$ wants to perform ranging with $RDEV\ B$. Then, the sequence of events for the TW-TOA protocol would be as follows.

(i) $RDEV\ A$ sends a range request frame, $RFRAME_{req}$, to $RDEV\ B$, and records the time the frame departs from its antenna, $t = T_1$, according to its local clock.

(ii) B receives the $RFRAME_{req}$ and sends back to A a range reply frame, $RFRAME_{rep}$.

(iii) A records the time that it receives the $RFRAME_{rep}$ at its antenna input according to its local clock. This time instant is denoted as $t = T_2$.

(iv) Finally, A calculates the difference between the two recorded times, $T_r = T_2 - T_1$, and computes the range using the formula $d = c\ T_r/2$, where c is the speed of radio wave in the transmission medium.

(v) By performing TW-TOA via at least three devices, A can compute its relative position with respect to them.

In reality, there is a non-zero processing delay at B. Therefore, in the above algorithm d would contain a positive bias. After incorporation of this processing delay, T_r can be approximated as $T_r = 2T_t + T_{ta}^B$, where T_t is the one-way time of flight of the $RFRAME$ and T_{ta}^B denotes the time elapsed at $RDEV\ B$ between its reception of the $RFRAME_{req}$ and its transmission of the $RFRAME_{rep}$, which is commonly referred to as the turn-around time. The ranging performance depends on two important factors. The first one is accurate estimation of T_t. Note that, as studied in Chapter 5 (c.f. (5.42)), the Cramer–Rao lower bound for the estimate of T_t is inversely proportional to the effective bandwidth of the signal and the square root of the SNR. Apparently, using wide bandwidth signals and higher SNR helps reduce range errors. The second factor is minimizing T_{ta}^B or having an accurate estimate of it. In what follows, we study how ranging protocols help cancel out the range error caused by T_{ta}^B.

An effective way to minimize T_{ta}^B is to avoid interactions between the PHY and MAC layers at $RDEV\ B$, and make the peer PHY prepare and transmit an $RFRAME_{rep}$ immediately after it receives an $RFRAME_{req}$, as illustrated in Fig. 6.2.

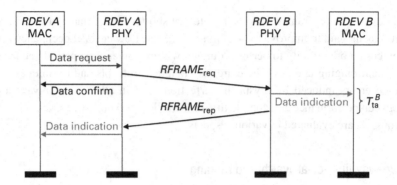

Fig. 6.2. Illustration of the TW-TOA ranging protocol without MAC layer involvement for minimum processing delay.

There are two drawbacks with this approach. First, in conventional communication systems, a MAC layer acknowledgment (ACK) needs to be sent (if required) for a successfully received packet. In Fig. 6.2, the $RFRAME_{rep}$ serves as an ACK, but the MAC layer has no control over the formation of the ACK and its content. Second, $RDEV\ B$ becomes vulnerable to malicious attacks. This is because the identification of the sender is typically carried in the MAC header of a frame. According to the protocol in Fig. 6.2, the $RFRAME_{rep}$ gets transmitted before the MAC header of the $RFRAME_{req}$ is processed. Thus, an $RFRAME_{req}$ sent by any device can trigger transmission of an $RFRAME_{rep}$, and the sender can easily figure out its range with B.

A better approach than excluding the MAC sublayer from ranging process just for the sake of minimizing T_{ta} is to encourage PHY and MAC layer interactions at B. As shown in Fig. 6.3, this would certainly prolong T_{ta}. However, note that in this scenario, B can easily compute T_{ta}, and send the result to A as a time-stamp report. The transmission

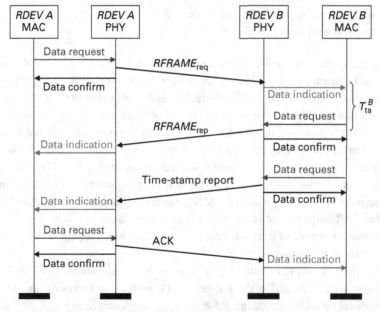

Fig. 6.3. Illustration of the TW-TOA ranging protocol with MAC layer involvement.

of the time-stamp report is subsequent to the $RFRAME_{\text{rep}}$. Only after receiving the time-stamp report, can A remove the range bias due to T_{ta}.

Note also that although the local device must record the timer value at which an RFRAME leaves its antenna, the local PHY can take a snapshot of its counter only after the frame leaves its modulator. Therefore, if not subtracted, this internal propagation delay between the timer circuitry and the output of the antenna induces a positive bias. Similarly, the receiving PHY can only take a snapshot of its timer when a frame enters its demodulator; and passing of the frame from its antenna to the demodulator also induces some delay. If internal propagation delays are in the order of nanoseconds, the resulting bias in the range estimate can reach tens of centimeters easily. It is possible to quantify these delays via test measurements or RF loop-back tests. When available, the MAC sublayer should use these biases to further correct its time-stamp values.

In the TW-TOA protocol, the $RFRAME_{\text{rep}}$ serves as an ACK for the $RFRAME_{\text{req}}$. Of course, MAC layer involvement increases the number of transmissions in the air from two to four. The additional two transmissions are the time-stamp report and its ACK. Indeed, the ACK is transmitted only if it is required. A single bit is dedicated typically in the MAC header of a frame to indicate whether the frame requires an ACK. In a dense network in which many devices frequently perform ranging, reducing the number of transmissions helps preserve computational and energy resources. Therefore, it is preferable not to mandate ACKs.

6.2.2 Differential two-way ranging protocols

Differential two-way ranging protocols (DTW-TOA) eliminate the need for a time-stamp report. Here the DTW-TOA protocols are classified as type I and type II. In [253] and [264], type I scheme is covered in detail, in which the source ($RDEV$ A) switches to the receive mode after a predetermined time interval T elapses following the transmission of an $RFRAME_{\text{req}}$. By this way, the peer device does not have to report T_{ta}^B.

Assume that A and B have no clock frequency offsets. As shown in Fig. 6.4, $RDEV$ A transmits an $RFRAME_{\text{req}}$ at local time $t_A = 0$. $RDEV$ B acquires the frame by locking onto the strongest multipath and then immediately resets its clock, $t_B = 0$. In some multipath channels, the first arriving signal component might not have the strongest energy. Therefore, in those channels the time difference between the first and strongest paths appears as a positive bias in the range estimate. This time offset is denoted in Fig. 6.4 as t_{off}^B. After acquisition is complete, B waits for T seconds and then transmits an $RFRAME_{\text{rep}}$. Similarly, assume that while receiving $RFRAME_{\text{rep}}$ at A, the time interval between the first and strongest signal paths is t_{off}^A. When the $RFRAME_{\text{rep}}$ is acquired, the local clock of A shows

$$t_A = 2T_t + T + t_{\text{off}}^B + t_{\text{off}}^A. \tag{6.1}$$

By running an off-line TOA refinement algorithm, A can have a reliable estimate of t_{off}^A. Let $\hat{t}_{\text{off}}^A$ denote this estimate. If the channel is assumed to be symmetric, and A and

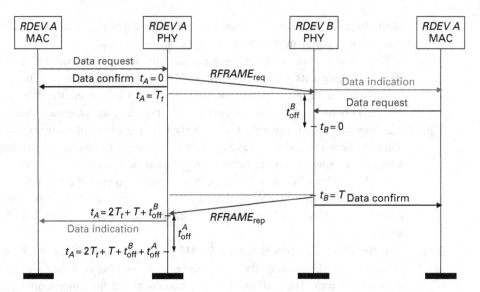

Fig. 6.4. Illustration of type I differential two-way ranging protocol, in which, following acquisition of an $RFRAME_{req}$, the peer device waits for time T to elapse before transmitting an $RFRAME_{rep}$.

B are identical devices, then $t_{off}^A \approx t_{off}^B$. After factoring this approximation into (6.1), the range estimate $\hat{d}$ can be obtained as

$$\hat{d} = \frac{c}{2}\left(t_A - 2\hat{t}_{off}^A - T\right). \tag{6.2}$$

Type II differential two-way ranging protocol is a slightly different version of type I. It puts additional burden on B for computing the estimate of t_{off}^B prior to transmission of $RFRAME_{rep}$ and then adjusting its transmission time to be ahead of T by $\hat{t}_{off}^B$ (Fig. 6.5); i.e.

$$t_B = T - \hat{t}_{off}^B. \tag{6.3}$$

Finally, A captures acquisition time of the $RFRAME_{rep}$ according to its local clock. Then, its clock shows

$$t_A = 2T_t + T + t_{off}^B - \hat{t}_{off}^B + t_{off}^A. \tag{6.4}$$

Note that A is still supposed to estimate t_{off}^A and factor it out. Finally, the range estimate becomes

$$\hat{d}_{AB} = c\left(T_t + \frac{t_{off}^A - \hat{t}_{off}^A}{2} + \frac{t_{off}^B - \hat{t}_{off}^B}{2}\right). \tag{6.5}$$

Both type I and type II protocols save two transmissions per ranging in comparison to the TW-TOA protocol, because the $RFRAME_{rep}$ serves as an ACK and there is no need to transmit time-stamp information. However, both protocols have a major drawback. In practical implementations, it is difficult to manage a PHY layer to transmit precisely at a preset time instant, especially if the required timing resolution is on the order of nanoseconds or sub-nanoseconds. Furthermore, any change

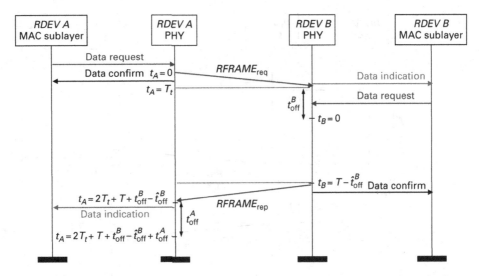

Fig. 6.5. Illustration of type II differential two-way ranging protocol, in which the peer device waits for time $T - \hat{t}^B_{\text{off}}$ to elapse after acquisition of a $RFRAME_{\text{req}}$ and then transmits a $RFRAME_{\text{rep}}$.

in channel access status might require deferring a transmission. In other words, channel may not be accessible when it is time to transmit an $RFRAME_{\text{rep}}$. Therefore, relying on a preset waiting interval T for ranging may carry the risk of a large error.

6.2.3 Symmetric double-sided ranging protocol

The symmetric double-sided (SDS) ranging protocol [265], which is illustrated in Fig. 6.6, aims to minimize range errors due to crystal imperfections. The SDS protocol consists of the following steps in order of execution.

(i) A transmits an $RFRAME_{\text{req}}$.
(ii) B replies with an $RFRAME_{\text{rep}}$.
(iii) A transmits a second $RFRAME_{\text{req}}$.
(iv) B replies with a time-stamp report that contains the measured T^B_{ta} and T^B_{round} values.

In order to quantify how the SDS protocol compensates for crystal frequency offsets and what its performance is in comparison with the TW-TOA protocol, assume that e_A and e_B denote clock frequency offsets of A and B, respectively. For a particular clock, its frequency offset can be considered to be constant during execution of a ranging protocol.

In the TW-TOA protocol, T^A_{round} is given by

$$T^A_{\text{round}} = 2T_{\text{t}} + T^B_{\text{ta}}, \qquad (6.6)$$

which implies

$$T_{\text{t}} = \frac{(T^A_{\text{round}} - T^B_{\text{ta}})}{2}. \qquad (6.7)$$

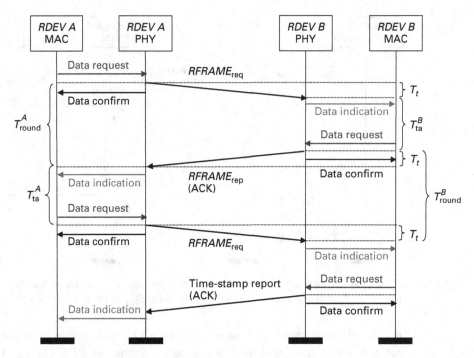

Fig. 6.6. Illustration of the symmetric double-sided ranging protocol.

After incorporating frequency offsets into (6.7), the estimate of T_t becomes

$$\hat{T}_t = \frac{T^A_{\text{round}}(1+e_A) - T^B_{\text{ta}}(1+e_B)}{2}. \tag{6.8}$$

Now define $\epsilon_{\text{tw}} = \hat{T}_t - T_t$ as the residual error for the time-of-flight estimate with the TW-TOA protocol. Then, the residual error is expressed as

$$\epsilon_{\text{tw}} = T_t e_A + \frac{(e_A - e_B)}{2} T^B_{\text{ta}}. \tag{6.9}$$

It is observed in (6.9) that ϵ_{tw} increases at longer ranges due to its dependency on $T_t e_A$. On the other hand, considering the fact that WPAN devices typically operate in short ranges (e.g. only tens of meters per link), and that T^B_{ta} varies from hundreds of microseconds to even milliseconds depending on hardware capabilities, the impact of $T_t e_A$ on ϵ_{tw} can be neglected. This leads to

$$\epsilon_{\text{tw}} \approx \frac{T^B_{\text{ta}}}{2}(e_A - e_B). \tag{6.10}$$

Minimizing T^B_{ta} and selecting a crystal with a low parts per million (ppm) help lower the residual error.

As for the SDS protocol,

$$\begin{aligned} T^A_{\text{round}} &= 2T_t + T^B_{\text{ta}}, \\ T^B_{\text{round}} &= 2T_t + T^A_{\text{ta}}. \end{aligned} \tag{6.11}$$

After extracting the time of flight T_t, we have

$$T_t = \frac{(T_{\text{round}}^A - T_{\text{ta}}^A) + (T_{\text{round}}^B - T_{\text{ta}}^B)}{4}. \qquad (6.12)$$

The estimate of T_t, with the inclusion of the crystal tolerances, becomes

$$\hat{T}_t = \frac{(T_{\text{round}}^A - T_{\text{ta}}^A)(1 + e_A) + (T_{\text{round}}^B - T_{\text{ta}}^B)(1 + e_B)}{4}. \qquad (6.13)$$

Then, the residual error $\epsilon_{\text{sds}} = \hat{T}_t - T_t$ is obtained as

$$\epsilon_{\text{sds}} = \frac{T_t}{2}(e_A + e_B) + \frac{(T_{\text{ta}}^B - T_{\text{ta}}^A)}{4}(e_A - e_B). \qquad (6.14)$$

It is plausible to assume that $T_{\text{ta}}^B \approx T_{\text{ta}}^A$, if A and B are peer devices. Then, (6.14) approximates to

$$\epsilon_{\text{sds}} \approx \frac{T_t}{2}(e_A + e_B). \qquad (6.15)$$

Finally, the comparison of (6.15) and (6.10) proves that the SDS protocol assures a smaller residual error.

Due to the fact that MAC layer ACK transmission is required in many instances, each of the SDS and TW-TOA protocols requires four message transmissions. Note that this is only under the assumption that the final ACK in the SDS protocol can be used to carry time-stamp data. However, traditionally an ACK is very short and cannot contain a long data such as a timestamp. For instance, the ACK in the IEEE 802.15.4b-2006 standard allocates only five octets for the PHY payload; and they are already in use for Frame Control, Sequence Number and FCS. In order to support the SDS protocol in an IEEE 802.15.4b-based network, the time-stamp report needs to be transmitted separately from the ACK. Thus, the SDS protocol is coerced to have the overhead of one extra transmission with respect to the TW-TOA. An alternative is to modify the IEEE 802.15.4 ACK frame structure and increase its payload size to avoid this extra transmission, but the devices that implement a modified ACK would not be inter-operable with the IEEE 802.15.4 devices.

6.2.4 Time-difference of arrival (TDOA)

Typically, a TDOA protocol requires deployment of an infrastructure with reference devices. What distinguishes the TDOA protocol from the TW-TOA protocol is that it relies on time synchronization between reference devices. Synchronization can be handled either via transmitting periodic beacons from a designated device so-called a *coordinator*, or wiring all the reference devices and driving them with a common clock [264, 266, 267].

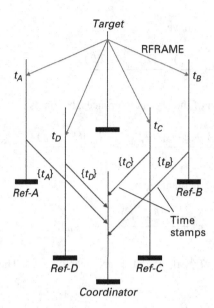

Fig. 6.7. Illustration of the centralized TDOA protocol with four reference devices and a coordinator as the central computing resource.

The message sequencing chart for the TDOA protocol is shown in Fig. 6.7. The operation of the protocol is as follows.

(i) The target transmits a broadcast RFRAME with its identification number (ID).
(ii) Each reference device that receives the broadcast computes its time of arrival.
(iii) Assume that the arrival times of the RFRAME at four reference devices A, B, C and D are t_A, t_B, t_C, t_D, respectively, according to the central clock they are synchronized to. Then, the reference devices report these time-stamps to their coordinator. These transmissions should be scheduled carefully to avoid collisions. Moreover, the target ID should also be encapsulated in the time-stamp report to distinguish multiple targets.
(iv) Given the reference device locations and the time-stamp data, the coordinator solves a non-linear optimization algorithm to estimate target coordinates.

6.3 Ranging in IEEE 802.15.4a standard

The IEEE 802.15.4a standard specifies PHY and MAC layers for IR-UWB and CSS communication systems. In this standard, ranging is optional and it is only enabled for the UWB PHY option. The preamble waveform is optimized for acquisition and synchronization, and the start of frame delimiter (SFD) is designed to support operations at low and high data rates. Supported data rates vary from 100 Kbps to 27 Mbps. In this section, packet structures and signal waveforms that are designed for ranging are discussed. Management of timing counters and preparation of time-stamp reports are also explained.

6.3 Ranging in IEEE 802.15.4a standard

The IEEE 802.15.4a uses the ALOHA protocol for channel access. In ALOHA, a device transmits a frame without sensing whether the channel is busy. If a transmission collides with another one, the frame is retransmitted after a random back-off. Achievable throughput, $\tilde{\eta}$, for the ALOHA with the assumption of a Poisson frame arrival rate λ is

$$\tilde{\eta} = \lambda \, e^{-2\lambda}. \tag{6.16}$$

For a detailed treatment of the ALOHA mechanism, the reader is referred to [268]. Note that RFRAMEs are very long in IEEE 802.15.4a. Especially, in the low rate option, the preamble and the SFD consist of 4096 and 64 symbols, respectively. Therefore, even a single frame may occupy the channel in the order of milliseconds, and retransmissions become very costly. A sparse network in which RDEVs perform ranging very often might experience as low throughput as a very dense network.

The main ranging protocol the standard adopts is the TW-TOA. However, it also enables the use of TDOA and SDS ranging protocols. To make decoding of ranging waveforms difficult for malicious devices and protect range information, the standard also describes a so-called private ranging protocol, which is optional. It enhances the integrity of ranging traffic in the case of a hostile attack. The reader is referred to Chapter 7 for a detailed study of the private ranging protocol.

Every IEEE 802.15.4a device communicates using the packet format illustrated in Fig. 6.8. The IEEE 802.15.4a packet consists of a synchronization header (SHR) preamble, a physical layer header (PHR) and a data field.

The SHR preamble is composed of a ranging preamble and an SFD. As summarized in Table 6.1, the ranging preamble is used for acquisition, channel sounding and leading edge detection. The SFD helps a receiver to synchronize to the beginning of the data portion of a frame. Only after establishing acquisition during the preamble, the receiver knows that it is receiving the preamble of a packet. However, it does not know when to expect the end of the preamble yet. It is the SFD[1] that flags the end of the preamble and the beginning of the PHY service data unit (PSDU). The PHR comes after the SHR and contains *data rate* and *frame length* information. Finally, the data field is the part that carries the communication data. In the following sections, the preamble and the SFD structures are described, and ranging related MAC and PHY tasks and capabilities are discussed, including counter management, crystal frequency offset management, and quantification of range measurements and their confidence level.

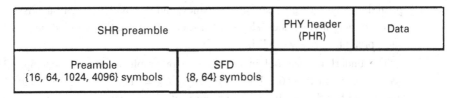

Fig. 6.8. Illustration of the IEEE 802.15.4a packet structure. The data part is BPM-BPSK modulated. (BPM-BPSK: Burst Position Modulation-Binary Phase Shift Keying) (After [91].

---

[1] Please refer to Section 6.3.6 for benefits of the SFD in ranging counter management.

Table 6.1. Functionalities of ranging preamble and SFD.

Preamble	Acquisition
	Channel sounding
	Leading edge detection
SFD	Frame synchronization
	Ranging counter management

Table 6.2. The basis preamble symbol set.

Index	Symbol
S_1	–1000010–1011101–10001–111100–110–100
S_2	0101–10101000–1110–11–1–1–10010011000
S_3	–11011000–11–11100110100–10000–1010–1
S_4	00001–100–100–1111101–1100010–10110–1
S_5	–101–100111–11000–1101110–1010000–00
S_6	1100100–1–1–11–1011–10001010–11010000
S_7	100001–101010010001011–1–1–10–1100–11
S_8	0100–10–10110000–1–1100–11011–1110100

6.3.1 Preamble structure

According to the standard, the ranging preamble may consist of one of {16, 64, 1024, 4096} symbols. The preamble length is specified by the application, and its selection criteria are based on channel multipath profiles, SNR, and receiving PHY capabilities (e.g., coherent/non-coherent reception, quality of search engine, and tracking capability). For instance, even the longest preamble does not help a receiving PHY with a poorly designed search engine. Longer lengths such as {1024, 4096} are preferred for low rate non-coherent receivers to improve SNR via more processing gain. Hence, they can have a better TOA estimate.

According to the standard, the PHY notifies its application how good each range measurement is via a parameter called the *figure of merit (FoM)*. By using this feedback, RDEVs can dynamically adapt the preamble length to channel conditions. Shortening preamble length lowers channel occupancy, and it provides more transmission opportunities for neighbor devices. However, it should be noted that acquisition is difficult with short preambles at low SNR links.

The underlying symbol in the ranging preamble is one of the length-31 ternary sequences, S_i, in Table 6.2. Each S_i of length $L_{ts} = 31$ contains 15 zeros and 16 non-zero codes, and has the much desired property of perfect periodic autocorrelation. In other words, periodic correlation side lobes are zero, and what is observed at the receiver between two consecutive correlation peaks is only the power delay profile of the channel (see Fig. 6.9). Thus, paths between autocorrelation peaks are ensured to be due to multipath channel, but not because of correlation side-lobes.

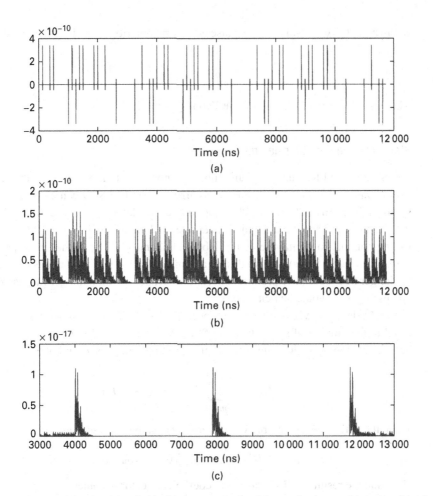

Fig. 6.9. (a) Preamble waveform $P_i(t)$; (b) the magnitude of the received preamble in a multipath channel, $h(t)$; (c) correlation of the received preamble with $P_i(t)$. Note that $T_{pri} = 125$ ns and $N_{sym} = 3$.

Similar to the signal parameter definitions in Chapter 5, $\omega(t)$ denotes the transmitted UWB pulse with unit energy, T_s is the symbol duration, E_s is the symbol energy, and N_f is the total number of pulses per symbol.

Let N_{sym} denote the number of symbol repetitions within the preamble and T_{pri} represent the pulse repetition interval. Then, for the ith basis symbol $\mathbf{S}_i$, the preamble symbol waveform $w_i(t)$ and the resulting preamble waveform $P_i(t)$ can be written as

$$w_i(t) = \sqrt{\frac{E_s}{N_f}} \sum_{j=0}^{L_{ts}-1} \mathbf{S}_i[j]\omega(t - jT_{pri}), \qquad (6.17)$$

$$P_i(t) = \sum_{n=0}^{N_{sym}-1} \mathbf{N}[n]w_i(t - nT_s), \qquad (6.18)$$

where $\mathbf{N} = \mathbf{1}_{1 \times N_{\text{sym}}}$. Note that the pulse waveform polarity vector for the entire preamble is then,

$$\mathbf{P}_i = \mathbf{N} \otimes \mathbf{S}_i, \qquad (6.19)$$

where the operation $\otimes$ stands for the Kronecker product.

6.3.2 Start of frame delimiter design

The SFD is used to establish frame timing. Upon detection of the SFD, the receiving device knows that the PHY header is to arrive next. The SFD can consist of 8 or 64 symbols. The IEEE 802.15.4a PHY supports a mandatory short SFD (8 symbols) for default mode (1 Mbps) and medium data rate, and an optional long SFD (64 symbols) for the nominal low data rate of 106 Kbps. The longer SFD provides more processing gain. Therefore, if one wants to design a communication system that has a long range, the longer SFD should be preferred, because SNR gets lower at longer ranges and more processing gain would be beneficial.

Let $\mathbf{M}$ denote a vector of ternary codes $\{-1, 0, +1\}$ and assume that its length is equal to the number of symbols in the SFD, N_{sfd}. Then, the SFD waveform $Z_i(t)$ can be generated by spreading the outer sequence $\mathbf{M}$ with the basis symbol $\mathbf{S}_i$:

$$Z_i(t) = \sum_{m=0}^{N_{\text{sfd}}-1} \mathbf{M}[m] w_i(t - mT_s). \qquad (6.20)$$

Then, the entire SHR preamble waveform $Y_i(t)$ can be expressed as

$$Y_i(t) = P_i(t) + Z_i(t - N_{\text{sym}} T_s). \qquad (6.21)$$

Note that the resulting SFD waveform coefficient vector becomes

$$\mathbf{Y}_i = \mathbf{M} \otimes \mathbf{S}_i. \qquad (6.22)$$

For generation of $\mathbf{P}_i$ and $\mathbf{Y}_i$, each 1 in the outer code $\mathbf{M}$ or $\mathbf{N}$ is spread as $\mathbf{S}_1$, -1 as $-\mathbf{S}_1$ and 0 as $0 \times \mathbf{S}_1$.

In what follows, let $\mathbf{M}_l$ and $\mathbf{M}_s$ indicate outer sequences for long and short SFDs respectively. They should have the following key properties.

- *Property I*: $\mathbf{M}_l[k] = \mathbf{M}_s[k]$, $\forall k, 0 \leq k \leq 7$. The correlation template for SFD detection in high data rate receivers should be equal to the short SFD itself. By making the first eight codes of $\mathbf{M}_l$ and $\mathbf{M}_s$ the same, the high data rate receivers are spared from running a second correlator to detect whether the SFD being received is the short one or the longer.
- *Property II*: $\mathbf{M}_l[k] = \mathbf{M}_l[k+8]$, $\forall k, 0 \leq k \leq 7$. By exploiting this feature, the high data rate receiver can identify the long SFD, because its correlation output fires twice while receiving the short SFD. This is due to repetition of the first eight codes of $\mathbf{M}_l[k]$. Hence, after the second firing, it can stop the correlator to save energy.

6.3 Ranging in IEEE 802.15.4a standard

Property III: $\sum_{k=0}^{7} \mathbf{M}_l[k] = 0$ and $\sum_{k=0}^{7} \mathbf{M}_s[k] = 0$. The first eight codes in $\mathbf{M}_l$ and $\mathbf{M}_s$ should be balanced. Therefore, when the correlation window is running through the preamble, its output returns zero (see Fig. 6.10). In other words, due to reduced number of correlation side lobes, detection performance for the SFD improves.

Property IV: $\sum_{k=0}^{N_{\text{sfd}}-1} (2 \times |\mathbf{M}_l[k]| - 1) = 0$ and $\sum_{k=0}^{7} (2 \times |\mathbf{M}_s[k]| - 1) = 0$. This requirement helps non-coherent receivers achieve a clear transition from the preamble into the SFD during correlation.

A coherent receiver correlates the received waveform $Y_i(t) + n(t)$ with a template matched to the SFD waveform. Then, assuming a single-path AWGN channel, the

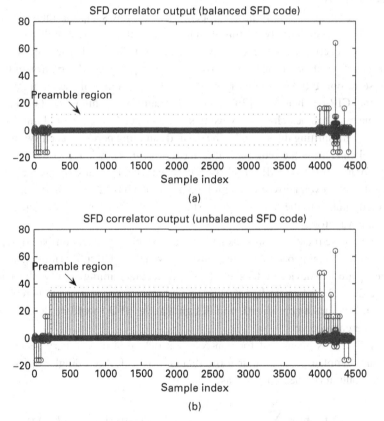

Fig. 6.10. (a) Illustration of the SFD correlator output in the case of a balanced SFD, which is $\mathbf{M}_s = [0, 1, 0, -1, 1, 0, 0, -1]$. Note that when the correlation window is in the preamble only region, the output is zero. When it starts moving into the SFD region, the correlator output returns side lobes until the SFD correlation peak. (b) The SFD correlator output when the SFD is unbalanced, $\mathbf{M}_s = [0, 1, 0, 1, 1, 0, 0, -1]$. Note that even during the preamble only part, the correlator output does not remain at zero, but returns side lobes. This deteriorates SFD detection performance at low SNRs.

correlation output $C_i(k)$ can be expressed as

$$C_i[k] = \sum_{k=0}^{\infty} \int_{kT_s}^{(k+1)T_s+N_{sfd}T_s} (Y_i(t) + n(t))Z_i(t - kT_s)dt. \qquad (6.23)$$

In Fig. 6.10(a), the correlator output for the short SFD case without noise is illustrated. When the entire correlation window overlaps with the preamble part of the received signal, no correlation side lobe is seen (for balanced SFDs only). This improves the SFD detection probability. When the correlation window partially overlaps with the preamble and the rest of it overlaps with the SFD field, side lobes occur. If any side lobe exceeds a detection threshold or the SFD peak happens to be smaller than the threshold, when noise is present, the SFD peak detection fails. This is called *frame synchronization error*, and it results in a packet loss, because the receiver will not be able to decode the packet header properly.

UWB receivers typically have to operate at very low sampling rates. This makes it very difficult to effectively capture the energy at each individual multipath component using Rake receivers, as it is extremely difficult to synchronize to each tap. A chip-spaced sampling of the channel can be used to detect the chip-spaced *observation* of the channel impulse response (CIR), which typically carries a fraction of the available energy of the actual CIR (such as 30% [269]). Note that sampling at high rates (such as chip rate or frame rate) can be achieved by using symbol-spaced sampling and multiple training symbols, and shifting the signal by the desired sampling period at each symbol. Another practical concern is the requirement to have *a-priori* knowledge of the received pulse shape for matched filter (MF) implementation, which may change from one environment to another and even between two multipath components [270]. Therefore, it is difficult to exactly match to the received pulse shape, especially in analog implementation of a template waveform.

Due to these practical concerns and limitations, energy detection-based implementation of UWB ranging becomes more feasible. Even though it suffers more from noise due to the square-law device, it does not require an accurate timing or pulse shape matching. The IEEE 802.15.4a uses BPSK-PPM modulation to support non-coherent reception. Then, it becomes a consideration in design that the SFD should be detectable also via non-coherent receivers.

In [271], it is given that for non-coherent detection of a ternary sequence S_i, the optimum template is its bipolar form, that is $2|S_i| - 1$. Accordingly, the non-coherent receiver output can be expressed as

$$NC_i[k] = \sum_{k=0}^{\infty} \int_{kT_s}^{(k+1)T_s+N_{sfd}T_s} (Y_i(t) + n(t))^2 \bar{Z}_i(t - kT_s)dt, \qquad (6.24)$$

where

$$\bar{Z}_i(t) = \sum_{m=0}^{N_{sfd}-1} \sum_{j=0}^{L-1} (2|M[m]S_i[j]| - 1) \omega(t - jT_{pri} - mT_s). \qquad (6.25)$$

Table 6.3. Autocorrelation peak-to-maximum side lobe (PMSL) and peak-to-average side lobe (PASL) levels (in dB) of the SFD codes in the IEEE 802.15.4a standard.

	Coherent		Non-coherent	
	PMSL	PASL	PMSL	PASL
M_l	7.27	17.6	8.06	20.9
M_s	6.02	13.2	6.02	18.0

In Table 6.3, peak-to-maximum side lobe (PMSL) and peak-to-average side lobe (PASL) values for both short and long SFDs in the IEEE 802.15.4a standard are given. According to those values, one might think that non-coherent detection of the SFD should have a better performance because of its higher PMSL and PASL. Even though it is true at very high SNRs, the squaring operation enhances noise at low SNRs, and SFD detection performance of the non-coherent detector suffers more than that of the coherent one.

6.3.3 PHR

The length of the PHR is 19 octets. The PHR consists of fields that indicate data rate, frame length, ranging flag, preamble length, and error correction and detection bits. Each of the data rate and preamble length is represented with two bits as shown in Fig. 6.11. A value 1 for the ranging flag indicates to the recipient PHY that it is an RFRAME.

The PHR is transmitted at the mandatory data rate. The data rate for the data field of the frame (see Fig. 6.8) is indicated within the *data rate* sub-field of the PHR. It is an exception that for the low data rate option (100 kbps) the PHR is transmitted at the low data rate. The extended SFD, which is 64 symbols long, is used as an indicator for the low rate.

6.3.4 Ranging-related PHY PIB attributes

Between the RF firmware and the MAC, the PHY functions as an interface, and it includes a management entity called the PHY layer management entity (PLME). The

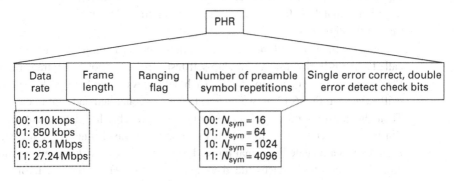

Fig. 6.11. Structure of the PHR.

interface provided by the PLME is used to invoke PHY layer management functions. The database of parameters or attributes that are essential for setting up the PHY and used for managing its operation is called the PHY PIB. For instance, maximum transmit power level of a radio can be specified as a PHY PIB attribute. In the IEEE 802.15.4a standard, there are only five ranging-related PHY PIB attributes. These are:

(i) *phyPreambleSymbolLength*
(ii) *phyRangingCapabilities*
(iii) *phyRFRAMEProcessingTime*
(iv) *phyTxRMARKEROffset*
(v) *phyRxRMARKEROffset*.

The *phyPreambleSymbolLength* is used for the UWB PHY only to specify the length of the basis preamble symbol. *0* indicates that length 31 ternary sequence is used to generate the preamble symbol, and *1* corresponds to length 127, which is only used in the private ranging mode.

The *phyRangingCapabilities* provides ranging capabilities of the UWB PHY. If the RDEV supports ranging, this attribute is set to *0x01*. If crystal offset characterization is supported, it is set to *0x02*. If the RDEV is capable of conducting private ranging, then its value should be *0x04*.

The *phyRFRAMEProcessingTime* refers to the processing time it takes for the PHY to handle an arriving RFRAME. The counting resolution is 2 ns. If two sequential RFRAMEs arrive with separation longer than the time interval indicated by this attribute, the PHY can keep up with the processing load.

The *phyTxRMARKEROffset* is used to specify the internal propagation time between the ranging counter and the transmit antenna. Four octets are allocated for this attribute. The resolution is 1/128th of a chip time at the mandatory chipping rate of 499.2 MHz.

The *phyRxRMARKEROffset* is used to specify the internal propagation time between the receive antenna and the ranging counter. Four octets are allocated for this attribute. The resolution is 1/128th of a chip time at the mandatory chipping rate of 499.2 MHz.

For accurate ranging, only the time-of-flight of a signal over the air should be considered. Therefore, an internal propagation delay between an antenna and its impedance-matching feed circuitry should be subtracted from the round trip propagation time of the signal.

Knowing the values of internal propagation times is a non-trivial task. One method is called a "loop-back" test. An on-chip self-test capability (not an RF transmission) for a PHY can provide a means to identify such delays. A typical loop-back test architecture is illustrated in Fig. 6.12, in which a test signal is generated and sent to the transmitter. Then, the transmitter feeds it back to the receiver branch. The internal propagation delay can be simply considered to be half the round trip travel time of the test signal. It is important to attenuate the signal power level before it enters the receiver branch. Note that a loop-back test induces added complexity to the PHY at the benefit of a calibrated range estimate.

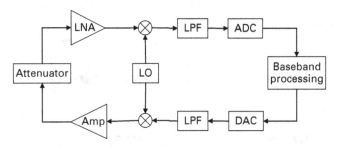

Fig. 6.12. A loop-back test architecture for RF transceivers [272].

In [273], an on-chip or on-wafer loop-back test technique for 5 GHz IC transceivers is explored. In [272], [274] and [275], techniques to improve loop-back tests are also studied.

In the IEEE 802.15.4a standard, a calibration mechanism that is controlled by the application is provided. The mechanism, when triggered by the application, causes its *PHY* to invoke implemented calibration capabilities to characterize the internal propagation time. Further details are given in Section 6.3.11.

6.3.5 MAC PIB attributes

Typically a list of MAC layer PIB attributes and variables that are accessible for the application is referred to as MAC PIBs. In 802.15.4a, there are three ranging-related MAC PIBs:

(i) *macAckWaitDuration*
(ii) *macRangingSupported*
(iii) *macMaxFrameTotalWaitTime*.

The attribute *macAckWaitDuration* describes, in number of data symbols, how long to wait for an acknowledgement of a transmitted data frame (e.g. $RFRAME_{req}$) (Fig. 6.13). Before it expires, the $RFRAME_{rep}$ shall be received by the local device. Let N_{ackw} denote *macAckWaitDuration*. Its value depends on the PHY settings, and it is calculated as

$$N_{ackw} = \left\lceil \frac{T_{ta}^B}{T_{sym}} + \frac{(N_{sfd} + N_{sym})T_s}{T_{sym}} + \frac{T_{phr}^{ackw} + T_{psdu}^{ackw}}{T_{sym}} \right\rceil, \quad (6.26)$$

where T_{phr}^{ackw} is the duration of the PHR of an ACK frame, which is only two octets long, T_{psdu}^{ackw} is the five-octet-long ACK payload, and T_{sym} is the duration of the data symbol. Note that the superscript of the turn-around time T_{ta}^B is omitted in (6.26) without losing generality.

The *macRangingSupported* attribute indicates whether ranging features are supported at the MAC sublayer. It takes boolean values; and FALSE means that ranging features are not supported.

The *macMaxFrameTotalWaitTime*, or N_{ftw}^{max}, describes, in number of preamble symbols, how long to wait for a frame intended as a response to a data request frame. For

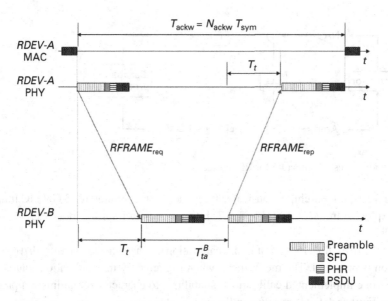

Fig. 6.13. Illustration of the *macAckWaitDuration*. Note that flight time T_t is negligible compared to the turn around time, T_{ta}^B. Delays between MAC and PHY interactions are omitted in the illustration.

instance, the response frame can be interpreted as the time-stamp report that follows $RFRAME_{rep}$, within the context of ranging. Then, the attribute value is given by

$$N_{ftw}^{max} = \frac{T_{fr}^{max} + LIFS}{T_s} + N_{ackw}, \quad (6.27)$$

$$T_{fr}^{max} = (N_{sym} + N_{sfd})T_s + T_{phr} + T_{psdu}^{max}, \quad (6.28)$$

where T_{fr}^{max} is the maximum frame duration, T_{phr} is the duration of the PHR of a frame, T_{psdu}^{max} is the maximum PSDU length, and $LIFS$ the frame spacing after a long frame. The $LIFS$ is set to $40T_s$ in the standard.

Let N_{proc} denote the time available for $RDEV\ B$ in number of preamble symbols to run leading edge detection algorithms and to prepare a precise time-stamp report. Assume that B can start its processing for finding the leading edge at the NHL, while its PHY starts transmitting the $RFRAME_{rep}$. Then,

$$N_{proc} < N_{ftw}^{max} - \frac{T_t + T_{ta}^B}{T_s}. \quad (6.29)$$

6.3.6 Counter management for ranging

Ranging counters are used to track elapsed time between two events. Their management is essential for high-ranging performance. From a local device perspective, the counter starts with the departure of the SFD of an $RFRAME_{req}$ from its antenna, and the counter stops when the SFD of an $RFRAME_{rep}$ arrives at its antenna as shown in Fig. 6.14. These instants are referred to as RMARKER. Similarly, the counter of a peer device

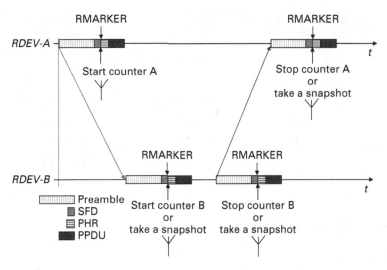

Fig. 6.14. Illustration of ranging counter start and stop events during packet transmission and reception.

starts when the SFD of an $RFRAME_{req}$ arrives at its antenna, and the counter stops when the SFD of an $RFRAME_{rep}$ leaves its antenna.

It is possible that the ranging counter happens to be already running when one of the above events occurs. This indicates that another ranging process is taking place. In this case, what to record as a start or stop time will be the instantaneous value of the counter.

While receiving a frame, the ranging flag can only be read by B only after it starts its ranging counter or captures a snapshot. Value 0 for the *ranging flag* in the PHR indicates that the arriving frame is not an RFRAME. Therefore, the captured counter value should be simply discarded.

The ranging counter is used as an abstraction in the standard; and its implementation is not described. Note that the time resolution at which counters are incremented directly impacts the achievable range resolution. Furthermore, in multipath channels counter instantiation may not correspond to the first arriving pulse of the PHR. Therefore, leading edge detection algorithms[2] should be run in order to estimate this offset. Then, it should be subtracted from the ranging counter value, before finalizing the range estimate.

6.3.7 Ranging figure of merit (FoM)

The figure of merit (FoM) indicates how accurate and reliable a range counter value is. Therefore, for every range counter value, an FoM is produced by an IEEE 802.15.4a device. The FoM consists of three subfields and an extension bit as shown in Fig. 6.15.

The confidence level is the probability that the detected arrival time of the leading edge of a signal will deviate from the true arrival time by at most the confidence interval.

---

[2] See Chapter 5 for detailed coverage of the leading edge detection techniques.

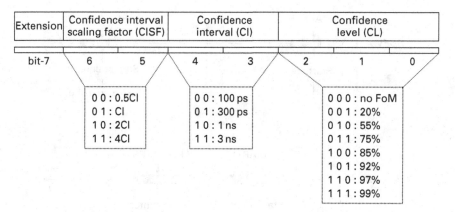

Fig. 6.15. Ranging figure of merit and its subfields.

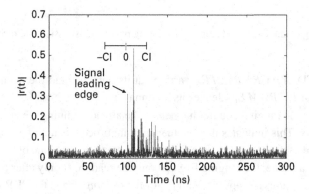

Fig. 6.16. Confidence interval with respect to the true arrival of the signal leading edge.

As exemplified in Fig. 6.16, assume that the leading edge of a signal arrives at $t = 100$ ns and also that the reported confidence interval (CI) and confidence level (CL) are 25 ns and 90% respectively. This means that with 90% probability the arrival time falls within [75, 125] ns. The confidence interval scaling factor (CISF) is used to extend the range of the CI to below 100 ps and above 3 ns. Then, the effective confidence interval CI_{eff} becomes $CI_{\text{eff}} = CISF \times CI$; clearly, it can be as low as 50 ps and as high as 12 ns. The standard does not mandate how to produce the CL, but considers it as an implementation issue.

Assume that an RDEV is receiving an IEEE 802.15.4a preamble with $N_{\text{sym}} = 128$. If it acquires the preamble signal at the second symbol, it can get processing gain via equal gain combining (EGC) from the remaining 126 symbols. Thus, it has an opportunity to improve the SNR by 21 dB, prior to making a judgment on the arrival time of the leading signal edge. If it achieves acquisition at the 100th preamble symbol, then the available processing gain becomes only 14 dB ($10 \log_{10} 26$). The correlator outputs after the EGC in these two cases are shown in Fig. 6.17.

As the SNR increases, finding the leading edge becomes easier and the ranging error gets smaller. With this in mind, one can simply relate the CL with the separation of two

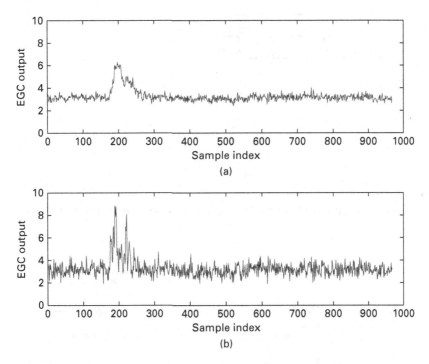

Fig. 6.17. Correlator output with EGC during reception of the IEEE 802.15.4a preamble with $N_{sym} = 128$, once acquisition is achieved. (a) Acquisition at the second preamble symbol allows processing gain of 21 dB, (b) acquisition at the 100th preamble symbol results in 14 dB processing gain for leading edge detection. Note that the CL would be higher in (a).

events: the acquisition instant and the ending of the preamble. In other words, the earlier the acquisition is within the preamble, the higher the CL is.

Finally, the FoM value of *0x80* is used to inform the upper layer that the ranging counter values are not correct. Moreover, the FoM value of *0x00* is used when there is no information about the quality of the corresponding range measurement. Untrustworthy measurements are reported as *0x79*.

6.3.8 Mitigation of clock frequency offsets

A local oscillator is used to generate time-reference for a device. A shift from the ideal frequency of an oscillator causes frequency offsets, and consequently it alters the clock cycle. Fig. 6.18 illustrates an ideal 500 MHz clock and two other clocks with $\mp 5\%$ frequency offsets with respect to the ideal clock. Any frequency offset would impact performance of processes and applications that heavily depend on an accurate time base, because devices observe time in number of their clock cycles. Ranging is one example. As given in (6.15) and (6.10), frequency offsets cause ranging errors. Especially for TW-TOA protocol where the turn-around time T_{ta}^B can reach milliseconds, frequency offsets can be very detrimental.

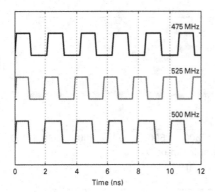

Fig. 6.18. Illustration of two clocks with $\mp 5\%$ frequency offsets with respect to an ideal clock of 500 MHz.

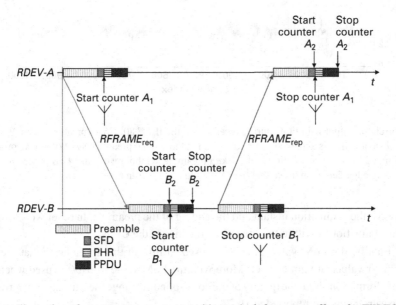

Fig. 6.19. Illustration of counter management to mitigate clock frequency offsets for TW-TOA protocol.

As discussed in Section 6.2.3, the SDS protocol mitigates ranging error that is induced by clock frequency offsets. It is also possible to improve ranging performance of the TW-TOA protocol by espousing the counter management scheme illustrated in Fig. 6.19.

Assume counters A_1 and A_2 are managed by A, and counters B_1 and B_2 by B. Let Δ_i denote the difference between the stop and start times of counter i. It is easy to see that Δ_{A_1} and Δ_{B_1} measure $T^A_{\text{round}}(1 + e_A)$ and $T^B_{\text{ta}}(1 + e_B)$, respectively. In addition, Δ_{A_2} measures the length of the PHR and PPDU of the $RFRAME_{\text{rep}}$ transmitted by B and Δ_{B_2} the length of the PHR and PPDU of the $RFRAME_{\text{req}}$. Assume that T_{frame} denotes the length of the PHR and PPDU, and that T_{frame} has the same value for $RFRAME_{\text{req}}$ and $RFRAME_{\text{rep}}$, when an ideal clock rate f_i is considered. Then, Δ_{A_2} and Δ_{B_2} are given by

6.3 Ranging in IEEE 802.15.4a standard

$$\Delta_{B_2} = f_i \frac{T_{\text{frame}}}{(1+e_A)} (1+e_B), \qquad (6.30)$$

$$\Delta_{A_2} = f_i \frac{T_{\text{frame}}}{(1+e_B)} (1+e_A). \qquad (6.31)$$

Note that B reports to A the values Δ_{B_2} and Δ_{B_1}. Define a correction factor υ, which is computed by A and given by

$$\upsilon \triangleq \sqrt{\frac{\Delta_{B_2}}{\Delta_{A_2}}}. \qquad (6.32)$$

Rewrite (6.8) as

$$\hat{T}_t = \frac{T_{\text{round}}^A (1+e_A)}{2} - \frac{\Delta_{B_1}}{2}. \qquad (6.33)$$

After Δ_{B_1} in (6.33) is substituted by Δ_{B_1}/υ, it can be easily shown that

$$\epsilon_{\text{tw}} = T_t e_A. \qquad (6.34)$$

Note that ϵ_{tw} does not depend on T_{ta}^B anymore and that it is distance dependent.

Example 6.1 *Devices A and B are 10 m apart from each other, and they are to perform TW-TOA based ranging by exchanging IEEE 802.15.4a packets. Assume that instantaneous clock rates of A and B are 495 MHz and 510 MHz, respectively. In other words, $e_A = -0.01$ and $e_B = 0.02$.*

(i) *If $T_{\text{frame}} = 100\,\mu s$ and $T_{\text{ta}}^B = 1\,ms$ when measured with a 500 MHz clock, calculate the turn around time and Δ_{B_2} measured by B in number of clock cycles.*

B runs its clock at 2% faster rate with respect to the 500 MHz clock. Therefore, for the true turn around time of 1 ms, B counts 510 000 clock cycles.

$$\Delta_{B_1} = T_{\text{ta}}^B (1+e_B) f_i$$
$$= 1 \times 10^{-3} (1+0.02) 500 \times 10^6$$
$$= 510\,000. \qquad (6.35)$$

With the 500 MHz clock, 100 μs long T_{frame} corresponds to 50 000 clock cycles. Note that A transmits the frame according to its own clock cycle, which is 1% longer than the ideal. Therefore, T_{frame} put into air by A is indeed 101.01 μs. B counts 51 515 clock cycles during this T_{frame} of 101.01 μs.

$$\Delta_{B_2} = f_i \frac{T_{\text{frame}}}{(1+e_A)} (1+e_B)$$
$$= 51\,515.15. \qquad (6.36)$$

(ii) *Compute* ΔA_1 *and* ΔA_2 *in number of clock cycles.*

$$\Delta A_1 = (2T_t + T_{ta}^B)(1 + e_A)f_i$$
$$= (2 \times 33.3 \times 10^{-9} + 1 \times 10^{-3})(1 - 0.01)500 \times 10^6$$
$$= 495\,033 \qquad (6.37)$$

and

$$\Delta A_2 = f_i \frac{T_{\text{frame}}}{(1 + e_B)}(1 + e_A)$$
$$= 500 \times 10^6 \frac{100 \times 10^{-6}}{(1 + 0.02)}(1 - 0.01)$$
$$= 48\,529.4. \qquad (6.38)$$

(iii) *What is the correction factor?*

$$v = \sqrt{\frac{\Delta B_2}{\Delta A_2}} = \sqrt{\frac{51\,515.15}{48\,529.4}} = 1.0303031. \qquad (6.39)$$

(iv) *Compute the estimated time of flight in seconds,* $\hat{T}_t$.

$$\hat{T}_t = \frac{1}{f_i}\left(\frac{\Delta A_1}{2} - \frac{\Delta B_1}{2v}\right)$$
$$= \frac{1}{500 \times 10^6}\left(\frac{495\,033}{2} - \frac{510\,000}{2 \times 1.0303031}\right)$$
$$\approx 33.03 \text{ ns}. \qquad (6.40)$$

(v) *Compute the residual time of flight error.*

$$\epsilon_{\text{tw}} = T - \hat{T}_t$$
$$= 33.3 \times 10^{-9} - 33.03 \times 10^{-9}$$
$$= 0.27 \text{ ns}. \qquad (6.41)$$

The residual time of flight error in (6.41) is caused by the round-off errors in the computations.

6.3.9 Time-stamp reports

Each time-stamp report contains the ranging counter start and stop values and the *FoM*. Management of counter values is explained in Section 6.3.6, and the *FoM* is discussed in Section 6.3.7.

6.3.10 Ranging-related service primitives

This section is intended more for practicing engineers and implementers rather than students. As stated in Section 6.1, primitives are used to pass attributes between layers and to trigger events. In this section, ranging-related primitives and attributes defined in the IEEE 802.15.4a standard and their use cases are discussed.

MCPS-Data.request

This primitive is generated at the NHL and issued to the local MAC sublayer, when ranging needs to be done with a peer entity. It consists of four ranging parameters:

(i) *msdu*
(ii) *msduLength*
(iii) *UWBRanging*
(iv) *UWBPreambleSymbolRepetitions*.

The *msdu* contains the MSDU to be transmitted by the MAC sublayer entity, and the *msduLength* shows its length. The *UWBRanging* is an indicator for ranging support, and it can take one of three values: *OFF*, *ALL* and *HEADER*. The *OFF* indicates that either ranging is not supported or not to be used for this particular transmission. *ALL* means ranging operation is supported and ranging counters should be enabled at the local PHY entity. *HEADER* denotes ranging support only by using the ranging flag without enabling the ranging counters. The *UWBPreambleSymbolRepetitions* shows the number of symbol repetitions the local PHY entity should use in the preamble of the corresponding frame.

PD-Data.request

This primitive is generated by a local MAC sublayer entity upon reception of an *MCPS-Data.request* primitive, and issued to the local PHY entity to initiate ranging transaction with a peer device. It consists of the following parameters:

(i) *psdu*
(ii) *psduLength*
(iii) *UWBRanging*
(iv) *UWBPreambleSymbolRepetitions*.

The *psdu* is the PSDU to transmit, and *psduLength* is its length. If the *UWBRanging* is set to *ALL*, a ranging capable PHY would start its ranging counters as soon as the RMARKER leaves its antenna (see Fig. 6.14). The support for ranging operation is later announced by setting the ranging flag in the PHY header of the frame.

Upon reception of this primitive, the local PHY entity forms a PPDU and attempts to transmit it to a peer entity. When the local PHY entity completes the transmission of the PPDU, it issues the *PD-Data.confirm* primitive.

PD-Data.confirm

This primitive is generated by the local PHY entity and issued to the local MAC sublayer entity in response to a *PD-Data.request*. It passes three attributes:

(i) *status*
(ii) *UWBRangingCounterStart*
(iii) *UWBRangingCounterStop*.

The *status* is set to *SUCCESS* if the PPDU is successfully transmitted. Otherwise, it takes a value that refers to an error code. Handling of the ranging counters is somewhat tricky. Only if the frame to be transmitted in response to the *PD-Data.request* primitive is an $RFRAME_{req}$, then the value of the *UWBRangingCounterStart* corresponds to the local time at which the RMARKER of the $RFRAME_{req}$ leaves the transmit antenna. If the frame to be transmitted is an $RFRAME_{rep}$, then the time at which the RMARKER of the $RFRAME_{rep}$ leaves the transmit antenna is stored in *UWBRangingCounterStop*.

MCPS-Data.confirm

This primitive is generated by the local MAC sublayer in response to an *MCPS-Data.request* primitive, upon reception of a *PD-Data.confirm* primitive. It takes the following parameters from the *PD-Data.confirm* primitive and passes them to the NHL:

(i) *status*
(ii) *UWBRangingCounterStart*
(iii) *UWBRangingCounterStop*.

Hence, the NHL is notified of the result of its request to transmit.

PD-Data.indication

This service primitive is generated at the PHY entity and issued to the local MAC entity. Its function is to notify the MAC sublayer that a PSDU is received from a peer PHY. The primitive includes the following parameters:

(i) *UWBPreambleSymbolRepetitions*
(ii) *UWBRangingReceived*
(iii) *UWBRangingCounterStart*
(iv) *UWBRangingCounterStop*
(v) *UWBRangingFOM*.

The *UWBPreambleSymbolRepetitions* refers to the preamble symbol repetitions, N_{sym}, of the received PPDU. The value '*ON*' for *UWBRangingReceived* indicates that the received frame is an RFRAME. If particularly it is an $RFRAME_{req}$, the local time that the RMARKER of the frame arrives at the receive antenna should be recorded and passed by parameter *UWBRangingCounterStart*. Thus, *UWBRangingCounterStart* points to the beginning of the turn around time T_{ta}^{B}. If the received frame is an $RFRAME_{rep}$, then local time for the RMARKER arrival should be passed by *UWBRangingCounterStop*. The *UWBRangingFOM* shows the one-octet-long *FoM* report for the received RFRAME.

MCPS-Data.indication

This primitive is generated by the MAC sublayer on receipt of a data frame, and is issued to the NHL. This primitive takes the parameters listed for the *PD-Data.indication* primitive and passes them to the NHL.

6.3.11 Ranging-related management primitives

These primitives are typically used for management of local tasks by local entities. They do not trigger over-the-air message exchange with peer entities.

MLME-Sounding.request
This primitive is used to request channel sounding information from the PHY entity. It is generated by the NHL and issued to the MAC layer management entity (MLME). It does not pass any parameters.

PLME-Sounding.request
It is generated by the MLME and issued to the PHY layer management entity (*PLME*). The request is for the PHY to pass sounding information to the NHL. Thus, leading edge search can be performed by the application instead of PHY itself.

PLME-Sounding.confirm
This primitive is generated by the PLME and issued to the MLME. It reports the result of a channel-sounding request from the PHY. The report consists of three parameters:

(i) *status*
(ii) *SoundingSize*
(iii) *SoundingList*.

The *status* indicates the status of the channel-sounding attempt. The *SoundingSize* indicates the number of sounding points being reported. The *SoundingList* consists of the amplitudes and times of all the sounding points. The amplitudes are reported relatively, and therefore they do not have a unit.

MLME-Sounding.confirm
It is generated by the MLME and issued to the NHL in response to a *MLME-Sounding.confirm* primitive. It relays the parameters passed by the *PLME-Sounding.confirm* primitive.

MLME-Calibrate.request
A set of primitives is used to calibrate for internal propagation delays (e.g. propagation delay between a modulator and an antenna, etc.). The *MLME-Calibrate.request* primitive is generated by the NHL and issued to the MLME. It does not pass any parameter, but requests RMARKER offset information from the MLME.

PLME-Calibrate.request
It is generated by the MLME upon receipt of the *MLME-Calibrate.request* primitive, and issued to the PLME. It causes the local PHY to return RMARKER offset information, if it has calibration capability.

PLME-Calibrate.confirm

This primitive is generated by the PLME and issued to the MLME. It reports the result of the calibration request from the PHY. The report consists of three parameters:

(i) *status*
(ii) *CalTxRMARKEROffset*
(iii) *CalRxRMARKEROffset*.

The *status* indicates the status of the calibration attempt. If calibration information is available, it is set to *SUCCESS*. The *CalTxRMARKEROffset* is the propagation time from the modulator, which triggers the ranging counter, to the transmitter antenna. Similarly, the *CalRxRMARKEROffset* is the propagation time from the receiver antenna to the demodulator.

MLME-Calibrate.confirm

This primitive is generated by the MLME and issued to the NHL. It simply relays the PHY response given by the *PLME-Calibrate.confirm* primitive to the NHL.

6.3.12 Exemplified use of the primitives via the TW-TOA protocol

This section provides an example use of primitives, and illustrates how primitive parameters at the NHL, MAC and PHY protocol layers are managed during the TW-TOA protocol-based ranging.

Let *local device* and *peer device* denote devices that initiate ranging and respond to the ranging request, respectively. Furthermore, assume that the local and peer device PHY entities are ranging capable.

In the IEEE 802.15.4a standard, as explained in Section 6.3.10, the *MCPS-Data.request* primitive is generated at the local NHL and issued to the local MAC to initiate a ranging process. In the example illustrated in Fig. 6.20, *UWBRanging* is set to *ALL* to let the PHY use all of its ranging capabilities and *UWBPreambleSymbolRepetitions* to 16 to perform ranging with the shortest preamble. The local MAC sublayer entity passes these parameters via the *PD-Data.request* primitive to the local PHY entity.

The counter snapshot shows *100* when the $RFRAME_{req}$ leaves the antenna of the local device. The transmission status and the counter start value are reported back to the NHL via *PD-Data.confirm* and *MCPS-Data.confirm* primitives. Upon reception of the $RFRAME_{rep}$, the counter snapshot is *1567*. The *FoM* field indicates that the confidence interval and confidence level of the leading edge detection are 1 ns and 99%, respectively.

In Fig. 6.21, transactions between protocol layers of the peer device are illustrated. Counter start and stop values are given as 617 and 1917, respectively. The difference, which is 1300, is the turn around time in number of clock cycles (or chip intervals). Assume that it is reported to the local device within the time-stamp message.

Ultimately, the local device knows that the total round-trip time equals 1467 clock cycles and also that 1300 clock cycles are due to processing delay at the peer device. Then, the resulting one-way flight time corresponds to 83 chip intervals. With a 500 MHz clock, the chip interval is 2 ns. Thus, the distance estimate becomes 49.8 m.

6.3 Ranging in IEEE 802.15.4a standard

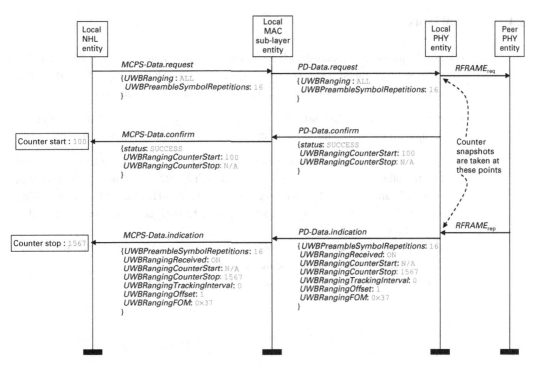

Fig. 6.20. Exemplified use of the TW-TOA protocol with primitive transactions and parameter settings between entities of the local device.

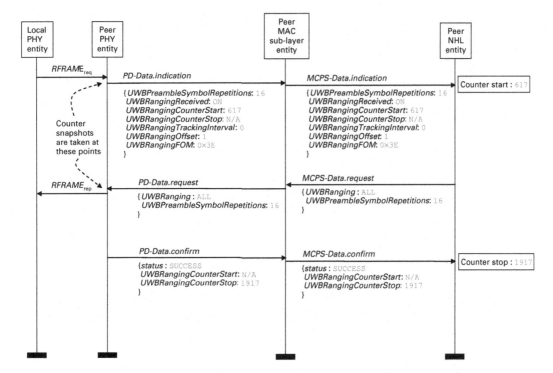

Fig. 6.21. Exemplified use of the TW-TOA protocol with primitive transactions and parameter settings between entities of the peer device.

6.4 Problems

(1) For the problem in Example 6.1, assume that T_{frame} parameter for the $RFRAME_{req}$ is three times longer than T_{frame} parameter for the $RFRAME_{rep}$. How does this change the correction factor? Does ϵ_{tw} change? Justify your answers.

(2) How will ϵ_{tw} in Example 6.1 change, if fixed point arithmetic is used in computations? Compute all the counter values and the correction factor.

(3) Devices A and B are 15 m apart from each other, and they are to carry on the type I differential two-way ranging protocol with IEEE 802.15.4a packets. Assume that $e_A = 0.02$ and $e_B = 0.01$ with respect to an ideal 500 MHz clock. The preset waiting time is $T = 100$ ms, and it is known *a priori* to both devices. Also assume that $t_{off}^B = 1$ ms and $t_{off}^A = 0$.
  (a) How would clock frequency offsets degrade the range estimation performance? Derive an expression for the range estimate.
  (b) Solve the same problem for the type-II ranging protocol.

7 Special topics in ranging

This chapter discusses three special topics related to ranging. First, techniques to mitigate various types of interference are presented. Second, carrier sensing methods that can be used to improve ranging performance for IEEE 802.15.4a networks are briefly reviewed. Finally, an overview of mechanisms that provide privacy and security for ranging signals and range information is given.

In this chapter, it is assumed that ranging is performed via frames that consist of preamble, start of frame delimiter (SFD), physical layer header (PHR) and payload, and also that the preamble is used for ranging (similar to the IEEE 802.15.4a systems studied in the previous chapter). Frames with longer preambles provide a higher processing gain for ranging due to improved SNR and lead to better ranging accuracy. This is because at high SNRs, detection of the direct path signal is easier. On the other hand, employing a longer preamble induces a drawback that the preamble becomes more vulnerable to interference and jamming attacks. In case that acquisition of a frame fails, the frame needs to be retransmitted.

Interference can be detrimental to ranging accuracy, even if it does not cause acquisition failure. At times the leading signal path gets buried under interference, so that it may be quite difficult to determine its arrival time. Remember from Chapter 6 that performance of ranging protocols is very sensitive to timing. This mandates rapid handling of all ranging related transmissions. If retransmissions of ranging frames were scheduled with high priority, regular data traffic would be penalized, and throughput and latency for the data traffic would degrade. Furthermore, each retransmission may potentially interfere with transmission of peer devices in the same network. It is not energy and throughput efficient to retransmit very long frames. All these factors suggest that a ranging receiver should deploy interference mitigation techniques to sustain its ranging performance.

Regulating multiple-access via carrier sensing in UWB networks may improve throughput and ranging accuracy, because carrier sensing prevents collisions and lowers the number of retransmissions. In [276] some novel approaches are introduced to improve carrier-sensing capability in IEEE 802.15.4a networks. A part of this chapter provides an overview of these standardized techniques.

In communication networks, packet preambles are very structured and unprotected. Therefore, during ranging signal exchanges, snooper and imposter devices can exploit the publicly known signal structures to figure out their relative distances with legitimate ranging devices. Furthermore, they can be deceptive to ranging receivers by replaying preamble waveforms. Therefore, it is essential to develop privacy and security mechanisms against malicious devices.

In summary, the first part of this chapter is allocated to discussion of various interference types and interference mitigation techniques. It is followed by a section that provides an overview of so-called coded payload modulation (CPM) techniques for carrier sensing. The CPM provides clear channel assessment to improve throughput and range accuracy without impacting frame lengths. Finally, privacy issues for ranging are addressed and practical defense mechanisms are discussed.

7.1 Interference mitigation

Interference may corrupt different parts of a received frame. Figure 7.1 illustrates four different interference scenarios. Each interference scenario is denoted with a letter. As shown in case A, if an interference signal overlaps with the preamble of a frame, some preamble symbols may get corrupted and those symbols cannot be used for SNR improvement. This type of interference does not necessarily cause acquisition failure, but it impacts range accuracy because the achievable processing gain decreases.

If interference affects the SFD part of a frame (case B), SFD detection may fail and frame synchronization is lost. Consequently, the PHR and payload data cannot be decoded. Furthermore, the ranging counter values that entirely rely on timing of the SFD detection become meaningless. Hence, retransmission is required.

The PHR contains data rate and frame length information according to the IEEE 802.15.4a standard. If the data rate field of the PHR gets corrupted (case C) irrecoverably, it becomes impossible to make use of the payload data. Even if the receiver successfully acquires and performs leading edge detection by processing the preamble and the SFD parts of the frame, the source of the frame cannot be identified and TOA information will be of no importance. Then, retransmission becomes necessary.

The payload consists of MAC header and payload. If interference damages the MAC header (case D), the sender of a frame cannot be identified. In this case, even after successful acquisition and synchronization, the MAC payload becomes useless.

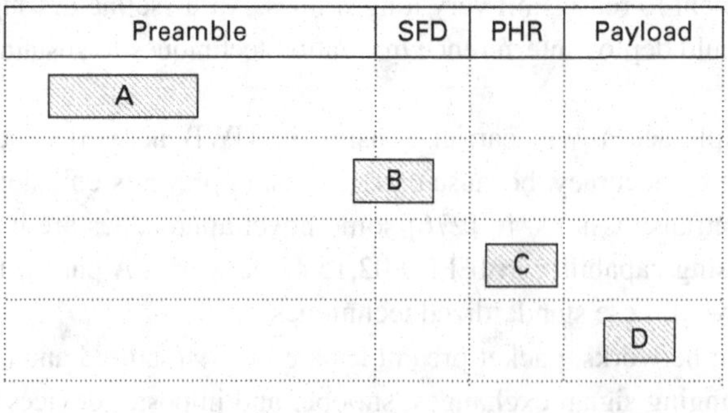

Fig. 7.1. Illustration of cases in which interference overlaps with various parts of a received frame.

7.1 Interference mitigation

Two types of interference for UWB signals are multiuser interference (MUI) and narrowband interference (NBI). Both MUI and NBI have been well studied in the literature for UWB data communication. On the other hand, interference mitigation for UWB acquisition and ranging still remains an active research area.

A UWB receiver can process the preamble by one of template matching (coherent), energy detection (ED) or transmitted-reference (TR) schemes. Although coherent ranging is superior in general, ED receivers offer advantages such as simplicity, operability at sub-Nyquist sampling rates (which determines the range resolution), and low cost, as pointed out in Chapter 5. They are also more resilient to pulse shape distortion, but they are less favored in practice because of their poor performance in the presence of MUI and NBI. This is because interference suppression techniques such as those for CDMA systems are not readily applicable to simple non-coherent receivers.

Typically, UWB ranging receivers obtain processing gain by combining a number of signal samples according to a transmitted waveform pattern [258]. However, in the presence of interference, the interfering signal may be regarded as a leading edge. Therefore, prior to any combining operations, it is crucial to remove NBI and MUI to the utmost extent for accurate ranging.

7.1.1 Narrowband interference mitigation

There exist legacy or other devices that operate in the UWB spectrum. For instance, IEEE 802.11a WLAN systems operate at 5.2 GHz and may interfere with IEEE 802.15.4a devices. Therefore, it is of great importance to limit interference between UWB and narrowband systems.

Assume that a sinusoidal signal $i(t)$ with frequency f_i interferes with UWB signal $s(t)$. Let f_L and f_H denote the lower and higher cutoff frequencies of the UWB signal, respectively, such that $f_L < f_i < f_H$, and

$$i(t) = A \sin(2\pi f_i t + \phi). \qquad (7.1)$$

In coherent processing according to Fig. 7.2, the received signal $r(t)$ is first fed into a bandpass filter (BPF) with frequency response $H(f)$. If we consider an ideal BPF at the

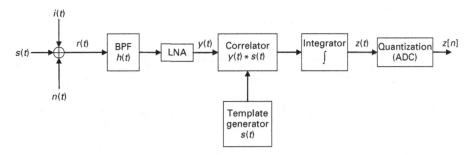

Fig. 7.2. Coherent processing of received UWB signal.

receiver RF front-end, $H(f)$ can be expressed as

$$H(f) = U(f - f_\mathrm{L}) - U(f - f_\mathrm{H}), \qquad (7.2)$$

where $U(f)$ is the unit step function defined as

$$U(f) = \begin{cases} 1, & \text{if } f > 0 \\ 0, & \text{otherwise} \end{cases}. \qquad (7.3)$$

Next, the filter output is amplified, and then correlated with a template $s(t)$. The correlation output is integrated and quantized. Typically, NBI is much stronger than a received UWB signal. Therefore, if the maximum power of $i(t)$ is outside the dynamic range of the LNA, the NBI can easily saturate it. One practical solution is, if the center frequency f_i of the interference is known, to add a notch filter with a null at f_i before the LNA.

The frequency spectrum $H_\mathrm{no}(f)$ of the ideal notch filter $h_\mathrm{no}(t)$ is given in [277] as

$$H_\mathrm{no}(f) = 1 - \left[U\!\left(f - f_i + \frac{B_\mathrm{no}}{2} \right) - U\!\left(f - f_i - \frac{B_\mathrm{no}}{2} \right) \right], \qquad (7.4)$$

where B_no denotes the width of the spectral notch.

A typical approach to suppress NBI with multiple tones is to divide the UWB spectrum into subbands and then deactivate the subband that corresponds to the frequency band of a narrowband device [74, 278]. This can be achieved using a receiver that employs a filter-bank as shown in Fig. 7.3.

There are also other methods to generate spectral nulls at all interfering frequencies. In [279], the template waveform $q(t)$ is constructed as a linear combination of N_ort orthogonal waveforms to be able to detect several types of UWB pulses by simply adjusting scaling coefficients of each waveform, i.e.

$$q(t) = \sum_{j=1}^{N_\mathrm{ort}} \varpi_j s_j(t), \qquad (7.5)$$

where $\int s_k(t) s_l(t) \mathrm{d}t = 0$ for $k \neq l$ and ϖ_j is the scaling coefficient for the jth waveform. In [280], a multicarrier template waveform that consists of several subband pulses is

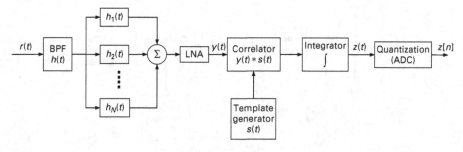

Fig. 7.3. A receiver architecture to suppress known multitone NBI.

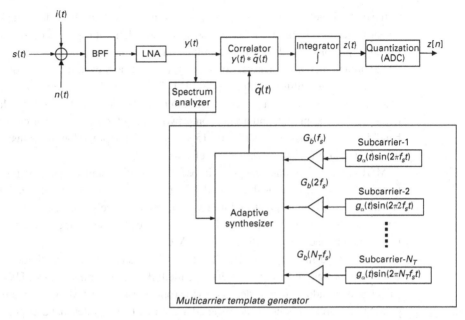

Fig. 7.4. A receiver architecture that uses a multicarrier template generator to mitigate NBI (After [280]). Note that $N_{\text{init}} = 1$ and $f_s = \Delta f$.

used to mitigate NBI. The template is correlated with the received signal (see Fig. 7.4). Assume that the template is formed from N_T subcarriers with a frequency spacing of Δf, starting at frequency $N_{\text{init}} \Delta f$. Then, the approximate template can be expressed as

$$\tilde{q}(t) = \frac{1}{N_T} \sum_{n=N_{\text{init}}}^{N_{\text{init}}+N_T-1} G_b(n\Delta f) g_\alpha(t) e^{j2\pi n \Delta f t} df , \qquad (7.6)$$

where $g_\alpha(t) = e^{-\frac{\alpha t^2}{\hat{\tau}^2}}$, with $\hat{\tau}$ representing half the subcarrier pulse width, and $G_b(n\Delta f)$ is the Gabor transform of $q(t)$, given by

$$G_b(n\Delta f) = \int q(t) g_\alpha(t-b) e^{-j2\pi n \Delta f t} dt . \qquad (7.7)$$

Another approach to mitigate NBI is to design a transmit pulse shape that generates notches at target interference frequencies. A pulse design-based scheme that creates a null at a particular frequency is presented in [281]. This approach is not very feasible due to transmit/receive antenna effects that may distort pulse shape and frequency-dependent fading in UWB channels.

7.1.2 Multiuser interference models

Users of a UWB network share a common spectrum. Therefore, management of multiple-access within the same spectrum is important. One multiple-access approach is to assign a unique time-hopping (TH) or direct sequence (DS) code to each user. If orthogonality of the codes is achieved, users can transmit simultaneously without degrading detection performance of one another.[1] Otherwise, cross-correlation properties of the codes determine the performance loss when simultaneous transmissions occur. In mesh networks this approach increases receiver complexity, because now each receiver should be capable of acquiring multiple unique codes. This burden on the receiver can be lowered if codes are assigned not on a per user basis, but per network. The IEEE 802.15.4a standard takes the latter approach and assigns only two unique ternary codes for each frequency band. In other words, in an IEEE 802.15.4a network, the preamble of a transmission can have one of at most two different waveforms.

MUI during preamble reception impacts acquisition and ranging. If the interferer transmits the same code as the desired user, the receiver may lock onto the interference signal. This leads to acquisition failure. If the interference buries the direct path signal, range accuracy also suffers as explained earlier. The statistics of the MUI determine how much ranging accuracy is affected by the MUI.

As discussed in Chapter 2, in a TH IR-UWB system, each user divides each of its symbol durations into N_f time intervals called *frames* and transmits one UWB pulse in each frame. In addition, in order to provide robustness against MUI, the positions of the pulses in different frames are determined according to TH codes that are specific to each user. In such a system, the received signal from K users can be expressed as

$$r(t) = \sum_{k=1}^{K} \sum_{l=1}^{L^{(k)}} \alpha_l^{(k)} s^{(k)}\left(t - \tau_l^{(k)}\right) + n(t), \qquad (7.8)$$

where $n(t)$ is AWGN with two-sided power spectral density $\mathcal{N}_0/2$, $\alpha_l^{(k)}$ and $\tau_l^{(k)}$ are the channel coefficient and delay of the lth path for the signal received from user k, respectively, and $L^{(k)}$ is the number of multipath components for user k. In addition, the signal $s^{(k)}(t)$ related to user k is given by

$$s^{(k)}(t) = \sqrt{\frac{E_s}{N_f}} \sum_{j=-\infty}^{\infty} a_j^{(k)} \omega\left(t - jT_f - c_j^{(k)} T_c\right), \qquad (7.9)$$

where the notations are adopted from (5.6); i.e. E_s represents the energy of a ranging symbol, $a_j^{(k)} \in \{-1, +1\}$ is the polarity code (or a ternary code, $a_j^{(k)} \in \{-1, 0, +1\}$, in general) for user k, T_f is the frame duration, $c_j^{(k)} \in \{0, 1, \ldots, N_h - 1\}$ is the TH code for user k, with N_h denoting the number of chips per frame, $\omega(t)$ represents the received UWB pulse with unit energy, and T_c is the chip duration.[2]

Note that the use of ternary codes for $a_j^{(k)}$ makes it easier to generate signals with perfect autocorrelation property,[3] which is useful for accurate range estimation, as studied in Chapter 6.

---

[1] In practice, due to asynchronism and multipath propagation, the orthogonality of signals from different users cannot commonly be maintained at the receiver.
[2] Note that N_f, T_f, and T_c can also depend on the user index k in general.
[3] For perfect autocorrelation property, the TH codes are selected as $c_i^{(k)} = c_j^{(k)}$ $\forall i, j$ for the preamble.

In order to investigate the effects of MUI in UWB systems, assume that $c_j^{(k)} = 0$ $\forall k, j$ and consider a correlation receiver that obtains the decision variables for ranging as[4] [282]

$$Z_\tau = \sum_{j=0}^{N_f-1} a_j^{(1)} \int_{\tau+jT_f}^{\tau+(j+1)T_f} r(t)\,\omega(t - jT_f - \tau)\,dt \qquad (7.10)$$

$$= E_\tau + I_\tau + N_\tau, \qquad (7.11)$$

where E_τ is the captured signal energy, N_τ is the output noise distributed as $\mathcal{N}(0, \mathcal{N}_0 N_f/2)$, and I_τ is the MUI, given by

$$I_\tau = \sqrt{\frac{E_s}{N_f}} \sum_{k=2}^{K} \sum_{m=0}^{N_f-1} \sum_{j=-\infty}^{\infty} \sum_{l=1}^{L^{(k)}} a_j^{(k)} \alpha_l^{(k)} R_\omega \left((j-m)T_f + \tau_l^{(k)} - \tau\right), \qquad (7.12)$$

with $R_\omega(\cdot)$ representing the autocorrelation function of $\omega(t)$. Quantification of the impact of MUI on ranging performance depends on characteristics of I_τ. While the distribution of I_τ is assumed to be Gaussian in many studies [283, 284], the observation in [285] shows that the Gaussian approximation is not always accurate for TH IR-UWB systems. This suggests that the Gaussian distribution may not be a good model of I_τ for certain scenarios. In [286], distribution of MUI in TH-PPM UWB systems is studied assuming that interference within different frames is statistically independent. Thus, the probability density function of overall MUI, $p_I(x)$, for $K - 1$ interfering users is modeled as a weighted sum of Gaussian PDFs with different mean and variance, and is given by

$$p_I(x) = \sum_{i=0}^{n} \frac{\tilde{c}_i}{\sqrt{2\pi}\,\sigma_i} e^{-\frac{(x-\mu_i)^2}{2\sigma_i^2}}, \qquad (7.13)$$

where $n = N_f(K - 1)$, $\tilde{c}_i$ is the weight coefficient, and μ_i and σ_i^2 are the mean and variance of the ith PDF, respectively. This is also referred to as the *Middleton Class A noise* [287].

In [282], MUI is modeled by the Laplacian distribution, which provides better characterization of MUI for MF receiver structures than the Gaussian distribution does.

7.1.3 Multiuser interference in IEEE 802.15.4a networks

The MUI distribution models in the previous section are not directly applicable to IEEE 802.15.4a networks. Some of the reasons for this include a limited support for simultaneously operating piconets (SOPs), assignment of different waveform structures for preamble and payload, and a wide dynamic range for the preamble length to payload length ratio.

The frequency band plan of the IEEE 802.15.4a standard specifies 16 frequency channels, and use of only two different preamble sequences is allowed within each band,

---

[4] For modeling MUI in the text, an MF-based decision variable is considered.

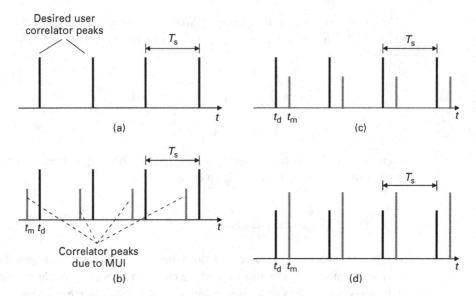

Fig. 7.5. Illustration of the impacts of MUI on ranging. The single user preamble interference case in a single-path channel without noise is considered. (a) Preamble correlator output when the input is only from the desired user's preamble. (b) Preamble correlator output when weaker interfering preamble arrives earlier than the desired user's preamble. (c) Preamble correlator output when weaker interfering preamble arrives later than the desired user's preamble. (d) Preamble correlator output when stronger interfering preamble arrives later than the desired user's preamble.

one for each SOP. For example, for channel 0, only sequences S_1 and S_2 are assigned (please see Table 6.2). Thus, multiuser preambles are constructed only from either S_1 or S_2.

Assume that the MUI comes from the same code user. In other words, both the desired user and multi-user preambles are constructed from symbol S_1 with duration T_s. Remember from Chapter 6 that a key property of the ranging symbols is their perfect periodic autocorrelation. When no MUI is present, the receiver correlator output would return peaks with period T_s as shown in Fig. 7.5(a). The MUI, when present, also generates periodically repeated peaks at the correlator output as illustrated in Fig. 7.5(b)–(d). Assume that the energies of the desired user and MUI correlator peaks are E_d and E_m, respectively and also that the first desired user peak arrives at time t_d and the first MUI peak at t_m. When $t_m < t_d$ and $E_m < E_d$ (see Fig. 7.5(b)), the receiver may lock onto the desired signal peaks for acquisition, but the MUI peak is subject to be detected as the leading path. When $t_m > t_d$ and $E_m < E_d$, the receiver is more likely to lock onto the desired signal and the MUI would look like another multipath component following the strongest path (see Fig. 7.5(c)). Of course, if the channel delay spread of the MUI in multipath channels is quite long such that it causes interference to the next correlator peak of the desired user, finding the leading path will be quite challenging. When $t_m > t_d$ and $E_m > E_d$, the receiver may simply lock onto the MUI peaks for acquisition, and consequently ranging with the desired user may fail. In practice, receiver behavior for acquisition depends on threshold settings and peak selection algorithms.

When the MUI and desired user symbols are not from the same code, the cross-correlation properties of the codes impact ranging and acquisition. Codes assigned to an IEEE 802.15.4a channel are not orthogonal. Therefore, in addition to the cross-correlation peaks, now the receiver has to deal with multiple side lobes. It would be almost impossible to distinguish multipaths of the desired user channel from the cross-correlation side lobes. If MUI corrupts only a portion of the desired user preamble, it may be possible to mitigate the impact of interference, although it requires computationally complex algorithms.

While perfectly balanced ternary sequences (PBTS) are used for the preamble, the data modulation adopts the BPM-BPSK scheme. Remember from Chapter 2 that the data symbol interval T_{sym} is divided into two halves in the IEEE 802.15.4a standard. Each half further consists of an active interval and a guard interval as shown in Fig. 7.6(a). Polarity of the burst and whether it is transmitted in the active interval of the first half or that of the second determine the values of the two bits for the BPM-BPSK modulation. Note that a burst is not allowed to be transmitted within guard intervals. For accurate modeling of the interference between the payload of the interfering user and the preamble of the desired user, one must analyze the cross-correlation properties of the PBTS waveforms with burst-hopping waveforms. By properly designing burst-hopping sequences, impacts of payload interference on the IEEE 802.15.4a preamble can be minimized. Fig. 7.6 illustrates an IEEE 802.15.4a data symbol structure in (a) and relative positioning of a preamble pulse with respect to the data burst in (b), (c) and (d) for $T_{sym} = T_f$. In Fig. 7.6(b), the preamble pulse falls onto the guard interval of the data symbol. Since the pulse repetition interval (PRI) is equal to T_{sym}, in this particular instance no preamble pulse collides with a data burst for a single-path channel. Figure 7.6(c) illustrates a case

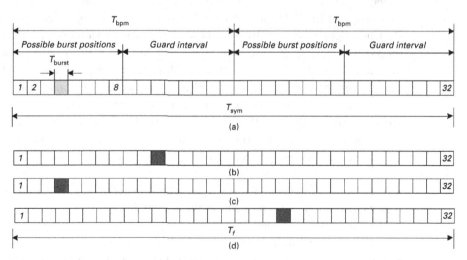

Fig. 7.6. Illustration of the structure of the IEEE 802.15.4a data symbol. (a) One data symbol. (b) Preamble frame interval T_f is equal to T_{sym} and the interfering preamble pulse arrives in the guard interval of the data symbol. (c) A preamble pulse overlaps with the burst. (d) A preamble pulse arrives within the fourth bin of the second half of the data symbol, while the burst is received in the fourth bin of the first half of the symbol.

Table 7.1. IEEE 802.15.4a implementation options.

Code length	T_{sym} (ns)	T_f (ns)	R_{pp} for $N_{sym} = \{16, 64, 1024, 4096\}$
31	8205.13	32.05	{0.0019, 0.0075, 0.1205, 0.4820}
31	1025.64	32.05	{0.0151, 0.0603, 0.9640, 3.8560}
31	128.21	32.05	{0.1205, 0.4820, 7.7117, 30.8470}
31	64.10	32.05	{0.2410, 0.9640, 15.4247, 61.6987}
31	8205.13	128.21	{0.0075, 0.0301, 0.4820, 1.9282}
31	1025.64	128.21	{0.0603, 0.2410, 3.8563, 15.4253}
31	512.82	128.21	{0.1205, 0.4820, 7.7126, 30.8506}
31	64.10	128.21	{0.9641, 3.8565, 61.7035, 246.8142}
127	8205.13	8.0128	{0.0019, 0.0077, 0.1234, 0.4937}
127	1025.64	8.0128	{0.0154, 0.0617, 0.9874, 3.9495}
127	128.21	8.0128	{0.1234, 0.4937, 7.8986, 31.5944}
127	32.05	8.0128	{0.4937, 1.9748, 31.5969, 126.3876}

in which catastrophic collisions are inevitable between preamble pulses and data bursts, if no burst hopping is used. In Fig. 7.6(d), the preamble pulse and the data burst arrive at different halves of the T_{sym}, and they do not collide.

It is important to know that the IEEE 802.15.4a standard offers settings in which preamble and payload lengths vary a lot. Most implementation options of the IEEE 802.15.4a standard are shown in Table 7.1. The code length indicates the length of the underlying PBTS used for the preamble. R_{pp} denotes the ratio of the preamble length to the payload length for preamble symbol repetitions N_{sym} of 16, 64, 1024 and 4096. Payload length is taken as 1029 symbols, which is the longest allowed. Even with the longest payload size, R_{pp} can be as high as 246.81. This means that in many cases the preamble of a multi-user transmission would interfere with the preamble of the desired user. *A-priori* probabilities of the *preamble-to-preamble* and *payload-to-preamble* interference cases should be considered in conjunction with their corresponding interference models in MUI analysis for IEEE 802.15.4a networks. The *preamble-to-preamble* interference can be seen in two forms, as explained earlier, due to code diversity. Key steps for a complete MUI analysis consist of the following items.

- Modeling MUI of the preambles at the acquisition correlator output, when the other users transmit the same preamble as the desired user.
- Modeling MUI of the preambles at the acquisition correlator output, when the other users transmit a different preamble from the desired user.
- Modeling MUI of the multi-user payload at the acquisition correlator output.
- Calculating R_{pp}.
- Determining *a-priori* probabilities for preamble-to-preamble interference and payload-to-preamble interference from R_{pp}.
- Modeling the total MUI by factoring *a-priori* probabilities of the three events.

7.1 Interference mitigation

Typically, a receiver combines correlator outputs separated by $N_{\text{sep}} = T_s/T_{\text{smp}}$ samples to improve SNR prior to leading path detection, where T_{smp} is the sampling interval. Even though such a combining operation improves the SNR, without any pre-filtering it can carry the interference into the correlator output. This is unfavorable, because clearly the interference may bury the leading path or be detected as the leading path itself. In either case, ranging performance gets degraded. One practical approach to mitigate transient interference is to apply a non-linear filter to the correlator output samples $z[n]$. One example for a non-linear filter is a median filter. The median filter is a special case of stack filters that have been widely used in digital image and signal processing [288, 289] to remove singularities caused by noise. The median filter replaces the center sample with the median of the samples within the filter window. A major drawback with non-linear filters is that they may degrade the desired signal at low SNR levels. Therefore, they must be used carefully.

Example 7.1 *Consider a preamble constructed from symbol sequence S_1 with $T_f = 32$ ns and $N_{\text{sym}} = 8$, where each UWB pulse (Gaussian monocycle with a bandwidth of 500 MHz) in the preamble has unit energy. This preamble is passed through a realization of CM-1 according to the IEEE 802.15.4a channel models [100], and corrupted by AWGN (SNR = 10 dB). Then, the received signal is fed into a square-law device and the output is sampled at $T_{\text{smp}} = 2$ ns intervals. After that, a template is generated from sequence $2|S_1| - 1$ using the same UWB pulse shape, and is integrated and sampled over 2 ns intervals. When the square-law output signal is correlated with the non-coherent template designed for S_1 at 2 ns sampling resolution, the correlator output consists of the channel profile repeated $N_{\text{sym}} = 8$ times with time resolution of T_{smp} as shown in Fig. 7.7(a). This is due to the perfect periodic autocorrelation property of the sequence S_1.*

In order to simulate the impact of transient interference at the correlator output, first a single preamble symbol is generated by using sequence S_2 from Table 6.2. Then, the amplitudes of the pulses in the interference signal are scaled such that they are three times stronger than the pulses in the desired symbol. After that, the generated symbol is passed through another realization of CM-1, and the resulting waveform is added, with a random delay, to the received signal obtained in the first step before the square-law device. After integrating and sampling the square-law device output at 2 ns intervals, the samples are correlated with the same template. In the presence of interference, the correlator would carry interference energy onto its output as shown in Fig. 7.7(b).

In order to mitigate the effects of this interference, consider the non-linear filtering operation on the correlation outputs $z[n]$, as shown in Fig. 7.8, where $N_{\text{sep}} = T_s/T_{\text{smp}}$. This filter takes a set of three correlator samples with a separation of N_{sep} samples and then outputs the median sample of the set. In Fig. 7.7c, the correlator output is illustrated after this filtering operation. The non-linear filter effectively removes the transient interference.

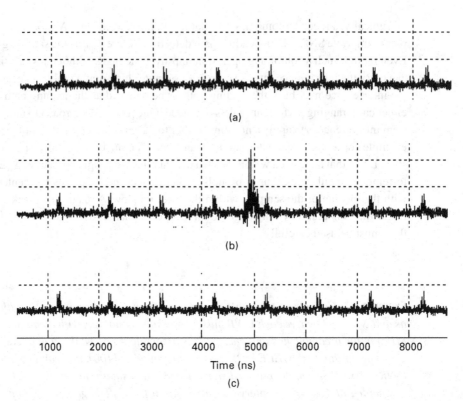

Fig. 7.7. Correlator output in a multipath channel when interference corrupts an IEEE 802.15.4a preamble symbol: (a) interference-free correlator output; (b) correlator output with transient interference; (c) after median filtering the correlator output that is corrupted by interference.

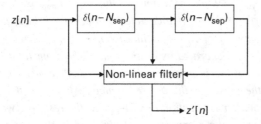

Fig. 7.8. A non-linear filter architecture that incorporates a length-3 median filtering.

7.1.4 Multiuser interference mitigation in TH-IR UWB

A TH-IR UWB receiver commonly combines energies over multiple symbols from multiple pulse positions using the TH sequences of the desired user to improve the signal-to-noise ratio (SNR) [258]. In the presence of MUI, such an energy-combining operation has a major drawback, as noted in the previous section. One solution simply lies in considering the collected energy samples from a different view, a two-dimensional

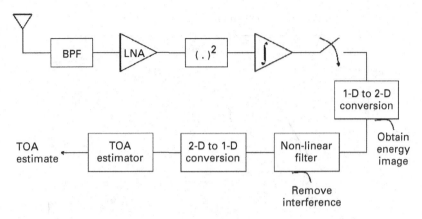

Fig. 7.9. Ranging with MUI cancelation using a non-linear filter (After [201]).

energy matrix as suggested in [201, 290]. A block diagram that represents the basic operations of the energy matrix with non-linear filtering approach is illustrated in Fig. 7.9. This work was later extended in [291] to mitigate narrowband interference.

In the energy matrix approach, first an empty matrix $\mathbf{Z}$ of size $N_f \times N_s$ is generated where N_f is the number of frames to be processed and N_s the number of energy samples to be collected from each frame. Assume that the signal $r(t)$ in (7.8) is received and also that the energy matrix is to be formed over a symbol duration at sampling rate $1/T_{\text{smp}}$. The collected energy samples at the ED receiver would be

$$z[n] = \int_{(n-1)T_{\text{smp}}}^{nT_{\text{smp}}} |r(t)|^2 dt . \tag{7.14}$$

Energy samples given in (7.14) are grouped together according to the transmitted TH code, and samples of the same group are used to populate a column of the energy matrix $\mathbf{Z}$; that is,

$$\mathbf{Z}[j,i] = z\left[i + (j-1)N_s + c_j^{(1)}\right] , \tag{7.15}$$

where $j \in \{1, 2, \ldots, N_f\}$, $i \in \{1, 2, \ldots, N_s\}$ and the chip interval T_c is assumed to be equal to T_{smp}. A typical energy matrix of a TH-IR signal plus interference after passing through separate IEEE 802.15.4a CM-1 channels is given in Fig. 7.10. Note that MUI and self-interference causes short discrete lines. The actual TOA corresponds to the left-most continuous vertical line in $\mathbf{Z}$.

A cause of self-interference is the imperfect auto-correlation of the TH codes. Note that the energy samples of a column are grouped according to the desired user's TH code. It is possible that only some of the grouped samples contain energy from the received signal due to a partial overlap with the signal's TH pattern. Especially if the uncertainty region for the TOA is larger than T_f, the energy collection process would cause more self-interference.

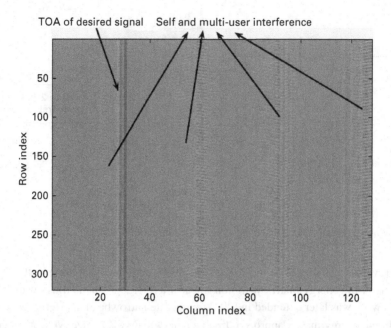

Fig. 7.10. Energy matrix for a TH IR-UWB system, where $E_d/\mathcal{N}_0 = 16$ dB (E_d is the desired signal energy), $E_m/\mathcal{N}_0 = 10$ dB (E_m is the interfering signal energy), $T_c = 4$ ns, $N_s = 4$, $T_{sym} = 512$ ns, $T_f = 128$ ns, and $T_{sym}/T_{smp} = 128$ [201].

To remove outliers in $\mathbf{Z}$, a moving length N_w median filter can be applied column-wise. Then, the elements of the new energy matrix $\mathbf{Z}^{(\text{med})}$ become

$$\mathbf{Z}^{(\text{med})}[j, i] = \text{med} \{\mathbf{Z}[j, i], \mathbf{Z}[j+1, i] \ldots \mathbf{Z}[j + N_w - 1, i]\}. \quad (7.16)$$

Once interference is removed, $\mathbf{Z}^{(\text{med})}$ is converted to a vector by the column-sum operation as follows.

$$\tilde{\mathbf{Z}}^{(\text{med})} = \mathbf{1}_{N_f - N_w + 1} \mathbf{Z}^{(\text{med})}. \quad (7.17)$$

Then, detection techniques can be applied onto $\tilde{\mathbf{Z}}^{(\text{med})}$ to locate the leading path.

Example 7.2 *Consider a TH-IR UWB system in which the ranging symbol consists of four frame intervals, and each frame interval is further divided into four chip intervals. The TH codes of the desired signal and the interference are $\{1, 1, 1, 0\}$ and $\{0, 0, 1, 1\}$, respectively (see Fig. 7.11(a)–(b)). Assume that the received signal is integrated and sampled at a period such that four samples are collected per frame interval, and also that no noise is present.*

(i) *Write an analytical expression for the correlator output after energy combining.*
   Energy combining requires energy samples $z[n]$ of the received signal to be combined in accordance with the corresponding TH code. The combined energy values $E_c[n]$ can be formulated as

$$E_c[n] = z[n + 0] + z[n + 4] + z[n + 4 + 4] + z[n + 4 + 4 + 3], \quad (7.18)$$

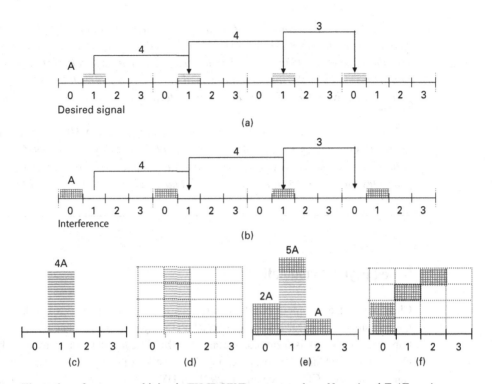

Fig. 7.11. Illustration of energy combining in TH-IR UWB systems, where $N_f = 4$ and $T_f/T_c = 4$. (a) Transmission pattern of the desired user's signal. (b) Transmission pattern of a TH-IR interference. (c) Energy vector after applying energy combining to the desired user's signal without interference. (d) Energy matrix of the desired user's signal without interference. (e) Energy vector after applying energy combining to desired user's signal with interference. (f) Energy matrix of the interference alone. Note that for simplicity of illustrations, both desired user signal and interference energy levels are assumed to be A in each frame (After [201]).

for $0 \le n \le 3$, assuming that TOA ambiguity is less than the frame interval T_f.

(ii) *If there was no interference, what would be the TOA index? Assume that the smallest index with non-zero energy is selected as the TOA.*
It can be calculated that $E_c[1] = 4A$ and $E_c[n] = 0$ for $n \ne 1$ therefore, the TOA index is 1 (see Fig. 7.11(c)–(d))

(iii) *In presence of the specified interference, what is the TOA index?*
The TOA index is 0, because $E_c[0] = 2A$ due to interference (see Fig. 7.11(e)).

(iv) *The size of $\mathbf{Z}$ would be 4×4. How would it be populated with the samples of the received signal?*

$\mathbf{Z}$ would be populated as follows

$$\mathbf{Z} = \begin{bmatrix} z[0+11] & z[1+11] & z[2+11] & z[3+11] \\ z[0+8] & z[1+8] & z[2+8] & z[3+8] \\ z[0+4] & z[1+4] & z[2+4] & z[3+4] \\ z[0+0] & z[1+0] & z[2+0] & z[3+0] \end{bmatrix}. \quad (7.19)$$

*After filling out each column of **Z** from samples separated according to the received signal's TH pattern, straight lines would be formed in the matrix whenever the samples contain signal energy (Fig. 7.11e). The detection of the left-most vertical line corresponds to the time index of the first arriving signal energy. If the MUI follows a different TH pattern, intuitively the energy matrix of the interference would not form a straight line (Fig. 7.11f).*

*Conventional energy combining does not account for the MUI, because it directly aggregates the energy samples. This is equivalent to summing the rows of **Z** along each column, yielding an energy vector. Note that the column-sum of the matrix in Fig. 7.11e generates the energy vector in Fig. 7.11c, and column-sum of (e) and (f) results in Fig. 7.11d.*

7.2 Coded payload modulation

Clear channel assessment (CCA) determines the current state of a wireless medium for collision avoidance. It is an important PHY layer function, because scheduling of transmissions for collision avoidance and other MAC layer protocol behaviors rely on the CCA. In narrowband systems, detecting the presence of energy at the carrier frequency can be used as a CCA mechanism [292]. In UWB systems, preambles that consist of periodically repeated sequences with minimum or near zero autocorrelation side lobes are used for time synchronization. Preamble structure in the IEEE 802.15.4a is a good example for wideband systems. The correlation peaks are used to detect the preamble; and these peaks are indicative of signal presence for CCA.

In [276], the authors develop a TDMA-type multiplexed preamble scheme that enables preamble-detection-based CCA for UWB systems. In this scheme, preamble symbols are multiplexed with the entire PPDU, by periodically inserting them into the header and payload parts of the IEEE 802.15.4a packet after every k-symbol-long interval as illustrated in Fig. 7.12. The inserted preamble symbols do not interfere with the PPDU data. This multiplexing should not impact the PPDU length. Therefore, increasing the number of inserted preamble symbols reduces the data-carrying capability of the frame.

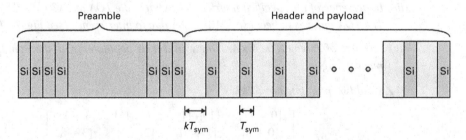

Fig. 7.12. TDMA style multiplexing of preamble and PPDU to support CCA for heavily loaded networks (After [276]).

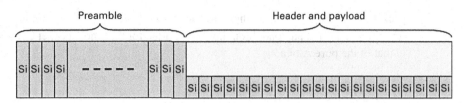

Fig. 7.13. CDMA style multiplexing of preamble and PPDU to support CCA (After [276]).

In the CDMA type multiplexed preamble scheme, periodically repeated preamble symbols are superimposed onto the entire PPDU (see Fig. 7.13). Power level of the overlayed preamble symbols are set to be much lower than the data symbols, because preamble symbols can cause interference to data demodulation. Another drawback is implementation complexity due to the need for generating pulses with two different power levels in the transmitter. The CDMA type multiplexing does not reduce data carrying capability of the frame, unlike the TDMA type.

As shown in Fig. 7.14, using CCA during the entire packet greatly improves throughput, compared to using CCA only for the preamble and the Aloha scheme. Aloha is the simplest channel-access technique, in which a node transmits without sensing the channel. If its transmission fails, the node backs-off by a random time interval, and then retries. When a CCA mechanism is deployed only for the preamble of the packets, a better payload throughput efficiency than Aloha is obtained. Using CCA during an entire

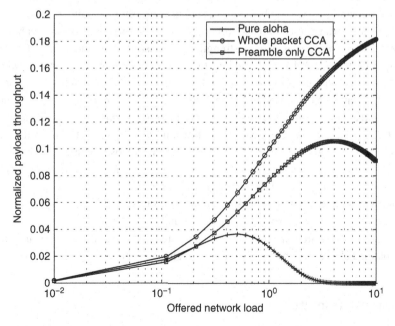

Fig. 7.14. Normalized payload throughput performance of CCA and pure Aloha (CS: Carrier sense). Preamble length is 1 ms, $k = 4$, data rate is 1 Mbps and the data size is 32 bytes (After [276]).

packet further improves throughput, as shown in Fig. 7.14. Furthermore, in [293], up to four times maximal payload throughput is reported for the whole packet CCA compared to that of the pure Aloha.

7.3 Private ranging

Most previous studies in the field have assumed that the transmitter–receiver pair involved in ranging know and trust one another, and devices use an agreed upon ranging waveform [243, 253]. Network (NWK) layer techniques for secure localization have been studied recently [294, 295]. In fact, existing communication standards such as WiFi, IEEE 802.15.4 and 802.11 enable security features such as authentication and encryption within the data link or NWK layers, but the PHY layer has very little, if any, responsibility to implement features that enhance network security. In what follows, security techniques applicable at the PHY layer will be highlighted for ranging systems.

There are typically two motivations behind location-related attacks. First, an intruder intends to figure out the location of sensor devices in protected areas and tries to tamper with and disable them. In the latter, it tries to prevent ranging devices from obtaining correct range information. Relative positioning information in a network can be used to optimize high-layer network operations such as route discovery and maintenance, multicasting and broadcasting. If falsified position information is passed around the network, location-based network functionalities can be easily subverted.

A malicious device can behave as a snooper, impostor or jammer. A snooper observes or listens to transmitted signals in secret to obtain information on whereabouts of other devices. By measuring signal strengths of transmissions and delays between range request frame $RFRAME_{req}$ and range reply frame $RFRAME_{rep}$, the snooper can have a coarse knowledge about its range to other devices. An impostor device, on the other hand, engages in deception under an assumed name or identity to cause distance reduction and enlargement in victims' range calculations. By simply replaying an originator's $RFRAME_{req}$, the impostor can trigger transmission of $RFRAME_{rep}$ by the target node. Hence, it can figure out its range to the target. An impostor can also impersonate a target by transmitting an $RFRAME_{rep}$ in response to an $RFRAME_{req}$. A jammer device simply emits interference to stall communication in a network. The most effective way to deal with a jammer is to back off until jamming stops. Also, some signal processing techniques can be effective to mitigate jamming interference.

Various protection mechanisms are suggested in the literature against the attack types explained above. In what follows these mechanisms are briefly discussed.

7.3.1 Challenge–response

In all ranging protocols, it is essential to prevent impersonation attacks so that the $RFRAME_{req}$ and $RFRAME_{rep}$ are ensured to be from legitimate devices. This can be

7.3 Private ranging

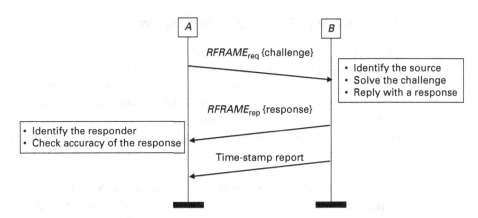

Fig. 7.15. Modification of the TW-TOA to deal with distance enlargement and reduction attacks.

achieved by incorporating a *challenge–response* phase into the TW-TOA as illustrated in Fig. 7.15. In [296], the authors suggest that $RFRAME_{req}$ can carry a secret *challenge* and $RFRAME_{rep}$ a *response* to the challenge. If the response is accurate, A can ensure B is not malicious.

An attacker may attempt to replay $RFRAME_{req}$ to get a response, but it is generally simple to deal with replay attacks at the protocol layers. Disallowing retransmissions with the same sequence number is the simplest way to defend against such attacks.

7.3.2 N_m-distance

The N_m-distance algorithm given in [296] requires N_m distance measurements between any two ranging devices. It then applies a median filter onto the measurement set to weed out large errors. It would be difficult for the attacker to cause enlargement in multiple measurements. Therefore, median filtering can prove to be effective unless more than half of the measurements are enlarged.

7.3.3 Randomizing turn-around time

It has already been discussed in Chapter 6 that turn-around time T_{ta}^B induces a large bias in the range estimate. In the TW-TOA protocol, T_{ta}^B is reported in a time-stamp report in an encrypted form to factor the bias out. Therefore, it becomes difficult for an observing malicious device to figure out the value of T_{ta}^B. In the type I and type II differential TOA protocols given in Chapter 6, a preset waiting interval T dominates the turn-around time. It may be possible for an eavesdropper to determine the value of T from statistical inference by listening to multiple ranging exchanges between A and B. As a precaution against such an attack, A and B can select a different T for each time they perform ranging. One suggestion is to vary T per ranging according to a pseudo-random sequence known to both A and B. Its value can be conveyed within the challenge frame described in Section 7.3.1.

7.3.4 Cyclic shift of perfectly balanced ternary sequences

Having a large family of preamble symbol waveforms minimizes the probability that an impostor would identify the ranging waveform used by ranging devices and improve its acquisition hideously. One drawback of the networks that employ the IEEE 802.15.4a signal structure is that the standard publicly specifies only eight length-31 PBTSs for the preamble to exploit their perfect periodic autocorrelation property for ranging. Therefore, it would not be difficult for an impostor to acquire a transmitted IEEE 802.15.4a sequence out of a set of eight, especially if it is running eight correlators in parallel.

The PBTS sequences have another interesting property that can be beneficial to deceive impostors and make them work harder to determine the turn-around time accurately. Assume that a ternary symbol waveform $\tilde{s}_i^{(l)}(t)$ is expressed as

$$\tilde{s}_i^{(l)}(t) = \sqrt{\frac{2E_s}{(N_f + 1)}} \sum_{j=1}^{N_f} \tilde{a}_{i,j}^{(l)} \omega(t - (j-1)T_f), \qquad (7.20)$$

where l indicates that the underlying sequence $\mathbf{S}_i^{(l)}$ is obtained by cyclic shifting of sequence $\mathbf{S}_i$ (or $\mathbf{S}_i^{(0)}$) by l times and $\tilde{a}_{i,j}^{(l)} \in \{-1, 0, 1\}$ is the ternary coefficient for the jth element in sequence $\mathbf{S}_i^{(l)}$, as illustrated in Fig. 7.16. Then, the transmitted ranging waveform can be expressed as

$$\tilde{r}_i^{(l)}(t) = \sum_{i=1}^{N_{\text{sym}}} \tilde{s}_i^{(l)}(t - iT_s). \qquad (7.21)$$

Let $\tilde{\phi}_i^{(m,n)}(t)$ denote the correlation of $\tilde{r}_i^{(m)}(t)$ and $\tilde{s}_i^{(n)}(t)$. The peak instants of $\tilde{\phi}_i^{(m,n)}(t)$ and $\tilde{\phi}_i^{(m+k,n)}(t)$ would differ by kT_s with respect to each other, as illustrated in Fig. 7.17. In other words, cyclic shift operation preserves the perfect periodic autocorrelation property, but it circularly shifts the correlation peaks.

This motivates the use of the cyclic-shift property of ternary sequences as a security tool. Simply, impostors or snoopers that are unaware of the cyclic shift index k can still observe the ranging waveform, but their range would be in error by ckT_s meters, where $c = 3 \times 10^8$ m/s. Note also that by incorporating a random cyclic shift in the transmitted waveform, an impostor is forced to search for the right sequence out of a set of 240 sequences.

7.3.5 Dynamic preamble selection

In the IEEE 802.15.4a standard, dynamic preamble selection (DPS) is used as a privacy mechanism for ranging. The preamble symbol to be used for ranging is communicated between ranging parties via an authentication message. In this message, the originator device informs the target of the indices of the ranging symbols to be used in $RFRAME_{\text{req}}$ and $RFRAME_{\text{rep}}$ frames. Afterwards, the TW-TOA ranging protocol is run.

7.3 Private ranging

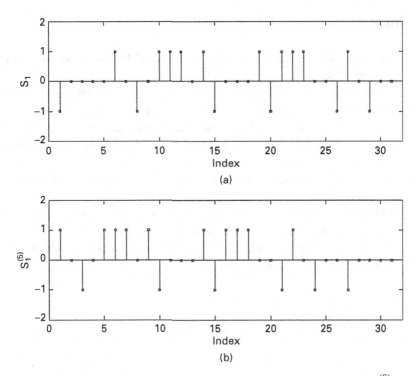

Fig. 7.16. Illustration of the cyclic shifting of a length-31 PBTS. (a) Ternary sequence $\mathbf{S}_1^{(0)}$; (b) $\mathbf{S}_1^{(5)}$.

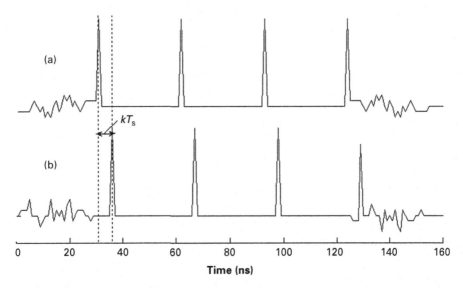

Fig. 7.17. Illustration of the change of peak locations when a cyclic shift is introduced onto the PBTS, for $T_s = 1$ and $k = 5$. (a) $\tilde{\phi}_1^{(0,0)}(t)$; (b) $\tilde{\phi}_1^{(5,0)}(t)$.

7.4 Problems

(1) How does interference affect the preamble, SFD, PHR, and payload parts of a received ranging frame in IEEE 802.15.4a systems? Explain the differences.
(2) Name three techniques that can be used to mitigate NBI in IEEE 802.15.4a systems.
(3) Name three methods that are used to model the MUI in TH-IR UWB systems.
(4) Explain why and how MUI degrades the ranging accuracy in an IEEE 802.15.4a system. How does the impact of interference change with respect to R_{pp} (ratio of the preamble length to the payload length)?
(5) Why and how do non-linear filters help in removing the MUI? Explain with an example.
(6) Discuss the advantages and disadvantages of coded payload modulation for transmission of IEEE 802.15.4a packets.
(7) Name and briefly summarize three protection mechanisms for improving the security of IEEE 802.15.4a ranging systems.

8 Practical considerations for UWB system design

In this chapter, practical issues for UWB system design are investigated. First, design of UWB signals for ranging applications is considered, and selection of various signal parameters, such as pulse repetition interval and bandwidth, is studied. Then, link budget calculations are performed in order to determine signal quality as a function of distance. Compared to narrowband systems, the large bandwidth of UWB systems introduces additional challenges for the design of certain system components. Therefore, hardware issues for UWB transmitters and receivers are investigated, and the design of power amplifiers, antennas, low-noise amplifiers and analog-to-digital converters is studied for UWB systems.

8.1 Signal design for ranging

Design of a ranging system requires careful selection of system parameters in order to optimize system performance under practical and regulatory constraints. In a ranging system, the main performance criteria are the ranging accuracy and the amount of time to perform ranging; i.e. *ranging time*.

8.1.1 Performance metrics

The ranging accuracy can be quantified by means of mean absolute error (MAE), which is defined as the expected value of the absolute value of the error; i.e.

$$\text{MAE} = \text{E}\left\{\left|\hat{d} - d\right|\right\}, \tag{8.1}$$

where $\hat{d}$ is the range estimate and d is the true range. In a practical scenario, the expected value in (8.1) is approximated by the sample mean of the absolute error:

$$\text{MAE} = \frac{1}{N} \sum_{i=1}^{N} \left|\hat{d}_i - d_i\right|, \tag{8.2}$$

where d_i and $\hat{d}_i$ are, respectively, the true range and the range estimate for the ith measurement, for $i = 1, \ldots, N$, with N denoting the number of measurements.

Another common metric for the ranging accuracy is root mean square error (RMSE), which is defined as the square root of the average value of the squared error; i.e.

$$\text{RMSE} = \sqrt{E\left\{(\hat{d} - d)^2\right\}}. \qquad (8.3)$$

Similar to (8.2), the expected value in (8.3) is approximated by the sample mean of the squared error in practice; i.e.

$$\text{RMSE} = \sqrt{\frac{1}{N}\sum_{i=1}^{N}\left(\hat{d}_i - d_i\right)^2}, \qquad (8.4)$$

where d_i and $\hat{d}_i$ are, respectively, the true range and the range estimate for the ith measurement, for $i = 1, \ldots, N$.

The main difference between the MAE and RMSE metrics is that the RMSE gives higher weights to large errors than the MAE does, as the errors are squared before the averaging operation in the calculation of RMSE. In other words, the RMSE is more sensitive to outliers.

For various node configurations, large deviations of ranging errors can be observed among different measurements, and the MAE or RMSE may not give sufficient information about the performance of the ranging system. For example, the ranging error can be very small for most measurements, but a few measurements with very large errors may dominate the MAE/RMSE. In such cases, a more meaningful performance metric is the cumulative distribution function (CDF) of the ranging error. In other words, the probability that the ranging error is smaller than a certain threshold can be specified for all threshold values; that is,

$$F(x) = P\left\{|\hat{d} - d| \leq x\right\}. \qquad (8.5)$$

Note that the CDF metric in (8.5) implicitly contains information about the *confidence level* metric studied in Section 6.3.7 of Chapter 6.

In addition to ranging accuracy, ranging time is another performance criterion for a ranging system. The ranging time is calculated as the amount of time that passes from the initial observation of the ranging signal until a ranging estimate is obtained. Commonly, it is the duration of the ranging signal that dominates the ranging time (i.e. the duration of the ranging signal is considerably larger than the amount of time for ranging calculations). A small ranging time is desirable since it means fast range estimation. This is important especially in communications systems that also perform ranging by utilizing a certain amount of each communications packet (for example, the preamble of each packet is used for ranging purposes in the IEEE 802.15.4a systems, as studied in Chapter 6). For large ranging times, a significant amount of each packet should be used for ranging purposes, which decreases the percentage of the packet that carries communications data; i.e. data rate decreases. Decreasing the duration of the ranging signal decreases the ranging time, but it also increases ranging errors in general. In other words, there is a trade-off between ranging time and ranging accuracy, and the aim is to obtain the best ranging accuracy in the shortest time interval.

8.1.2 Selection of signal parameters

In order to observe the relation between ranging accuracy and ranging time, consider the following UWB signal

$$s(t) = \sum_{j=-\infty}^{\infty} a_j \omega(t - jT_f), \qquad (8.6)$$

where $\omega(t)$ is a UWB pulse with a bandwidth of B MHz, T_f is the pulse repetition interval (PRI) in seconds, which is larger than the pulse width, and a_j is a randomization sequence uniformly distributed on $\{-1, +1\}$. As studied in Exercise 3 of Chapter 2, the average power spectral density (PSD) of $s(t)$ can be obtained as

$$\Phi_{ss}(f) = \frac{|\Phi(f)|^2}{T_f}, \qquad (8.7)$$

where $\Phi(f)$ is the Fourier transform of $\omega(t)$.

Assume that there is a regulatory constraint on the average PSD of UWB signals, such as the FCC's -41.3 dBm/MHz limit (see Chapter 2), and that the average PSD of the UWB signal in (8.6) is very close to that limit for all frequencies over its B MHz band. In other words, the UWB signal is assumed to transmit at the maximum power level allowed by the regulatory limit. For such a scenario, the maximum pulse energy can be obtained, for example from the FCC limit (-41.3 dBm/MHz), as

$$E \approx B \, T_f \, 10^{-7.13} \quad (\text{J}). \qquad (8.8)$$

For a ranging system that employs N_f pulses for range estimation, the ranging time is equal to $N_f T_f$ seconds.[1] As studied in Chapter 4, the ranging accuracy is inversely proportional to SNR. For a single-path AWGN channel, the CRLB expression in Section 4.1.3 can be expressed as

$$\text{RMSE} \geq \frac{c}{2\sqrt{2\pi}\sqrt{\text{SNR}}\,\beta}, \qquad (8.9)$$

where RMSE is as in (8.3) (with $\hat{d}$ representing an unbiased range estimate), c is the speed of light, β is the effective bandwidth and $\text{SNR} = E N_f / \mathcal{N}_0$ with $\mathcal{N}_0$ representing the spectral density of zero mean white Gaussian noise. Since the pulse energy is proportional to T_f as shown in (8.8), SNR increases as $N_f T_f$ increases. In other words, SNR is proportional to the ranging time. Therefore, the ranging accuracy increases with the ranging time; i.e. longer ranging times result in better ranging accuracy.[2]

In addition to the regulatory constraints on the maximum average power, there are also practical constraints on the peak power that can be used in an integrated circuit [297]. For a pulse-based UWB system, the average power constraint determines the total energy of UWB pulses that can be transmitted over a given time interval, whereas the peak

---

[1] Amount of time for ranging calculations is not taken into account.
[2] Although the analysis is performed for the signal model in (8.6), the inverse relation between ranging accuracy and ranging time holds true in general.

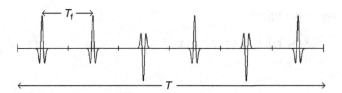

Fig. 8.1. A binary-coded ranging signal with a pulse repetition interval (PRI) of T_f seconds.

power constraint limits the energy of each UWB pulse. These constraints are important for signal design in UWB ranging systems.[3]

Example 8.1 *Consider a ranging signal of T seconds that consists of N_f UWB pulses, as shown in Fig. 8.1. The time interval between consecutive pulses (i.e., PRI) is T_f seconds, and the bandwidth is B Hz. Note that the pulse sequence is also polarity (binary) coded in order to enhance its correlation properties.*

For such a signal, the peak power can be calculated as

$$P_{\text{peak}} = \frac{E}{T_c} = \frac{P_{\text{avg}} T_f}{T_c}, \qquad (8.10)$$

where E is the pulse energy, T_c is the pulse width and P_{avg} is the average power. Assuming $T_c \approx 1/B$, (8.10) can be expressed as

$$P_{\text{peak}} = \frac{P_{\text{avg}} B}{PRF}, \qquad (8.11)$$

where $PRF = 1/T_f$ is the pulse repetition frequency.

From (8.11), it is observed that a constraint on the peak power imposes a lower bound on PRF; i.e.

$$PRF \geq \frac{P_{\text{avg}} B}{\bar{P}_{\text{peak}}}, \qquad (8.12)$$

where $\bar{P}_{\text{peak}}$ is the maximum peak power allowed in the system.

For a CMOS implementation of 90 nm process, the peak power limits are given by 15.6 mW and 2.5 mW for 2.5 V and 1 V peak-to-peak voltages, respectively [297]. For those peak powers, the lower bounds on the PRF can be obtained as in Fig. 8.2 for various bandwidths.

As can be observed from (8.8) and (8.12), the lower bound on the PRF ($= 1/T_f$) is proportional to the square of the bandwidth for a given value of P_{peak}. This is because as the bandwidth increases, the pulse energy (the average power) increases (cf. (8.8)) and the pulse duration T_c decreases. Therefore, in order to maintain the same peak power, the distance between the pulses, T_f, should be decreased in proportion to the square of the bandwidth (cf. (8.10)).

---

[3] In addition to practical limitations of integrated circuits, there might also exist regulatory constraints on the peak power of UWB signals [67].

8.1 Signal design for ranging

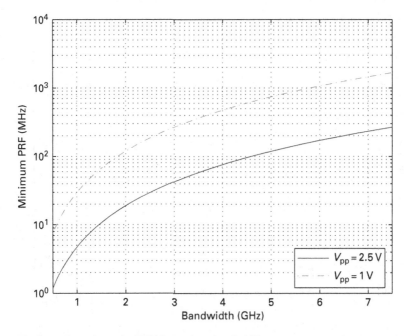

Fig. 8.2. The lower bounds on the PRF for various bandwidths.

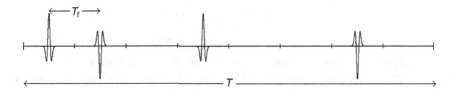

Fig. 8.3. A ternary-coded ranging signal with a pulse repetition interval (PRI) of T_f seconds.

In the previous example, binary coding is employed in order to enhance correlation properties of the pulse sequence for TOA estimation purposes. Another approach is to use ternary-coded pulses [298–300], which means transmitting a positive polarity pulse for $+1$, a negative polarity pulse for -1 and no pulse for 0.

Example 8.2 *In this example, the same scenario as in Example 8.1 is considered except that ternary coding is employed for the pulse sequence (Fig. 8.3). It is assumed that the number of zeros in the ternary code is equal to the number of non-zero values.*

In this case, each pulse can get twice the energy of a pulse in the previous example, since there are $N_f/2$ pulses with non-zero codes. Therefore, the peak power is twice the peak power in (8.11); i.e.

$$P_{\text{peak}} = \frac{2 P_{\text{avg}} B}{PRF}, \tag{8.13}$$

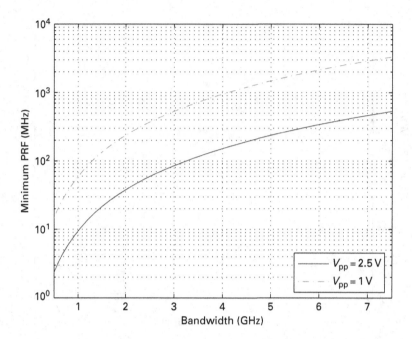

Fig. 8.4. Minimum PRF versus bandwidth for a ternary-coded UWB pulse sequence.

Then, the PRF is bounded as follows:

$$PRF \geq \frac{2P_{\text{avg}}B}{\bar{P}_{\text{peak}}}, \qquad (8.14)$$

where $\bar{P}_{\text{peak}}$ is the maximum peak power allowed in the system.

By using the same peak power values as in Example 8.1, the lower bounds on the PRF can be obtained as in Fig. 8.4 for various bandwidths.

Note that the PRF in Example 8.2 is defined as $1/T_f$ similar to the binary coded case. In fact, for ternary-coded pulse sequences, there are no pulses at the positions corresponding to zeros in the ternary code. Therefore, $PRF = 1/T_f$ is also called *peak PRF*[4] [91]. In addition, *mean PRF* is defined as the total number of pulses transmitted during a signal interval divided by the length of the signal duration. For example, the mean PRF is equal to $1/(2T_f)$ for the signal in Example 8.2.

In the design of a ranging signal, the minimum PRF is calculated as in the previous examples.[5] In order to determine a specific PRF value, the trade-offs between high and low PRFs need to be considered. Low PRFs are well suited for operation in environments with high delay spreads, particularly for non-coherent receivers. Non-coherent reception favors low PRFs since energy per pulse, or the peak power, is high for such systems.

---

[4] In this book, the term PRF is used to refer to peak PRF.
[5] In the case of regulatory constraints on the peak power of UWB signals, the minimum PRF calculations should take into account both the integrated circuit limitations, as considered in Examples 8.1 and 8.2, and those regulatory constraints.

However, for environments with short delay spreads and for coherent receivers, operation at high PRFs is preferable to have high data rates and shorter packet duration [91]. In addition, low PRF systems can facilitate low cost/power implementations since some units in the receiver, such as ADC, can be run only when pulses arrive [301].

In order to benefit from the advantages of low and high PRFs and facilitate the use of various receiver types in different environments with widely varying delay spreads, multiple PRFs can be specified in a system. For example, the IEEE 802.15.4a standard specifies PRFs of 7.8, 31.2 and 124.8 MHz. Note that the PRF values are selected as the multiples of each other so that a single pulse generator can be used at the transmitter (by exciting it less frequently for low PRFs) and sub-sampling of the signal with a high PRF can be used for low PRF cases [91].

After determining the length of the ranging signal and PRF, another important issue is pulse coding. Coded pulses in a ranging signal can provide robustness against multipath and multiple-access interference. As studied in Section 5.2, the autocorrelation properties of a code determine its robustness against multipath interference, whereas its cross-correlation properties are effective in mitigating multiple-access interference. Another important criterion is the length of the code; better correlation properties can be obtained with longer codes, but shorter codes ease the acquisition process.

As studied in Chapter 6, in the IEEE 802.15.4a standard, length-31 and length-127 (optional) ternary codes are employed in the synchronization preamble. As shown in Fig. 8.5, the preamble consists of N_{sym} repetitions of a preamble symbol,

$$P_i(t) = \sum_{n=0}^{N_{\text{sym}}-1} w_i(t - nT_s), \quad (8.15)$$

where T_s is the symbol interval and $w_i(t)$ is the preamble symbol given by

$$w_i(t) = \sqrt{\frac{E_s}{N_f}} \sum_{j=0}^{L_{\text{ts}}-1} \mathbf{S}_i[j]\omega(t - jT_f), \quad (8.16)$$

with E_s and N_f denoting, respectively, the symbol energy and the number of pulses per symbol, T_f (also called T_{pri} in Chapter 6) being the time interval between consecutive

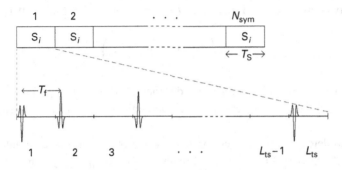

Fig. 8.5. Preamble structure in the IEEE 802.15.4a standard, where $N_{\text{sym}} \in \{16, 64, 1024, 4096\}$ and $L_{\text{ts}} \in \{31, 127\}$.

pulses (including the ones with zero codes), and S_i representing a ternary code of length L_{ts}.

The length-31 ternary codes ($L_{ts} = 31$) used in the standard are as shown in Table 6.2 of Chapter 6. These codes have the perfect periodic autocorrelation property; i.e. their periodic correlation side lobes are zero, which makes sure that the received signal paths between autocorrelation peaks are due only to the multipath channel.

In order to obtain processing gain, the coded ranging (preamble) symbol is repeated a number of times as shown in (8.15). In the IEEE 802.15.4a standard, the preamble consists of 16, 64, 1024 or 4096 symbols. As discussed before, longer ranging signals increase the ranging accuracy but also decrease the data rate in communications systems that perform ranging by using a specific section of each packet. Therefore, the preamble length is selected according to communications and ranging requirements of a given application. This selection is based on various system parameters such as propagation environment and receiver type (coherent/non-coherent). For example, longer preambles are preferable for low-rate, non-coherent receivers, in order to provide more processing gain for such systems to achieve reasonable SNR values.

8.2 Link budget calculations

As shown in (8.9), the received SNR is an important factor in determining the accuracy of a ranging system. As the SNR gets higher, ranging errors decrease since noise becomes less effective. In order to calculate the received SNR in a given system, the effects of transmit and receive antennas and the propagation channel should be taken into account.

Let $P_{tx-amp}(f)$ represent the output power spectrum of the transmit amplifier (at the antenna connector) and $\eta_{tx-ant}(f)$ denote the efficiency of the transmit antenna. Then, the transmit power spectrum can be expressed as

$$P_t(f) = P_{tx-amp}(f)\eta_{tx-ant}(f). \tag{8.17}$$

As the signal propagates in the air, the power density decreases with the distance as studied in Chapter 3. In addition, for UWB signals, this path loss is frequency dependent. Power density at distance d and frequency f can be expressed as follows [91]

$$P(f,d) = K_0 \frac{P_t(f)}{4\pi d_0^2} \left(\frac{d}{d_0}\right)^{-n} \left(\frac{f}{f_c}\right)^{-2\kappa}, \tag{8.18}$$

where K_0 is a constant, d_0 is the reference distance, f_c is the center frequency, n is the path loss exponent and κ is the exponent that determines the frequency dependency of the path loss.

In order to determine the power spectrum at the output of a receive antenna, the spectrum in (8.18) is multiplied by the antenna efficiency $\eta_{rx-ant}(f)$ and the antenna area (effective aperture of the antenna) $A_{rx}(f)$. Since the antenna area is given by $\lambda^2(G_{rx}(f))/4\pi$, with $G_{rx}(f)$ denoting the receive antenna gain [302] and λ representing the wavelength, the received power spectrum can be obtained from (8.18) as

$$P_r(d, f) = K_0 \frac{P_t(f)}{4\pi d_0^2} \left(\frac{d}{d_0}\right)^{-n} \left(\frac{f}{f_c}\right)^{-2\kappa} \frac{\lambda^2}{4\pi} G_{rx}(f)\eta_{rx-ant}(f). \quad (8.19)$$

For $G_{rx}(f) = 1$, (8.19) can be expressed, using (8.17) and $\lambda = c/f$, as

$$P_r(d, f) = PL_0 \, P_{tx-amp}(f)\eta_{tx-ant}(f)\eta_{rx-ant}(f)\frac{(f/f_c)^{-2(\kappa+1)}}{(d/d_0)^n}, \quad (8.20)$$

where

$$PL_0 = \frac{K_0 c^2}{(4\pi d_0 f_c)^2} \quad (8.21)$$

is the path-loss at the reference distance d_0 and reference frequency f_c for an ideally efficient isotropic antenna [91].

For a specific UWB pulse shape (at the output of the transmit amplifier) and transmit and receive antenna efficiencies, the received power at a given distance can be calculated from (8.20) by using empirically obtained values for PL_0, n and κ. Note that the total average received power should be calculated by integrating $P_r(d, f)$ over the frequency range of the system. After calculating the received power, SNR ($= E_s/N_0$) can be obtained for a given ranging symbol interval T_s and noise spectral density N_0 as follows[6]

$$SNR(d) = 10\log \frac{T_s \int_{f_L}^{f_H} P_r(d, f) df}{N_0} - L_{imp} - NF, \quad (8.22)$$

where L_{imp} accounts for the implementation loss (in dB), NF is the receiver noise figure (in dB),[7] $f_L = f_c - B/2$ and $f_H = f_c + B/2$, with B denoting the system bandwidth.

Example 8.3 *Consider a UWB ranging system that utilizes 1 GHz bandwidth with a center frequency of $f_c = 5$ GHz. Assume that the transmit and receive antennas are ideally efficient and isotropic. For a noise spectral density of -174 dBm/Hz, the aim is to plot SNR versus distance d between the transmitter and the receiver for various lengths of the ranging interval and for various propagation environments. It is assumed that the implementation loss is 3 dB and the receiver noise figure is 7 dB.*

For the system model as described above, $\eta_{tx-ant}(f) = \eta_{rx-ant}(f) = 1$ and $G_{rx}(f) = 1$. Then, (8.20) simplifies to

$$P_r(d, f) = PL_0 \, P_{tx-amp}(f)\frac{(f/f_c)^{-2(\kappa+1)}}{(d/d_0)^n}. \quad (8.23)$$

---

[6] In the presence of a person close to the antennas, there occurs additional attenuation of the received power [91, 303]. This effect is not considered in this analysis.

[7] The receiver noise figure is defined as the contribution of the receiver itself to the thermal noise at its output.

Table 8.1. PL_0, κ and n for various IEEE 802.15.4a channels.

	PL_0 (dB)	κ	n
Residential LOS (CM-1)	−43.9	1.12 ± 0.12	1.79
Residential NLOS (CM-2)	−48.7	1.53 ± 0.32	4.58
Office LOS (CM-3)	−35.4	0.03	1.63
Office NLOS (CM-4)	−59.9	0.71	3.07

In many cases, $P_{\text{tx-amp}}(f)$ approximates the FCC mask quite well. Therefore, it can be approximately modeled as a constant level of -41.3 dBm/MHz over the bandwidth of interest.

For a reference distance of $d_0 = 1$ m and $f_c = 5$ MHz, the values of PL_0 can be obtained from the IEEE 802.15.4a channel model. The values of PL_0, κ and the path loss exponent n are shown in Table 8.1 for various IEEE 802.15.4a channel models.[8]

From (8.23) and the assumption of constant $P_{\text{tx-amp}}(f)$ over the bandwidth of interest, (8.22) can be expressed as

$$SNR(d) = 10 \log \frac{PL_0 P_{\text{tx-amp}} T_s (f_L^{-2\kappa-1} - f_H^{-2\kappa-1})}{(2\kappa+1) \mathcal{N}_0 f_c^{-2(\kappa+1)} (d/d_0)^n} - L_{\text{imp}} - NF. \quad (8.24)$$

In Fig. 8.6, the ranging interval T_s is set to 1 μs, and received SNR is plotted versus distance for LOS and NLOS residential and office environments according to the IEEE 802.15.4a channel models.[9] It is noted that the NLOS channels have significantly lower SNR values than the LOS channels.

In Fig. 8.7, SNR is plotted versus distance for various durations of ranging symbols over residential LOS channels (CM-1). The symbol durations of $T_s = 1$ μs and $T_s = 4$ μs approximately correspond to the symbol duration options in the IEEE 802.15.4a standard [91]. It is observed that as the ranging symbol gets longer, higher SNR values are observed at the receiver. Similarly, combining signal energy from a number of ranging symbols increases the SNR. Similar observations are made from Fig. 8.8, which repeats the previous experiment for residential NLOS channels (CM-2).

Calculation of SNR as a function of distance provides important guidelines for the system design. However, the performance of a ranging system also depends heavily on the channel multipath profile. In other words, distribution of the total signal power among multipath components and arrival instants of those components affect the performance of the system significantly. For example, strength of first incoming signal component is an important parameter that determines error probabilities of the first path detection

---

[8] The path loss values in dB are usually represented as positive numbers, since they are already considered as "loss". However, the negative signs are explicitly shown in Table 8.1 in order to clarify the real values to be used in the equations (cf. (8.24)).
[9] $\kappa = 1.12$ and $\kappa = 1.53$ are used for CM-1 and CM-2, respectively.

8.2 Link budget calculations

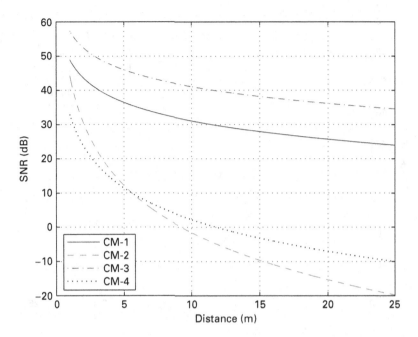

Fig. 8.6. SNR versus d for various IEEE 802.15.4a channel models.

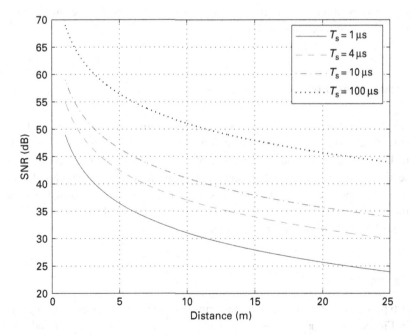

Fig. 8.7. SNR versus d for various durations of the ranging symbol (CM-1).

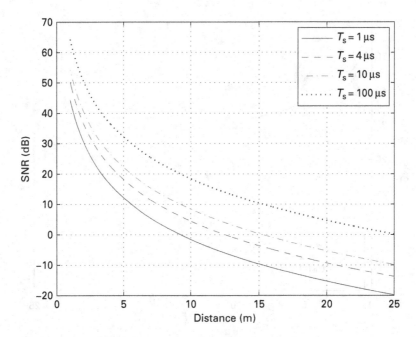

Fig. 8.8. SNR versus d for various durations of the ranging symbol (CM-2).

algorithms in Chapter 5. Therefore, in the design of a ranging system, various characteristics of the channel need to be considered in addition to the SNR characterization. In other words, small-scale fading and shadowing effects, which are studied in Chapter 3, should also be taken into account.

8.3 Hardware issues

After determining its parameters, implementation of a UWB system requires consideration of analog and digital hardware components for transmission and reception of UWB signals. Although conventional design techniques can be applied to certain sections of a UWB transmitter/receiver, the extremely large bandwidth of UWB signals introduces practical design challenges, especially for analog components. In this section, typical UWB transmitters and receivers are considered and various practical issues specific to UWB systems are investigated.

8.3.1 Transmitter

A typical UWB transmitter that performs both communications and ranging is shown in Fig. 8.9. For communications, input data is first coded (channel coding) and then modulated (symbol mapper). For example, a string of binary data can be coded by a convolutional encoder and then the coded bits can be mapped to BPSK symbols. In

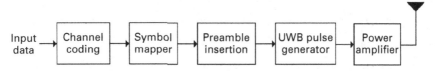

Fig. 8.9. Block diagram of a UWB transmitter.

addition to communications, a certain part of the transmission is allocated for ranging, which employs, for example, binary or ternary code sequences for TOA estimation (e.g. perfectly balanced ternary sequences (PBTSs) studied in Section 6.3 of Chapter 6). Commonly, transmission is performed in terms of packets, which contain both communications and ranging signals. Since ranging signals constitute the initial section of each packet, they are also called preambles.[10] In Fig. 8.9, communications signals are obtained first and then ranging signals are inserted at the beginning of the communications signals.

Digital communications and ranging signals at the output of the preamble insertion block are fed into the UWB pulse generation block, which converts the digital signal sequence into an analog UWB pulse sequence. Depending on the pulse generation technique, a local oscillator may be used in the UWB pulse generator to shift the frequency content of the signal into a specific UWB band. After that, a power amplifier (PA) can be used to increase the power of the signal to a certain level. Finally, an antenna or an antenna array transmits the UWB signal into the space.

UWB pulse generation

One of the most important components in a UWB transmitter is the pulse generator unit. Pulse generators can be classified in various ways depending on the use of an up-conversion unit in them, or on the amount of filtering needed for pulse generation.

One category of pulse generators does not require an up-conversion unit; i.e. there is no need for a mixer and a local oscillator for pulse generation. The reason for this is that the generated pulses occupy the UWB spectrum without any frequency translation. In addition, some of such pulse generators do not need any pulse shaping (filtering) either, since the generated pulses are already utilizing the UWB spectrum efficiently. For example, in [304] and [305], the fifth derivative of a Gaussian pulse (Fig. 8.10) is generated, which utilizes the spectrum effectively under the FCC regulations [305]. Therefore, no filtering or frequency translation is required. The technique in [304] and [305] for generating the fifth derivative of the Gaussian pulse is to first generate digital triangular pulses, which are similar to Gaussian shapes in practice, and then to combine a number of those triangular pulses with different delays and polarities, as shown in Fig. 8.11. The basic idea is to generate short duration triangular ("Gaussian-like") pulses by means of simple logic gates as shown in Fig. 8.12. For a stable input of 0 or 1, the output of the NAND gate is always equal to 1, since its inputs are complements of each other. When there is a transition from 0 to 1 at the input, the lower input of the NAND gate becomes

---

[10] In a system that performs both communications and ranging, preamble signals are used for timing acquisition, frequency recovery, packet and frame synchronization and channel estimation, in addition to range estimation.

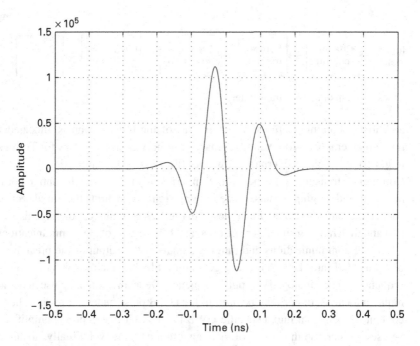

Fig. 8.10. Fifth derivative of a Gaussian pulse with unit energy.

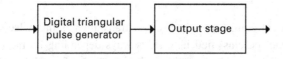

Fig. 8.11. Block diagram of a pulse generator that directly generates the fifth derivative of a Gaussian pulse.

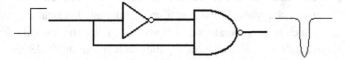

Fig. 8.12. Generation of a pulse with negative polarity.

1 before the upper input becomes 0 since the inverter induces a certain amount of delay. Therefore, for a short time interval, both of the inputs to the NAND gate become 1, resulting in an output of 0. When the upper input of the NAND gate becomes 0, the output becomes 1 again. By this way, a short duration pulse with negative polarity is generated at the output of the NAND gate as shown in the figure. Similarly, positive polarity pulses at different delays can be generated by using inverters, NOR and NAND gates, and then a combination of those short pulses can be obtained to generate a UWB pulse with the desired shape [304, 305].

Another approach for pulse generation without any frequency translation or filtering is to use antennas for shaping UWB pulses [306, 307]. In [306], very low power applications are considered, and UWB pulses are generated by means of a rectangular

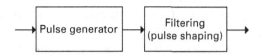

Fig. 8.13. Generation of a UWB pulse with filtering.

pulse generator that is connected to an antenna through an impedance-matching circuit. The differentiation effect of the transmitting antenna shapes the rectangular signal in a form that is similar to a Gaussian monocycle. In [307], a technique called direct antenna modulation (DAM) [308, 309] is considered for UWB pulse generation. In DAM, RF switches are integrated into a resonant antenna, such as a patch antenna, and are used to turn on and off the radiation slots of the antenna directly. In other words, the antenna is used as a switch-type pulse modulator.

In addition to pulse generators that directly generate the desired pulse shape, some pulse generators employ filtering for pulse shaping [310–315]. The structure of such pulse generators can be represented as in Fig. 8.13, where the first block is generating signals and the second block is shaping them for efficient use of the spectrum under regulatory constraints. Although various implementation approaches are possible, the common technique employed in the pulse generator block is to use combinatorial logic to generate pulses with specific pulse widths, which are determined by tunable or fixed delay elements (cf. Fig. 8.12). For the pulse-shaping block, a bandpass filter (BPF) is implemented to shape the pulses obtained by the pulse generator block. For UWB systems, especially the ones that utilize the full band (3.1–10.6 GHz), design of BPFs presents a number of practical challenges. UWB filters should have low insertion loss,[11] low and flat group delay and provide sufficient out-of-band suppression (for regulatory constraints) [312, 313].

Instead of generating the pulse directly in a specific band, some transmitters first generate a pulse at baseband and then up-convert it to a center frequency [316–318]. Up-conversion can be performed conventionally by a mixer and a local oscillator. However, it is also possible to design mixers that not only perform the up-conversion operation, but also implement pulse shaping. In [318], such a mixer is designed; it exploits the exponential behavior of BJTs in order to approximate Gaussian pulses by hyperbolic tangent (tanh) functions. Triangular pulses are input to this mixer as shown in Fig. 8.14,

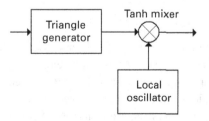

Fig. 8.14. A UWB pulse generator with tanh mixer, which performs pulse shaping as well as up-conversion.

---

[11] Insertion loss is defined as the power loss of a filter in the passband.

and the output of the mixer is approximately a Gaussian-shaped pulse that is up-converted to a desired center frequency.

Comparison of pulse generation techniques with and without up-conversion reveals that the transmitters with up-conversion circuitry consume more energy, since the local oscillator operates at the center frequency, which is quite high for UWB systems (e.g. above 3 GHz for operations in the 3.1–10.6 GHz band). On the other hand, the up-conversion operation provides better spectral control and is commonly preferred for high data rate systems. For ranging systems, architectures without up-conversion are preferable, since high data rate communication is not required and low-power architectures are quite desirable in most ranging applications.

Power amplifiers

After UWB pulse generation, a power amplifier (PA) can be employed to increase the power of the signal delivered to the antenna. For UWB systems operating under extremely low power regulations, transmitter design without a PA is also possible. For example, for unlicensed use of UWB systems according to the Japanese regulations, the electric field intensity permitted at 3 m is 35 V/m for the 0.322–10 GHz band, which facilitates designs without PAs [319].

Commonly, PAs consume a significant portion of transmit power. Therefore it is desirable to have efficient PAs in order to minimize the power consumed at a transmitter. *Efficiency* of a PA is defined as the ratio of the signal power delivered to the load to the total power consumed by the amplifier. Ideally, the PA should have an efficiency of 1, which means that all the power consumed by the PA is delivered to the load [320].

The main issues for UWB PA design are efficiency, bandwidth, and linearity. Specifically, a UWB PA should have high efficiency and provide good linearity and wideband matching [321]. Distributed amplifiers [322–325] are commonly employed for wideband applications since they provide linearity and wideband matching, but their efficiency is usually low. Similarly, shunt feedback amplifiers [326, 327] provide good wideband matching and flat gain, but their efficiency is not very high, either. In order to have high efficiency, BPF-based input matching amplifiers can be used, which employ BPFs and common-source amplifier topologies [321, 328–330]. In comparison with distributed and feedback amplifier topologies, such amplifiers consume very small amounts of power.

UWB antennas

The last component in the UWB transmitter in Fig. 8.9 is the antenna, which provides radiation of electromagnetic energy into the space. Although antenna design has been studied extensively for narrowband and wideband systems, UWB antenna design for communications and ranging applications under regulatory constraints presents additional challenges. In terms of antenna parameters, requirements for UWB antenna design can be summarized in the following manner.

- **Impedance bandwidth.** Ideally, the incoming signal towards the antenna should be completely radiated into the space, which requires perfect impedance matching at all frequencies. However, in practice, some of the incident signal is lost due to reflections. Impedance bandwidth of an antenna specifies a frequency band over which this signal loss is not very significant. Specifically, it is defined as the bandwidth over which at most 10% of the incident signal power is lost.

For UWB antenna design, the main challenge is to obtain very wide impedance bandwidths. In other words, impedance matching should be quite good for a very wide range of frequencies. For that reason, various bandwidth-broadening techniques are employed in UWB antenna design [331]. Common methods include using specific antenna geometries such as biconical, helix, or bow-tie structures [332], beveling or smoothing [333–336], resistive loading [337], slotting (or adding a strip) [338, 339], notching, and optimizing location or structure of the antenna feed [340–342].

- **Group delay.** Another desired feature of a UWB antenna is to have a near constant group delay (i.e., linear phase), which prevents pulse distortions during the transmission. For narrowband systems, this requirement is always satisfied since the input signal occupies a very small frequency band. However, for UWB systems, group delay should also be taken into account so that the transmitted pulse is not distorted significantly.[12]

- **Radiation efficiency.** Radiation efficiency quantifies the conductor and dielectric losses at the antenna, and is defined as the ratio of the radiated power to the input power at the terminals of the antenna [332]. Since UWB signals operating under regulatory constraints can transmit low-power signals only, radiation efficiency of UWB antennas should be quite high in order to facilitate ranging/communications at reasonable distances. For example, resistive loading, which is used for broadening the impedance bandwidth of an antenna [337], is not very desirable for UWB systems, since it reduces the radiation efficiency.

- **Radiation pattern.** Depending on their radiation patterns, antennas can be classified as directional and omnidirectional antennas. Whereas directional antennas radiate power in certain directions, omnidirectional antennas radiate power uniformly in a given plane. In the absence of any position information related to UWB nodes (terminals), omnidirectional antennas are preferred as signals can arrive in various directions. Directional antennas can be useful for base stations or access points when the signal is to be transmitted in a specific direction.

- **Size and geometry.** For most applications, small and planar UWB antennas are desirable. For example, when attaching UWB positioning devices to humans, antennas should be very small in order not to cause any disturbance. For that reason, antennas printed onto printed circuit boards (PCBs) are highly desirable [331].

Considering the impedance bandwidth requirement, there are various antennas that have very wide well-matched bandwidths, such as transverse electric magnetic (TEM) horns [343–347], biconical antennas [348, 349], cylindrical antennas (with resistive loading) [350, 351] and helical antennas [352]. However, these antennas are not commonly

---

[12] More generally, a UWB antenna should radiate a pulse that is very similar to the exciting pulse or its derivative [337].

preferred for UWB applications as they are non-planar and/or quite large. On the other hand, certain self-complementary, log-periodic antennas, such as planar log-periodic slot antennas, log-periodic dipole arrays and conical log-spiral antennas, also have wide impedance bandwidths,[13] but they have frequency-dependent group delays which cause distortion of radiated signals [331].

Planar antennas are well-suited for UWB systems as they are compact and can be printed on PCBs ([331] and [353], and references therein). In addition, they can have wide impedance bandwidths and near constant group delays if their geometries and feeding structures are designed appropriately [331]. As examples of planar *dipole* antennas, a bow-tie antenna, a diamond dipole antenna and a square dipole antenna are shown in Fig. 8.15. Planar *monopole* antennas can be mainly classified into polygonal and elliptical monopoles depending on their geometric structures [331] (Figs. 8.16 and 8.17). Polygonal monopoles can be used as UWB antennas after certain impedance-matching techniques, such as feeding adjustment, beveling, smoothing or slotting, as shown in Fig. 8.16. For an elliptical antenna, optimization of the axes of the ellipse and the smooth

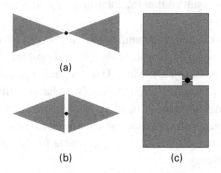

Fig. 8.15. Planar dipole antennas: (a) bow-tie antenna; (b) diamond dipole antenna; (c) square dipole (After [331]).

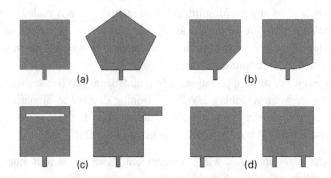

Fig. 8.16. Polygonal monopole antennas (a)–(d). Beveling/smoothing (b); slotting/strip adding (c); and feeding adjustment (d) techniques can be used for wideband impedance matching purposes (© 2006 IEEE) [331].

---

[13] Theoretically, they are frequency-independent.

Fig. 8.17. An elliptical monopole (a), and its modifications for small size and wide impedance bandwidth (b)–(d) (After [331]).

Fig. 8.18. A Vivaldi UWB antenna.

transition between the radiator and the feeding strip provide a wide impedance bandwidth [331]. In addition, various geometric modifications, such as beveling, can be made for further improvement of the impedance bandwidth, as shown in Fig. 8.17. Another planar UWB antenna is the Vivaldi antenna in Fig. 8.18, which can theoretically have an infinite bandwidth, and provides a very wide impedance bandwidth in practice [354].

Although planar antennas can achieve wide impedance bandwidths and facilitate compact designs, they may not provide omnidirectional radiation at all frequencies. For UWB applications requiring small omnidirectional antennas, roll monopoles can be employed, which are obtained by twisting a planar radiator to a roll shape [331]. These antennas not only provide wide impedance bandwidths like planar monopoles, but also have omnidirectional radiation patterns like cylindrical antennas [355, 356].

8.3.2 Receiver

Signals transmitted from UWB antennas are collected by UWB receivers in order to obtain ranging and/or communications information contained in those signals. UWB receivers can be broadly classified into two different classes depending on the amount of processing performed in the analog domain. Fig. 8.19 illustrates an "all-digital" UWB

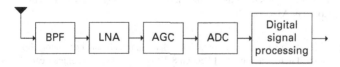

Fig. 8.19. "All-digital" UWB receiver. The signal is converted to digital as early as possible, and processing is performed in the digital domain.

receiver[14] structure, which converts the analog UWB signal at the antenna output into a digital signal as quickly as possible, and performs main signal processing operations in the digital domain [357–359]. This receiver structure is also referred to as a *direct sampling* receiver, as studied in Chapter 5. In an all-digital UWB receiver, the analog UWB signal collected by the antenna[15] is first passed through a BPF for out-of-band noise/interference mitigation, and then amplified by a low-noise amplifier (LNA). Then, the signal is scaled by an automatic gain control (AGC) unit, which adjusts the level of the UWB signal according to the analog-to-digital converter (ADC) specifications. The ADC converts the analog signal into a digital one, and is a key block in UWB receivers due to the large bandwidths of the UWB signals. As studied in [360], ADC performance can be improved by appropriate selection of AGC parameters, which can prevent suboptimal regimes of an ADC being dominated by quantization or saturation noise. After the ADC, signal processing operations can be performed in the digital domain, which provides flexibility to perform various estimation and/or decoding algorithms.

The second class of UWB receivers perform certain correlation operations in the analog domain, and then convert the signal into digital, as shown in Fig. 8.20 [361–363].[16] Depending on the receiver type, the analog correlator in the figure can be replaced by an energy detector or a delay-and-correlate operation, as studied in Chapter 5. Although a single analog correlator is shown in Fig. 8.20, a bank of correlators can also be employed to implement a Rake receiver in the analog domain [362], which facilitates collection of signal energy from multiple signal paths.

The main advantage of UWB receivers that perform certain operations in the analog domain is to relax ADC requirements. For such receivers, the analog signal at the ADC input can be sampled at considerably slower rates, such as frame or symbol rate, than the analog signal at the ADC input of an all-digital UWB receiver, which commonly requires Nyquist rate sampling. However, analog correlation causes certain performance loss compared to all-digital structures due to circuit mismatches and reduced receiver flexibility.

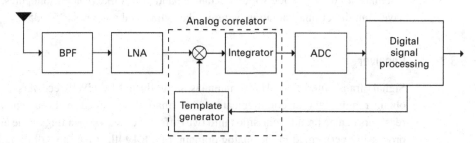

Fig. 8.20. UWB receiver with an analog correlator. Some signal processing is performed in the analog domain in order to relax ADC requirements.

---

[14] Of course, the receiver is not completely digital, but it has minimal amount of analog circuitry.

[15] UWB antennas studied in the subsection of Section 8.3.1 entitled "UWB antennas" can also be employed to receive UWB signals by utilizing the reciprocity relation for antennas [337].

[16] For both classes of UWB receivers, BPF outputs can be mixed with an oscillator output if the receiver performs a down-conversion operation [361].

8.3 Hardware issues

In the remainder of this section, two crucial components of a UWB receiver, LNA and ADC, are investigated in detail.

Low-noise amplifiers

LNAs are used to amplify weak signals captured by antennas. Although advanced LNA design techniques exist for narrowband systems, the very wide bandwidth of UWB signals makes many of those techniques unsuitable for UWB receivers [364]. The main challenges in UWB LNA design are related to providing low noise figure (NF), good impedance matching and sufficiently high and flat power gain over a wide frequency range.

The NF of an LNA is defined as the difference between the input SNR and the output SNR,[17] where all quantities are in dB; i.e.

$$NF = SNR_{in} - SNR_{out}. \qquad (8.25)$$

For UWB systems, defining SNR as a meaningful metric of performance requires careful consideration. Since the input signal occupies a wide range of frequencies and the noise can be colored, a higher ratio between signal power and total noise power may not always mean a better receiver performance [364]. For this reason, the NF of a UWB system can be defined by means of a linearly weighted average of single-tone NFs [365], or by using matched filter bounds instead of SNRs in NF calculations [364].

The flat gain requirement over an ultra-wide bandwidth of interest is another challenging issue for UWB LNA design [366–369]. Since imposing strict gain-flatness requirements limits design flexibility, it is also possible to introduce LNAs with slightly unflat gains in order to achieve low power and good noise performance [370].

Table 8.2 compares performance of various UWB LNAs proposed recently [328, 371–374]. It is observed that the current LNA structures based on CMOS technology can provide about 10–15 dB power gain with power consumptions of around 10 mW over the specified ultra-wide bandwidths. Note that the NF varies significantly in certain designs within the bandwidth, whereas it is more stable in others. In addition to the LNAs based on the CMOS technology, a UWB LNA is proposed in [375] based on the silicon–germanium heterojunction bipolar transistor (SiGe HBT) technology, which provides a higher power gain (16.4–18 dB) than the UWB LNAs in Table 8.2, but also consumes significantly more power (42.5 mW).

Analog-to-digital conversion

Conversion of an analog signal into a digital signal consists of sample-and-hold (S/H) and quantization operations. Depending on how fast the analog signal is sampled, how many bits are used to represent the digital signal, and the amount of power dissipated by the S/H and the quantization circuitry, performance of various ADCs can be analyzed. These three quantities, sampling rate, resolution, and power dissipation, are the main parameters in comparing various ADC structures.

---

[17] The *noise factor* F is defined as the ratio of the input SNR to the output SNR; i.e. $F = 10^{0.1NF}$, where NF represents the noise figure.

Table 8.2. Performance comparison of various UWB LNAs [371]. Input return loss is defined as the ratio of the reflected signal to the incident signal, which quantifies the matching characteristics. For perfect matching, the input return loss becomes zero.

	[371]	[328]	[372]	[373]	[374]
Technology (CMOS)	0.18 μm	0.18 μm	0.18 μm	0.18 μm	0.13 μm
Bandwidth (GHz)	3.4–11.4	2.3–9.2	1.3–10.7	1–11.6	3.1–10.6
Power gain (dB)	13.5–16	9.2	8.5	10.8–12	13.7–16.5
Noise figure (dB)	3.1–6	4–9	4.3–5.3	4.7–5.6	2.1–2.8
Power consumption (mW)	11.9	9.2	4.5	10.6	9
Input return loss (dB)	< −8	< −9	< −10	< −11	< −10
Supply voltage (V)	1.8	1.8	1.8	1.5	1.2
Chip area (mm^2)	1.2	1.1	1.0	0.66	0.87

Analog-to-digital conversion at a UWB receiver can be performed after LNA and AGC blocks as shown in Fig. 8.19, or after correlation operations in the analog domain as shown in Fig. 8.20. The main advantage of performing correlations in the analog domain is that ADCs can operate at much lower rates since sampling per frame or symbol becomes sufficient. In this way, low-power UWB receivers can be designed [319, 361].

Although analog correlations facilitate low-rate sampling, they reduce receiver flexibility and suffer from circuit mismatches. For example, the number of correlators is usually limited, and sophisticated narrowband interference (NBI) mitigation techniques cannot be employed for UWB receivers that perform analog correlation [357]. Therefore, it is desirable to move the ADC block as close to the antenna as possible, which yields an "all-digital" UWB receiver with improved flexibility [357–359].

For all-digital UWB receivers, very high speed ADCs are necessary since sampling UWB signals at the Nyquist rate requires obtaining a few billion samples per second (Gsps). Fortunately, resolution requirement is not as strict as the sampling rate requirement. As the received UWB signals are commonly immersed in strong AWGN and interference due to their low spectral density, an ADC with a few bits of resolution is sufficient [358, 360, 376–378]. Specifically, ADCs with more than four bits of resolution provide only marginal improvement over a four-bit ADC for UWB systems [358, 376]. Also, use of one-bit ADCs is possible in the absence of NBI [359, 379–381]. However, in the presence of NBI, around four bits are necessary to mitigate the effects of NBI [359, 376].

In order to meet the fast sampling rate requirement with the current ADC technology, various channelization techniques have been proposed [357, 381–385]. In time-domain channelization, N_A ADCs are time-interleaved in order to obtain an effective sampling rate of $N_A f_s$, where f_s is the sampling rate of each ADC [381–383]. Channelization can also be performed in the frequency domain by means of a filter bank implementation as shown in Fig. 8.21. The advantage of the frequency domain channelization over the time domain one is that the former has significantly reduced dynamic range requirements [357]. Both time-domain and frequency-domain channelization can be considered in the general framework of analog-to-digital conversion based on projection of input signal over a set of basis functions [384].

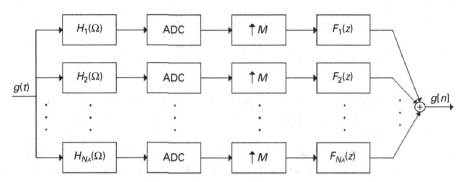

Fig. 8.21. Frequency domain channelization with an upsampling factor of M. The effective sampling frequency is $N_A f_s$, where f_s is the sampling rate of each ADC. $H_1(\Omega), \ldots, H_{N_A}(\Omega)$ and $F_1(z), \ldots, F_{N_A}(z)$ are the analysis and synthesis filters, respectively. The analysis filters can be implemented by means of mixers and low-pass filters in order to relax implementation requirements (After [357]).

In addition to channelization approaches, subsampling techniques can also be employed to relax the ADC sampling rate requirements [386, 387]. In a subsampling UWB receiver, no down-conversion is performed and the passband UWB signal is directly sampled at twice the signal bandwidth instead of the maximum signal frequency. For example, for a UWB signal that occupies the bandwidth from 4 to 5 GHz, a sampling rate of 2 Gsps is used. Subsampling can relax the sampling rate requirements significantly for UWB signals with relatively small bandwidths. However, it is not very effective for full-band UWB signals that occupy the 3.1–10.6 GHz band.

In addition to the techniques, such as channelization and subsampling, for relaxing the sampling rate requirements, the architecture of the ADC also carries significant importance for the performance of a UWB receiver. Among various types of ADC structures, such as flash, pipelined, successive approximation register (SAR), sigma-delta and folding [388], flash ADCs have the highest sampling rates due to their parallel structure for comparing input signals with successive reference signal levels [389]. However, their complexity increases exponentially with resolution; therefore, high-resolution flash ADCs dissipate a significant amount of power [359, 390]. On the other hand, SAR ADCs consume low power, but they are quite slow in comparison with flash ADCs. However, time-interleaving can be incorporated into the SAR ADC architecture for reasonably fast analog-to-digital conversion (such as 500 Msps) with low power consumption [391, 392].

8.4 Problems

(1) Consider a range estimate that is modeled as

$$\hat{d} = d + n, \tag{8.26}$$

where d is the true range and $n \sim \mathcal{N}(0, \sigma^2)$ is the measurement noise.
(a) Calculate the RMSE of the range estimate.
(b) Calculate the CDF of the ranging error.

(2) Consider a UWB ranging signal given by

$$s(t) = \sum_{j=0}^{N_f-1} a_j \omega(t - jT_f), \quad (8.27)$$

where $\omega(t)$ is the UWB pulse with a pulse width of T_c, N_f is the number of UWB pulses in the ranging signal, T_f is the PRI, and $\mathbf{a} = [a_0\ a_1 \cdots a_{N_f-1}]$ is a ternary sequence.

Calculate the (peak) PRF, the mean PRF and the peak power of the ranging signal for a pulse energy of 7.4×10^{-12} J, $T_c = 1$ ns, $N_f = 10$, $T_f = 0.1$ μs, and $\mathbf{a} = [1\ 0\ -1\ 1\ 0\ 1\ 0\ -1\ 1\ 0]$.

(3) (*programming exercise*) Repeat Example 8.3 for a UWB ranging system that utilizes 2 GHz bandwidth with a center frequency of $f_c = 5$ GHz.

(4) What are the advantages and disadvantages of an "all-digital" (direct sampling) UWB receiver in comparison with a UWB receiver that performs analog correlations?

9 Recent developments and future research directions

In this chapter, some of the recent interesting research work related to UWB ranging and positioning are briefly reviewed. The purpose of the chapter is not to describe these studies in detail, but rather to point out specific recent references that may yield further research.

9.1 Development of accurate ranging/positioning algorithms

As discussed in the previous chapters, ranging and localization via UWB radios have been investigated extensively in the literature. While CRLB and ZZLB provide lower bounds on the ranging/localization accuracy, low-complexity and efficient estimators that approach these bounds in practical scenarios are still needed.

There are numerous recent research studies that aim at improving UWB ranging/localization accuracy. One research direction is joint estimation of range and location. In [170], it is shown that a two-step approach that uses independent decisions in ranging and localization steps is asymptotically optimal at high SNRs. However, it requires perfect estimates of delays, attenuations, and pulse shapes related to the received multipath components (MPCs) in order to construct an optimum correlation template at the receiver, which is very difficult to achieve in practice. Without perfect *a-priori* information of the channel parameters, such a two-step method returns unreliable TOA estimates during the ranging step. Since the measurements are separately performed at each reference node, without a constraint that all the measurements correspond to the location of the same mobile terminal, such approaches are suboptimal [137]. A better approach would be to make *least commitment*, where intermediate information is preserved and propagated till the end [393]. In other words, the received channel responses should not be discarded until a final decision regarding the target node location is made. For example, in [394], the ranging thresholds are set so that the residual error is minimized. In another work [395], the ranging algorithm outputs several range estimates with associated likelihood values; these 'soft' range estimates are then utilized in the positioning algorithm which employs the projections onto convex sets (POCS) technique.

As discussed in the previous chapters, one way to address the transceiver complexity problem in MF receivers is to adopt the TR-UWB approach. Since a TR-UWB receiver does not require timings and amplitudes of received multipath components, it results in a simple structure. However, the required analog delay line is hard to realize in practice.

To address this problem, recently a frequency-shifted reference (FSR) UWB technique has been proposed [396, 397]. In FSR-UWB, two signals that are relatively shifted in the frequency domain are used as reference and data signals. When the frequency shift is smaller than the channel coherence bandwidth, the two signals propagate through the same channel, and experience about the same fading. At the receiver, the received waveform is multiplied by a frequency-shifted version of itself. The resulting signal yields the decision variable. Unlike TR-UWB, FSR-UWB does not need an analog delay line. However, like the TR-UWB, FSR-UWB suffers from the energy loss due to allocating half of the energy to the reference. In [398, 399], a digital multi-carrier differential (MCD) UWB is proposed, which employs a differential modulation/demodulation approach in the frequency domain for digital UWB receivers. In MCD-UWB, only the first carrier is used as the reference carrier, and a differential relationship is established between adjacent carriers. With FFT/DCT operations, MCD-UWB can flexibly realize variable data rates by transmitting more than one symbol in each block interval. Compared to the FSR-UWB, MCD-UWB sufficiently reduces the energy loss due to the reference carrier. It is proved that for a low-rate scheme when only one symbol is differentially modulated over all carriers, the similar simple receiver structure can be obtained as that of the FSR-UWB. Similar to the FSR-UWB, MCD-UWB also avoids the analog delay line.

Another research direction aims to improve the positioning accuracy by cooperative localization techniques. As discussed in [26], the cooperative localization problem can be represented by spools of thread interconnected with springs. This analogy is depicted in Fig. 9.1. Some of the spools (reference nodes) are nailed to their known positions whereas some of the others with unknown coordinates can move around. The lengths of the springs are determined by the measured distances and they can be compressed or stretched. Once the system is released, it converges to a minimum energy location estimate at the equilibrium point. In [206], Denis *et al.* further investigated the problem of distributed localization for UWB systems in NLOS scenarios and for realistic ranging error models. It is shown that cooperative/distributed maximization of log-likelihood of range estimates yields more accurate results compared to distributed weighted least squares (WLS) techniques.

The localization accuracy can be improved using hybrid techniques, such as hybrid TDOA/AOA [161], or hybrid TOA/fingerprint-based techniques [158]. However, this may come at the expense of increased complexity. Enhanced location trackers can

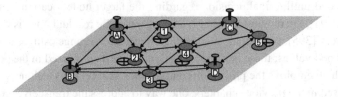

Fig. 9.1. Illustration of cooperative localization. It is analogous to finding resting point of spools of thread which are connected by a network of springs (represented by arrows) (After [26]). The actual locations of the target nodes are indicated by ⊕ and the figure illustrates the equilibrium point.

increase the positioning accuracy while tracking a target node in motion, such as by using Kalman filters and particle filters [400]. It is possible to improve the leading edge detection and ranging performance by using forward error correction coding [401].

As discussed in [402], real world localization accuracy may often be much worse than that predicted by simulations. A major factor in the prediction gap is assumed to be due to the differences between theoretical noise models (which is commonly assumed to be Gaussian distributed) and empirical noise characteristics. While [402] focuses on multi-hop localization, the prediction gap applies to localization systems in general if improper noise models are used. The ranging noise in a UWB system may have a significantly different probability distribution than a Gaussian noise depending on the employed ranging algorithm [206, 403]. If not properly accounted for, this may yield an inferior localization accuracy in practice. Further research may be needed to close the gap between theorical and practical results.

It is a non-trivial issue to design low cost and low complexity UWB chipsets that can be fitted into consumer devices (e.g., mobile handsets [404]). Enabling this is not only the job of the chip manufacturers, but also of the system designers. They should design low-complexity, yet efficient, ranging/localization techniques so that they can be implemented in a cost-effective way. The trade-off is that, with low-complexity implementations (e.g. using an energy detector that operates at sub-Nyquist sampling rates), ranging typically requires a longer air time. This is because the noise has to be averaged to improve the SNR.

9.2 Training-based systems and exploiting the side information

The positioning accuracy can be improved by making use of some *a-priori* information about the measured radio signals at different geographical locations. Such information are commonly referred to as fingerprint information. They should be location-sensitive and can be collected during a training (off-line) phase in a database. During the real-time (on-line) phase, the fingerprint information can be used to locate the mobile node. A block diagram of a fingerprint-based localization system is depicted in Fig. 9.2.

A fingerprint database can be simply composed of received signal strengths from/at different reference nodes and at different mobile (target node) locations [21, 405]. However, it may as well capture more detailed fingerprint information. For example, in [153], seven parameters of the magnitude of the received channel impulse response (CIR) are used as fingerprint information: mean excess delay, RMS delay spread, maximum excess delay, total received power, number of MPCs, power of the first path, and arrival time of the first path. During the off-line phase, this information is collected at different measurement locations and they are used for training an artificial neural network (ANN). In the on-line phase, this ANN is used for mapping the measured parameters to an estimate of the mobile's location.

Rather than using only certain parameters of the CIR, the estimated CIR itself can be used to estimate the mobile's location as in [151]. First, the CIRs at different locations are recorded in a database during the off-line phase. During the on-line phase, a cost function

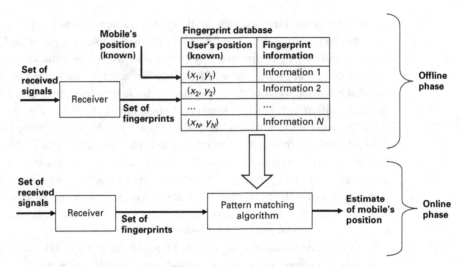

Fig. 9.2. Illustration of localization with a training-based system (After [153]).

(e.g., the Euclidean distance) is computed between the estimated CIR and each of the CIRs in the database. Then, the location corresponding to the smallest cost function is chosen as the mobile's location.

A similar concept is applied to UWB systems in [150, 406] using a single reference node. First, *a-priori* knowledge of the average power delay profile is obtained at different geographical locations. Then, a likelihood-based algorithm is used to determine if the CIR of a received signal belongs to a certain location or not. A drawback of the proposed algorithm is that with a single reference node, it only distinguishes between two positions. Hence, it may be more suitable for obtaining a rough estimate of the mobile's location, and more accurate techniques may be developed using multiple reference nodes.

One of the important drawbacks of fingerprinting techniques is that they may require a large number of training points for accurate localization. This is because the solution space is typically limited by the number of training locations. Moreover, the effectiveness of the fingerprint database may show variations in time due to changes in the propagation environment (e.g., due to movements of people, furnitures, opening/closing of doors, changes in humidity). In [154], a zero-configuration system that does not require an on-site survey is proposed to solve this problem. The key idea is that on-line signal strength measurements among different reference nodes (with known locations) are used to capture the variations in the RF channel. Then, this information is used while mapping the signal strength values to actual geographical distances,[1] which can then be used to estimate the location of the target node. Hence, adverse effects of changes in the RF channel are captured and mitigated simultaneously by making use of the infrastructure.

The map of the geographical region can be quite useful for tracking and NLOS mitigation applications. For example, while tracking a mobile, if the map of a building is available *a priori*, the location estimator may be tuned so that the estimate cannot jump

---

[1] In particular, singular value decomposition (SVD) techniques are used for mapping.

from one side of a wall to the other side of the wall. In [407], a ray-tracing algorithm is proposed which uses the available map information for mitigating NLOS effects. The results show significant improvements in localization accuracy.

The *a-priori* information may also be simply composed of the statistics of an unknown (location-specific) parameter (e.g. NLOS bias, RMS delay spread, and mean excess delay) within a certain geographical region. During the on-line phase, such *a-priori* knowledge of these statistics may be used to improve the positioning accuracy [119, 408]

9.3 NLOS mitigation

Many practical scenarios for UWB ranging and localization involve NLOS paths between the target node and a reference node, which degrade the positioning accuracy. A simple way to mitigate the NLOS effects is to identify the NLOS reference nodes and discard them during localization [17]. However, there is always the probability of wrong identification (i.e. detecting a LOS reference node as NLOS, or vice versa). Moreover, if an NLOS reference node is not used for localization, the remaining number of available reference nodes (and their geometry) may not be suitable for obtaining an accurate estimate of the target node's location.[2]

An NLOS reference node carries useful information about the location of the target node, which can be utilized to improve the localization accuracy. Since the NLOS bias in a TOA estimate is always positive (and is usually relatively larger than the error due to background noise), the location of the target node is bounded by a circle centered at the NLOS reference node. This circle has a radius equal to the range estimate of the NLOS reference node, and can be used as a constraint while calculating the location of the target node. In [19, 197], quadratic and linear programming techniques for NLOS localization are introduced where the constraints in the algorithms are obtained from the NLOS reference nodes.

Another way to mitigate the effects of the NLOS bias is to use a weighted least squares algorithm [4, 119]. The contribution of the NLOS measurements may be weighted appropriately and suppressed in the localization algorithm. The weights may be obtained from the variances of the range estimates [4], or from different statistics of the multipath components of the received signal [119].

In [409], a database technique based on universal kriging algorithm is introduced for mitigating the effects of NLOS bias.[3] First, during the off-line stage, the NLOS bias errors are recorded at a number of target node locations. Then, the universal kriging technique is used to interpolate the NLOS errors at untrained locations. The basic assumption is that the NLOS error is spatially correlated. During the online stage, the NLOS error

---

[2] For example, a minimum of three LOS reference nodes are required for TOA-based 2-D localization, and a minimum of two LOS reference nodes are required for AOA-based 2-D localization.

[3] For a more detailed discussion and definition of universal kriging, the reader is referred to [410].

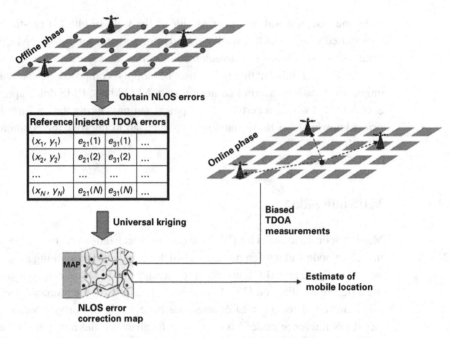

Fig. 9.3. Illustration of the kriging technique for NLOS mitigation (After [409]).

correction matrix obtained by kriging is used for improving the accuracy of the location estimate. Basic operation of the algorithm is depicted in Fig. 9.3.

In addition to different techniques that are explicitly designed to mitigate NLOS effects, cooperative localization techniques [206] and tracking algorithms [400, 411], if intelligently designed, may also improve the localization accuracy in NLOS environments. Despite the numerous NLOS identification and mitigation algorithms in the literature, there is still a large research space for designing efficient UWB ranging/localization techniques in practical NLOS scenarios.

9.4 Multiple accessing and interference mitigation

In order to accommodate multiple users in the same channel, it is essential to use efficient multiple access mechanisms for localization. As discussed in Chapter 5, orthogonal channels can be assigned in time, frequency, code, or space domains. A commonly used multiple access technique in ad-hoc and sensor network systems is the carrier sense multiple access (CSMA). In CSMA, only a single user is allowed to access the channel within a certain time period. However, due to the high *bandwidth to data-rate ratio* of UWB systems, such a single-channel approach for UWB multiple access may be wasteful [412]. Instead, in [412], a TH-CDMA-MAC protocol is introduced for improved scalability and faster convergence time of the location estimates.

Even with proper multiple access designs, there may still be interference from other users (e.g. from a simultaneously operating piconet (SOP)), and this may degrade the ranging accuracy. Hence, efficient interference mitigation algorithms may be required to achieve accurate localization. In [413], a generalized successive interference cancellation/matching pursuits (GSIC/MP) algorithm is proposed for mitigating the effects of multiple access interference in DS-CDMA systems. It provides reliable channel estimates for sparse multipath channels with weak direct paths. Since it can combat near/far effects, it can be used in scenarios where no power control is available.

A method for mitigation of the multi-user interference in time-hopping and direct-sequence non-coherent IR-UWB systems is introduced in [201, 290], which was discussed in detail in Section 7.1.4. A block diagram summarizing the basic principle of the algorithm is illustrated in Fig. 7.9. To summarize the concept, first, a 2-D image of the signal is obtained by de-spreading it with the desired user's spreading code and observing the samples over multiple ranging symbols. When the rows of the 2-D image are observed, the desired signal repeats itself at each row, while the interference has a random pattern (since it has a different spreading waveform). Hence, interference can be suppressed by applying a non-linear filter (e.g. a minimum filter or a median filter) on the columns of the 2-D image. Then, 2-D to 1-D conversion is performed followed by an appropriate TOA estimation algorithm.

While interference to the ranging system may be from other UWB transmitters, accuracy of UWB localization systems is also affected from narrowband interference. In a recent work, impact of narrowband and wideband interference on the accuracy of practical UWB ranging systems is evaluated in realistic multipath environments [291]. Further analysis of the issue and development of efficient narrowband interference cancelation algorithms may be an interesting research direction.

Another way of limiting the interference to other users in a network is to implement power control. In [414], it is shown that localization accuracy fluctuates or "fades" as a target node moves through the network of reference nodes. If power control is used, fluctuations in the localization error can be reduced and localization accuracy can be improved. Intelligent power control techniques may also be used to minimize the interference to other users while simultaneously maintaining a satisfactory localization accuracy.

9.5 Cognitive ranging and localization

Cognitive radio has recently been popular due to its promising characteristics and its capability to adapt to the changes in the environment [415]. In a recent work by Haykin, the concept of *cognitive radar* [416, 417] is introduced, where cognition is applied to radar systems. In a later work, Celebi *et al.* introduced the concept of cognitive positioning systems (CPSs) in which the positioning accuracy can be varied by adaptively changing the system parameters [418, 419]. The CPS is composed of two modes: (1) bandwidth determination, and (2) hybrid overlay and underlay enhanced dynamic spectrum management (H-EDSM) system. As illustrated in Fig. 9.4, the bandwidth determination step chooses the appropriate bandwidth required for the desired positioning accuracy. Then, it is consulted to the H-EDSM for the availability of the required effective bandwidth. If there is available bandwidth, the reference cognitive radio transmits the signal at the available spectrum. The target cognitive radio, upon receiving the signal, estimates the mobile location using adaptive TOA (A-TOA) estimation, where A-TOA is capable of performing ranging at different bandwidths.

In [249], ranging in dynamic spectrum access networks (DSANs) using cognitive radios is studied. More specifically, due to the dynamic nature of the DSANs, the effects of absolute bandwidth, operating center frequency, and frequency-dependent feature of multipath components (or the change in environment) on the ranging accuracy are investigated. Issues related to ranging in DSANs can be investigated further and ranging algorithms can be developed for such wireless networks. Location awareness engine architecture for cognitive radios and cognitive wireless networks is introduced in [420]. The proposed location awareness engine is responsible for handling all location information-related tasks such as location estimation and sensing, seamless positioning and interoperability, statistical learning and tracking, security and privacy, and mobility management.

The impact of bandwidth on the accuracy of UWB ranging systems is further analyzed in [421, 422]. The ranging error is analyzed for two different scenarios in [421]: (1) detected direct path (DDP) scenarios, and (2) undetected direct path (UDP) scenarios. For DDP scenarios, the ranging error consistently decreases with increasing bandwidth.

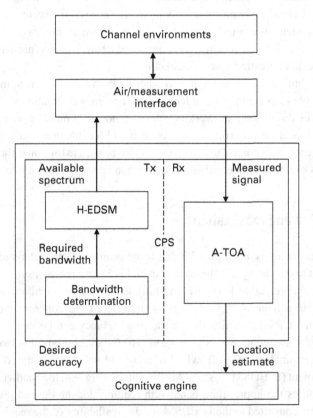

Fig. 9.4. Illustration of CPS block diagram (©2007 IEEE) [418].

However, as the bandwidth increases, it becomes more likely that the amplitudes of the first arriving MPCs may be smaller than the ranging threshold in UDP scenarios.[4] Since the total ranging error is the summation of the errors for DDP and UDP scenarios, the authors show that there actually exists an optimum bandwidth where the ranging error is minimized.[5] In [422], the authors analyze the relation between the bandwidth of UWB systems and the ranging accuracy using experimental data collected in three-bedroom and four-bedroom apartments. The bandwidths between 500 MHz and 7 GHz are considered with 500 MHz increments. While the ranging accuracy increases significantly with bandwidth until 2.5 GHz, improvements in the ranging accuracy after 5 GHz become insignificant.

Apart from the system bandwidth, sampling rate of the receiver[6] (beside other algorithm-specific parameters such as the integration window length, ranging threshold, and search window length) is another parameter that may affect the ranging accuracy. Therefore, even if the system bandwidth increases, improvement in the ranging accuracy may not be significant if the receiver capabilities/parameters are not also improved/modified.

9.6 Anchor placement

Placement of the reference nodes carries critical importance for the positioning accuracy. The Cramer–Rao lower bound (CRLB) and geometric dilution of precision (GDOP) are commonly used as tools for assessing the optimality of a certain sensor placement [423].

In [203], an iterative algorithm called RELOCATE is proposed for optimally placing the reference nodes. For a fixed position of the target node, it optimally places the reference nodes so as to minimize the CRLB. Extension of the algorithm for multiple locations of the target node (such as a walking path within a building) is also presented.

Practical aspects of 3-D placement of the reference nodes is evaluated in [424] using well known optimal solutions. Horizontal dilution of precision (HDOP) and vertical dilution of precision (VDOP) are used to describe errors in horizontal and vertical dimensions. An example scenario for placing four reference nodes within a cubic room is considered. Placing all the target nodes on a planar surface (e.g., four different corners of the room's ceiling) yields relatively low HDOPs but large VDOPs. On the other hand, if the target nodes are placed in an "as good as possible" tetrahedron configuration, the

---

[4] At small bandwidths, each observable MPC is composed of a number of paths that arrive close to each other. Hence, the amplitude of each observable MPC is proportional to the sum of the individual MPCs. As the system bandwidth increases, each of these individual MPCs become resolvable, and their amplitudes decrease.

[5] It should be noted that this finding is based on a specific approach for selecting the ranging threshold.

[6] As discussed in Chapter 5, the receiver may sample the received signal at smaller than Nyquist rate for a low-complexity implementation.

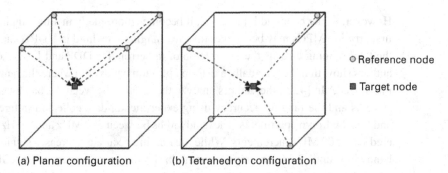

Fig. 9.5. A rectangular room with (a) planar, (b) tetrahedron placement of reference nodes. Placing the reference nodes on an "as good as possible" tetrahedron yields significantly better vertical accuracy.

HDOP is relatively smaller while the VDOP is significantly smaller compared to the planar configuration (see Fig. 9.5).

Optimum geometries of the reference nodes for different number N_m of reference nodes are derived in [425]. In general, the reference nodes are placed on a geometry whose corners are "equally" distributed on a unit spherical surface. The five solutions to this problem for $N_m = 4, 6, 8, 12, 20$ correspond to a tetrahedron, octahedron, cube, icosahedron, and dodecahedron, respectively, which are also referred to as platonic solids. Also, any superposition of centered platonic solids yields another optimum geometry [425].

9.7 UWB radar in health-care

Due to its high time resolution, UWB signaling is very suitable for short range radar type of applications, in which it is desirable to estimate the range, direction, and speed of a target object. In particular, UWB radars have a large number of potential applications in health-care and medicine. A nice tutorial on the applications of UWB radar in medicine can be found in [426].

A recently studied application of UWB radars is the estimation of vital signal parameters. Its use in the detection of chest cavity motion and in the estimation of respiration and heart-beat rates are described and analyzed in [427]. Considering a static environment (other than the subject under consideration), the channel impulse response can be modeled as

$$h(t, \tau) = \alpha_b \delta(\tau - \tau_b(t)) + \sum_i \alpha_i \delta(\tau - \tau_i), \qquad (9.1)$$

where the first term captures the respiratory variations and the other terms correspond to the static channel (see Fig. 9.6). In order to get rid of the static components, a motion filter can be used. First, the received signal is averaged over a large number of observations to capture only the static channel. Then, this is subtracted from the instantaneous signal.

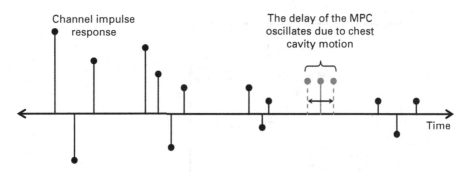

Fig. 9.6. The change in the delay of a certain MPC due to chest cavity motion.

The remaining signal basically captures the respiratory variations in the related signal component.

In another work [428], the variations of the radar return spectrum around the interferometric minima are used to detect chest cavity motion. For a breathing person, the variations are very large compared to a non-breathing person. Furthermore, fundamental lower bounds on the estimation of the vital signal parameters using UWB are derived in [429].

9.8 UWB for simultaneous localization and mapping

Simultaneous localization and mapping (SLAM) is the task of jointly and incrementally building a map of the environment while simultaneously estimating the target's own location. In [430], it is proven that the SLAM problem can be solved where the map uncertainty and position uncertainty can be improved up to a fundamental limit determined by the initial position uncertainty. In [431], it is shown that by using the a-optimal information measure, a more accurate map than existing approaches can be developed by using a greedy, closed-loop strategy.[7]

The use of UWB-IR receivers for indoor mapping and positioning is investigated in [432–434]. The proposed technique is capable of positioning and mapping without using any fixed references, and simultaneously constructs a map of the room for simple two-wall and four-wall scenarios. Figure 9.7 illustrates a simple scenario where two mobiles (target nodes), M_1 and M_2, may receive LOS or reflected signals from each other (or reflections of their own signals). By processing different echoes, the mobiles can obtain the distances to each other as well as to the walls. As the number of walls increases, the complexity of the algorithm gets larger.

SLAM can be used in many interesting practical scenarios. For example, the European Union-funded project EUROPCOM [435] envisions a scenario in which UWB radio is

---

[7] *D-optimality* and *a-optimality* are two different optimality criteria in experimental design theory. While the d-optimal information measure uses the product of eigenvalues, the a-optimal information measure uses the sum of eigenvalues [431]. For a more detailed discussion on these information measures, the reader is referred to [431].

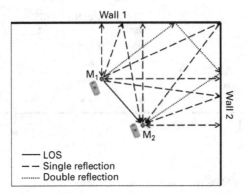

Fig. 9.7. Two-dimensional indoor mapping without any infrastructure (After [432]).

used in emergency situations (particularly within large buildings) where locations of the personnel are displayed in a control vehicle. Since the building map may not be available *a priori*, or it may change due to damaged walls etc., SLAM can be used to obtain an up-to-date map of the building. In [436], audio signals are used to emulate the behavior of the UWB transmitter/receiver and the CLEAN algorithm is used to obtain a 3-D map of the environment.

9.9 Secure ranging and localization

Until recently, ranging and localization have been mainly studied in secure environments. However, many of the traditional localization techniques are susceptible to different attacks (e.g., Sybil attack [437], wormhole attack [20, 437], jamming attack [20], distance enlargement/reduction [438][8]) in hostile environments. Recognizing this vulnerability, few recent research works address the issue of secure localization in adversarial settings [437–441].

Secure localization can be achieved to some extent by using longer ranging codes during TOA estimation as discussed in Chapter 7. For example, using an MTOK sequence of length 127, an imposter has to search a larger number of possible ranging sequences to detect the correct sequence compared to that when an MTOK sequence of length 31 is used. This decreases the chances that the attack becomes successful before the ranging process is completed. Moreover, compared to shorter length preambles (e.g. 16 or 64 repetitions), longer length preambles (e.g. 1024 or 4096 repetitions), even though potentially yielding higher accuracy levels, may be more susceptible to attacks due to a larger number of repetitions of the ranging preamble.

On the other hand, as discussed in [20], many of the localization-specific attacks may be *non-cryptographic* where conventional security mechanisms are unlikely to remove the threats, and higher-level security mechanisms are required. For example, a malicious user may alter the signal strength or TOA from a particular reference node

---

[8] For a detailed review of different attacks specific to localization systems, the reader is referred to [20, 438].

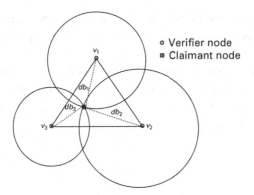

Fig. 9.8. Verifiable multilateration (VM) with three verifiers. If the claimant node enlarges the measured distance to one of the verifiers, it has to prove that the measured distance to at least one of the other verifiers is decreased (which cannot happen due to the distance bounding property) (After [438]).

by placing attenuators between the target node and the reference node.[9] This produces outliers in the measurement data from different reference nodes, and hence degrades the localization accuracy. In [20], robust estimators (in particular, least median of squares estimators) are used for making the location estimate attack-tolerant for such distance enlargement/reduction attacks. In that case, the position estimate is obtained as

$$(\hat{x}, \hat{y}) = \arg \min_{(x,y)} \mathrm{med}_i \left[\sqrt{(x_i - x)^2 + (y_i - y)^2} - \hat{d}_i^2 \right]^2, \tag{9.2}$$

where (x_i, y_i) is the coordinate of the ith reference node, $(\hat{x}, \hat{y})$ is the estimate of the target node position, and $\hat{d}_i$ is the distance measurement related to the ith reference node. This position estimation scheme can tolerate up to 50% outliers among all measurements, in the absence of noise.

In [438], the verifiable multilateration (VM) algorithm is introduced, which enables secure positioning in the presence of attackers who may maliciously modify the measured distances. The VM algorithm is based on the *distance bounding* property, which states that a claimant (e.g., a mobile that wishes to spoof its position to the network) may only claim that it is more distant from a verifier (e.g., a reference node) than it actually is. If a claimant increases its measured distance to one of the verifiers in order to spoof its position, it also has to prove that at least one of the measured distances to one of the other verifiers should decrease for consistency (see Fig. 9.8). However, this is not possible due to the distance bounding property.

In [437], the authors consider and examine specific adversarial models (e.g., wormhole attack, Sybil attack, and compromise of network entities) and develop a secure localization method called high-resolution range-independent localization (HiRLoc), which combines communication range constraints with cryptographic primitives. The

---

[9] Or, the malicious user may also serve as a reference node and purposefully transmit a manipulated signal with a different signal strength and/or delay for misleading the target node.

basic idea in HiRLoc is the variation of the transmission parameters at the reference points, such as the antenna orientation and the communication range (via power control), or both.

Another secure localization technique that is based on transmission range variation is presented in [441]. For each transmission power level, a unique nonce (random number) is transmitted by the anchors, and only the intended sensor is able to decipher the nonces due to the employed encryption key.

9.10 Concluding remarks

Ranging and localization via UWB radios will enable numerous exciting applications for next generation wireless systems. Despite the recent research and development on UWB ranging and localization, there are still many research areas and directions to explore. This chapter has summarized only a few of the possible research directions. Together with the release of the IEEE 802.15.4a standard, we will observe an increasing number of UWB ranging/localization chip sets and devices, which will trigger new research and development. Maybe we will soon see UWB localizers as an essential component of our cellular handsets. Maybe, in few years from now, we will never have to search for our lost keys again due to centimeter-accuracy UWB localizers. Regardless, it is for sure that UWB localizers will be there to make our daily lives easier.

References

1. J. Pagonis and J. Dixon, Location awareness and location based services – Part I, 2004. [Online]. Available: http://www.symbian.com
2. K. Muthukrishnan, M. E. M. Lijding and P. J. M. Havinga, Towards smart surroundings: Enabling techniques and technologies for localization. In *Proc. Int. Workshop on Location- and Context-Awareness*, Oberpfaffenhofen, Germany, May 2005, pp. 350–362.
3. J. Roth, A decentralized location service providing semantic locations. Ph.D. dissertation, University of Hagen, 2005.
4. J. J. Caffery and G. L. Stuber, Overview of radiolocation in CDMA cellular systems. *IEEE Commun. Mag.*, **36**: 4 (1998), 38–45.
5. K. Pahlavan, P. Krishnamurthy and J. Beneat. Wideband radio propagation modeling for indoor geolocation applications. *IEEE Commun. Mag.*, **36**: 4, (1998), 60–65.
6. C. C. Chong, F. Watanabe and H. Inamura, Potential of UWB technology for the next generation wireless communications. In *Proc. IEEE Int. Symp. Spread Spectrum Techniques and Applications*, Manaus, Brazil, Aug. 2006, pp. 422–429.
7. Loki website, (2006) [Online]. Available: http://www.loki.com
8. S. J. Ingram, D. Harmer and M. Quinlan, Ultra-wideband indoor positioning systems and their use in emergencies. In *Proc. IEEE Position Location and Navigation Symp. (PLANS)*, Monterey, CA, Apr. 2004, pp. 706–715.
9. Idtechex website. [Online]. Available: http://www.idtechex.com
10. J. Hightower and G. Borriello, Location systems for ubiquitous computing. *IEEE Computer*, **34**: 8, (2001), 57–66.
11. I. Guvenc, Enhancements to RSS based indoor tracking systems using Kalman filters. M.S. Thesis, University of New Mexico, 2002.
12. J. J. Caffery and G. L. Stuber, Subscriber location in CDMA cellular networks. *IEEE Trans. Veh. Technol.*, **47**: 2, (1998), 406–416.
13. K. W. Cheung, H. C. So, W. K. Ma and Y. T. Chan, Least square algorithms for time-of-arrival-based mobile location. *IEEE Trans. Sig. Processing*, **52**: 4, (2004), 1121–1128.
14. Y. T. Chan and K. C. Ho, A simple and efficient estimator for hyperbolic location. *IEEE Trans. Sig. Processing*, **42**: 8, (1994), 1905–1915.
15. Y. T. Chan, H. Y. C. Hang and P. C. Ching, Exact and approximate maximum likelihood localization algorithms. *IEEE Trans. Veh. Technol.*, **55**: 1, (2006), 10–16.
16. P. C. Chen, A non-line-of-sight error mitigation algorithm in location estimation. In *Proc. IEEE Int. Conf. Wireless Commun. Networking (WCNC)*, vol. 1, New Orleans, LA, Sep. 1999, pp. 316–320.
17. Y. T. Chan, W. Y. Tsui, H. C. So and P. C. Ching, Time of arrival based localization under NLOS conditions. *IEEE Trans. Veh. Technol.*, **55**: 1, (2006), 17–24.

18. W. Kim, J. G. Lee and G. I. Jee, The interior-point method for an optimal treatment of bias in trilateration location. *IEEE Trans. Veh. Technol.*, **55**: 4, (2006), 1291–1301.
19. X. Wang, Z. Wang and B. O. Dea, A TOA based location algorithm reducing the errors due to non-line-of-sight (NLOS) propagation. *IEEE Trans. Veh. Technol.*, **52**: 1, (2003), 112–116.
20. Z. Li, W. Trappe, Y. Zhang and B. Nath, Robust statistical methods for securing wireless localization in sensor networks. In *Proc. IEEE Int. Symp. Information Processing in Sensor Networks (IPSN)*, Los Angeles, CA, Apr. 2005, pp. 91–98.
21. P. Bahl and V. N. Padmanabhan, RADAR: An in-building RF-based user location and tracking system. In *Proc. IEEE Int. Conf. on Computer Commun. (INFOCOM)*, Tel Aviv, Israel, Mar. 2000, pp. 775–784. [Online]. Available: http://citeseer.nj.nec.com/bahl00radar.html
22. Ekahau inc. website. [Online]. Available: http://www.ekahau.com
23. J. Hightower, C. Vakili, G. Borriello and R. Want, Design and calibration of the SpotON ad-hoc location sensing system. 2001. [Online]. Available: http://citeseer.nj.nec.com/hightower01design.html
24. D. Niculescu and B. Nath, DV based positioning in ad-hoc networks. *Telecomms. Syst.*, **2**: 2, (2004), 133–151.
25. K. Siwiak and J. Gabig, IEEE 802.15.4IGa informal call for application response, contribution#11, Doc.: IEEE 802.15-04/266r0, July 2003. [Online]. Available: http://www.ieee802.org/15/pub/TG4a.html
26. N. Patwari, J. N. Ash, S. Kyperountas, A. O. Hero, R. L. Moses and N. S. Correal, Locating the nodes: Cooperative localization in wireless sensor networks. *IEEE Sig. Processing Mag.*, **22**: 4, (2005), 54–69.
27. K. Romer and F. Mattern, The design space of wireless sensor networks. *IEEE Wireless Commun. Mag.*, **11**: 6, (2004), 54–61.
28. Habitat monitoring on Great Duck Island. [Online]. Available: http://www.greatduckisland.net
29. P. Juang, H. Oki, Y. Wong, M. Martonosi, L. S. Peh and D. Rubenstein, Energy-efficient computing for wildlife tracking: Design tradeoffs and early experiences with ZebraNet. In *Proc. Int. Conf. Architectural Support for Programming Languages and Operating Syst. (ASPLOS-X)*, San Jose, CA, Oct. 2002, pp. 96–107.
30. G. W. Allen, K. Lorincz, M. Ruiz, *et al.*, Deploying a wireless sensor network on an active volcano. *IEEE Internet Comput., Special Issue on Data-Driven Applications in Sensor Networks*, **10**: 2, (2006), 18–25.
31. A. Baggio, Wireless sensor networks in precision agriculture. In *ACM Workshop on Real-World Wireless Sensor Networks (REALWSN)*, Stockholm, Sweden, June 2005.
32. F. Michahelles, P. Matter, A. Schmidt and B. Schiele, Applying wearable sensors to avalanche rescue. *Comput. Graphics*, **27**: 6, (2003), 839–847.
33. J. D. Lundquist, D. R. Cayan and M. D. Dettinger, Meteorology and hydrology in Yosemite national park: A sensor network application. *Lect. Notes Comput. Sci.*, **2634**, (2003), 518–528.
34. G. Tolle, J. Polastre, R. Szewczyk, *et al.*, A macroscope in the redwoods. In *Proc. ACM Conf. on Embedded Networked Sensor Syst. (SenSys)*, San Diego, CA, Nov. 2005, pp. 51–63.
35. Z. Butler, P. Corke, R. Peterson and D. Rus, Networked cows: Virtual fences for controlling cows. In *Proc. Workshop on Applications of Mobile Embedded Syst. (WAMES)*, Boston, MA, June 2004.
36. G. Simon, A. Ledeczi and M. Maroti, Sensor network-based countersniper system. In *Proc. ACM Conf. on Embedded Networked Sensor Syst. (SenSys)*, Baltimore, MD, Nov. 2004, pp. 1–12.

37. W. Merrill, L. Girod, B. Schiffer, *et al.*, Defense systems: self healing land mines. In *Wireless Sensor Networks: A Systems Perspective*. (Norwood, MA: Artech House Publisher, 2005), chapter 18.
38. D. A. Lawrance, R. E. Donahue, K. Mohseni and R. Han, Information energy for sensor-reactive UAV flock control. In *Proc. AIAA Unmanned Unlimited Technical Conf., Workshop and Exhibit*, Chicago, IL, Sep. 2004.
39. G. Virone, A. Wood, L. Selavo, *et al.*, An advanced wireless sensor network for health monitoring. In *Transdisciplinary Conf. on Distributed Diagnosis and Home Healthcare (D2H2)*, Arlington, VA, Apr. 2006.
40. S. Consolvo, P. Roessler, B. Shelton, A. LaMarca, B. Schilit and S. Bly, Computer-supported coordinated care: Using technology to help care for elders. In *Intel Research Internal Report, IR-TR-2003-131*, Dec. 2003.
41. C. Kidd, R. J. Orr, G. D. Abowd, *et al.*, The aware home: A living laboratory for ubiquitous computing research. In *Proc. Int. Workshop on Cooperative Buildings*, Mar. 1999, pp. 191–198. [Online]. Available: http://www.awarehome.gatech.edu
42. M. Srivastava, R. Muntz and M. Potkonjak, Smart kindergarten: Sensor-based wireless networks for smart developmental problem-solving environments. In *Proc. ACM SIGMOBILE Int. Conf. on Mobile Computing and Networking*, Rome, Italy, July 2001, pp. 166–179.
43. X. Wang, F. Silva and J. Heidemann, Follow-me application–active visitor guidance system. In *Proc. ACM Int. Conf. Embedded Networked Sensor Syst. (SenSys)*, Baltimore, MD, Nov. 2004, p. 316.
44. Multispectral solutions, Sapphire DART positioning system. [Online]. Available: http://www.multispectral.com/sapphiredart.html
45. Ubisense website. [Online]. Available: http://www.ubisense.net
46. Time domain corporation website. [Online]. Available: http://www.timedomain.com/products/P350.pdf
47. Pal650ek positioning technology. [Online]. Available: www.multispectral.com/pdf/PAL650EK.pdf
48. Thales research website. [Online]. Available: http://www.thalesresearch.com
49. Aetherwire website. [Online]. Available: http://www.aetherwire.com/CDROM/General/AWL/execsum.html
50. IMEC website. [Online]. Available: http://www.imec.be/wwwinter/mediacenter/en/UWBISSCC2007.shtml
51. Fujitsu website. [Online]. Available: http://http://pr.fujitsu.com/jp/news/2007/01/9-1.html
52. Argos website. [Online]. Available: http://www.argosinc.com
53. Polaris Wireless website. [Online]. Available: http://www.polariswireless.com
54. Aeroscout website. [Online]. Available: http://www.aeroscout.com
55. Pango networks website. [Online]. Available: http://www.pangonetworks.com
56. R. Want, A. Hopper, V. Falcao and J. Gibbons, The active badge location system. *ACM Trans. Information Syst.*, **10**: 1, (1992), 91–102.
57. A. Harter, A. Hopper, P. Steggles, A. Ward and P. Webster, The anatomy of a context-aware application. In *Proc. ACM Int. Conf. on Mobile Computing and Networking (Mobicom)*, Seattle, WA, Aug. 1999, pp. 59–68.
58. N. Priyantha, A. Chakraborty and H. Balakrishnan, The Cricket location-support system. In *Proc. Annual Int. Conf. on Mobile Computing and Networking (Mobicom)*, New York, NY, 2000, pp. 32–43.
59. RF technologies web site. [Online]. Available: http://www.rftechnologies.com/pinpoint

60. Axcess website. [Online]. Available: http://www.axcessinc.com
61. InnerWireless Website. [Online]. Available: http://www.innerwireless.com
62. Metris website. [Online]. Available: http://www.metris.com
63. R. Fontana, A brief history of UWB communications. [Online]. Available: http://www.multispectral.com
64. G. F. Ross, The transient analysis of certain TEM mode four-port networks. *IEEE Trans. Microw. Theory and Tech.*, **MTT-14**: 11, (1966), 528–547.
65. C. L. Bennett and G. F. Ross, Time-domain electromagnetics and its applications. *Proc. IEEE*, **66**: 3, (1978), 299–318.
66. G. F. Ross, Transmission and reception system for generating and receiving base-band duration pulse signals without distortion for short base-band pulse communication system. US Patent No. 3,728,632, April 17, 1973.
67. Federal Communications Commission, *First Report and Order 02-48*, Feb. 2002.
68. M. Z. Win and R. A. Scholtz, Impulse radio: How it works. *IEEE Commun. Lett.*, **2**: 2, (1998), 36–38.
69. F. Ramirez-Mireles and R. A. Scholtz, Multiple-access performance limits with time hopping and pulse-position modulation. In *Proc. IEEE Military Commun. Conf. (MILCOM)*, Boston, MA, Oct. 1998, pp. 529–533.
70. S. Gezici, Z. Sahinoglu H. Kobayashi and H.V. Poor, Ultra-wideband impulse radio systems with multiple pulse types, *IEEE Select. Areas Commun.*, **24**: 4, (2006), 892–898.
71. D. Barras, F. Ellinger, H. Jackel and W. Hirt, Low-power ultra-wideband wavelets generator with fast start-up circuit. *IEEE Trans. Microw. Theory and Techniques*, **54**: 5, (2006), 2138–2145.
72. L. Yu and L. B. White, Design of complex wavelet pulses enabling PSK modulation for UWB impulse radio communications. In *Proc. IEEE Int. Conf. on Wireless Broadband and Ultra Wideband Commun.*, Sydney, Australia, Mar. 2006.
73. X. Wu, Z. Tian, T. N. Davidson and G. B. Giannakis, Optimal waveform design for UWB radios. *IEEE Trans. Sig. Processing*, **54**: 6, (2006), 2009–2021.
74. Y. Wu, A. F. Molisch, S.-Y. Kung and J. Zhang, Impulse radio pulse shaping for ultra-wide bandwidth (UWB) systems. In *Proc. IEEE Personal, Indoor, and Mobile Radio Commun. (PIMRC)*, vol. 1, Beijing, China, Sep. 2003, pp. 877–881.
75. B. Parr, B. Cho, K. Wallace and Z. Ding, A novel UWB pulse design algorithm. *IEEE Commun. Lett.*, **7**: 5, (2003), 219–221.
76. P. Runkle, J. McCorkle, T. Miller and M. Welborn, DS-CDMA: The modulation technology of choice for UWB communications. In *Proc. IEEE Int. Conf. Ultra-wideband Syst. Technol. (UWBST)*, Baltimore, MD, Nov. 2003, pp. 364–368.
77. E. Saberinia and A. H. Tewfik, Multi-user UWB-OFDM communications. In *Proc. IEEE Pacific Rim Conf. on Commun., Computers, and Sig. Processing (PACRIM)*, vol. 1, Victoria, Canada, Aug. 2003, pp. 127–130.
78. Federal Communications Commission, Part 15 - Radio Frequency Devices, Feb. 2006. [Online]. Available: http://www.fcc.gov/oet/info/rules/part15/part15-2-16-06.pdf
79. G. Breed, A summary of FCC rules for ultra wideband communications. *High Frequency Electronics*, Jan. 2005. [Online]. Available: http://www.highfrequencyelectronics.com/Archives/Jan05/HFE0105_Tutorial.pdf

80. The Commission of the European Communities, Commission Decision of 21 February 2007 on allowing the use of the radio spectrum for equipment using ultra-wideband technology in a harmonised manner in the Community. Official Journal of the European Union, 2007/131/EC, Feb. 23, 2007. [Online]. Available: http://eur-lex.europa.eu/LexUriServ/site/en/oj/2007/l_055/l_05520070223en00330036.pdf

81. Electronic Communications Committee, The harmonised conditions for devices using ultra-wideband (UWB) technology with low duty cycle (LDC) in the frequency band 3.4-4.8 GHz. ECC/DEC/(06)LL, July 17, 2006. [Online]. Available: http://www.ero.dk/4FBCBB18-8C4A-41B1-831E-FF3AE19449CA

82. The Commission of the European Communities, Commission Decision of 8 July 2004 on the harmonisation of radio spectrum in the 79 GHz range for the use of automotive short-range radar equipment in the Community. Official Journal of the European Union, 2004/545/EC, July 13, 2004. [Online]. Available: http://eur-lex.europa.eu/LexUriServ/site/en/oj/2004/l_241/l_24120040713en00660067.pdf

83. The Commission of the European Communities, Commission Decision of 17 January 2005 on the harmonisation of the 24 GHz range radio spectrum band for the time-limited use by automotive short-range radar equipment in the Community. Official Journal of the European Union, 2005/50/EC, Jan. 25, 2005. [Online]. Available: http://eur-lex.europa.eu/LexUriServ/site/en/oj/2005/l_021/l_02120050125en00150020.pdf

84. R. Kohno, Interpretation and future modification of Japanese regulation for UWB. IEEE P802.15-06/261r0, May 16, 2006.

85. WiMedia Alliance. [Online]. Available: http://www.wimedia.org

86. UWB Forum. [Online]. Available: http://www.uwbforum.org

87. ECMA-368, High rate ultra wideband PHY and MAC standard, 1st edition. Dec. 2005. [Online]. Available: http://www.ecma-international.org/publications/files/ECMA-ST/ECMA-368.pdf

88. ECMA-369, MAC-PHY interface for ECMA-368, 1st edition. Dec. 2005. [Online]. Available: http://www.ecma-international.org/publications/files/ECMA-ST/ECMA-369.pdf

89. A. R. S. Bahai, B. R. Saltzberg and M. Ergen, *Multi-carrier Digital Communications: Theory and Applications of OFDM*, 2nd ed. (New York, NY: Springer; 2004).

90. IEEE standard for information technology, telecommunications and information exchange between systems, Local and metropolitan area networks specific requirements, Part 15.4: Wireless medium access control (MAC) and physical layer (PHY) specifications for low-rate wireless personal area networks (LR-WPANs), May 2003. [Online]. Available: http://standards.ieee.org/getieee802/download/802.15.4-2003.pdf

91. IEEE P802.15.4a/D4 (Amendment of IEEE Std 802.15.4), Part 15.4: Wireless medium access control (MAC) and physical layer (PHY) specifications for low-rate wireless personal area networks (LRWPANs), July 2006.

92. S. B. Wicker and V. K. B. (editors), *Reed–Solomon Codes and Their Applications*, 1st ed. (Piscataway, NJ: Wiley-IEEE Press; 1999).

93. R. Yao, G. Gao, Z. Chen and W. Zhu, UWB multipath channel model based on time-domain UTD technique. In *Proc. IEEE Global Commun. Conf. (GLOBECOM)*, vol. 3, San Francisco, CA, Dec. 2003, pp. 1205–1210.

94. R. Yao, W. Zhu and Z. Chen, An efficient time-domain ray model for UWB indoor multipath propagation channel. In *Proc. IEEE Veh. Technol. Conf. (VTC)*, vol. 2, Orlando, FL, Sep. 2003, pp. 1293–1297.

95. H. Sugahara, Y. Watanebe, T. Ono, K. Okanue and S. Yarnazaki, Development and experimental evaluation of RS 2000 - a propagation simulator for UWB systems. In *Proc. IEEE Conf. UWB Syst. and Technol. (UWBST)*, Kyoto, Japan, May 2004, pp. 76–80.
96. B. Uguen, E. Plouhinec, Y. Lostanlen and G. Chassay, A deterministic UWB channel model. In *Proc. IEEE Conf. UWB Syst. and Technol. (UWBST)*, Baltimore, MD, May 2002, pp. 1–6.
97. S. S. Ghassemzadeh, R. Jana, C. Rice, W. Turin and V. Tarokh, Measurements and modeling of an UWB indoor channel. *IEEE Trans. Commun.*, **52**: 10, (1996), 1786–1796.
98. J. R. Foerster and Q. Li, UWB channel modeling contributions from Intel, June 2002, IEEE P802.15-02/279r0-SG3a.
99. R. Qiu, A study of the UWB wireless propagation channel model and optimum UWB receiver design. *IEEE J. Select. Areas Commun.*, **20**: 9, (2002), 1628–1637.
100. A. F. Molisch, K. Balakrishnan, C. C. Chong, *et al.*, IEEE 802.15.4a channel model - final report. Sep., 2004. [Online]. Available: http://www.ieee802.org/15/pub/TG4a.html
101. S.-K. Yong, TG3c channel modeling sub-committee final report. Mar. 2007, doc: IEEE 15-07-0584-00-003c. [Online]. Available: http://www.ieee802.org/15/pub/TG3c.html
102. A. F. Molisch, Ultra-wideband propagation channels – Theory, measurement, and modeling. *IEEE Trans. Veh. Technol.*, **54**: 5, (2005), 1528–1545.
103. M. A. Richards, *Fundamentals of Radar Signal Processing*. (New York, NY: McGraw-Hill; Electronic Engineering Series, 2005).
104. M. I. Skolnik, *Introduction to Radar Systems*, 2nd ed. (New York, NY: McGraw-Hill, 1980).
105. R. E. Collin, *Antennas and Radiowave Propagation*. (Singapore: McGraw-Hill, Electronic Engineering Series, 1985).
106. O. S. Heavens, *Optical Properties of Thin Film Solids*. (Mineola, NY: Dover Publications; 1965).
107. A. Safaai-Jazi, S. M. Riad, A. Muqaibel and A. Bayram, Ultra-wideband propagation measurements and channel modeling, Nov. 2002, DARPA NETEX Program Report.
108. S. Hantscher, B. Praher, A. Reisenzahn and C. G. Diskus, Comparison of uwb target identification algorithms for through-wall imaging applications. In *Proc. European Radar Conf.*, San Diego, CA, Sep. 2006, pp. 104–107.
109. Y. Yunqiang and A. E. Fathy, See-through-wall imaging using ultra wideband short-pulse radar system. In *Proc. IEEE Antennas and Propag. Society Int. Symp.*, vol. 3B, Washington, DC, July 2005, pp. 334–337.
110. C.-C. Chong, Ultra-wideband channel modeling and its impacts on system design. In *Ultra-Wideband Wireless Communication*. (Hoboken, NJ: Wiley Interscience, 2006).
111. J. Kunisch and J. Pamp, Measurement results and modeling aspects for the UWB radio channel. In *Proc. IEEE Int. Conf. Ultra-wideband Syst. Technol. (UWBST)*, Baltimore, MD, May 2002, pp. 19–23.
112. A. Alvarez, G. Valera, M. Lobeira, R. Torres and J. L. Garcia, New channel impulse response model for UWB indoor system simulations. In *Proc. IEEE Veh. Technol. Conf. (VTC)*, May 2003, pp. 1–5.
113. A. Saleh and R. Valenzuela, A statistical model for indoor wireless multipath propagation. *IEEE J. Select. Areas Commun.*, **5**: 2, (1987), 128–137.
114. A. F. Molisch, K. Balakrishnan, C. C. Chong, *et al.*, A comprehensive model for ultra-wideband propagation channels. *IEEE Trans. Antennas Prop.*, **54**: 11, (2006), 3151–3166.
115. T. K. Sarkar, Z. Ji, K. Kim, A. Medouri and M. Salazar-Parma, A survey of various propagation models for mobile communication. *IEEE Antennas Propag. Mag.*, **45**: 3, (2003), 51–82.

116. I. Guvenc and H. Arslan, Comparison of two searchback schemes for non-coherent TOA estimation in IR-UWB systems. In *Proc. IEEE Sarnoff Symp.*, Princeton, NJ, Mar. 2006.
117. B. Alavi and K. Pahlavan, Bandwidth effect on distance error modeling for indoor geolocation. In *Proc. IEEE Personal, Indoor, and Mobile Radio Commun. (PIMRC)*, vol. 3, Beijing, China, Sep. 2003, pp. 2198–2202.
118. H. Celebi, I. Guvenc and H. Arslan, On the statistics of channel models for UWB ranging. In *Proc. IEEE Sarnoff Symp.*, Princeton, NJ, Mar. 2006.
119. I. Guvenc, C. C. Chong and F. Watanabe, NLOS identification and mitigation for UWB localization systems. In *Proc. IEEE Wireless Commun. Networking Conf. (WCNC)*, Hong Kong, Mar. 2007, pp. 3488–3492.
120. M. Win, Ultra-wideband spread spectrum techniques for wireless multiple access communications. Ph.D. Dissertation, University of South California, May 1998.
121. D. Cassioli, M. Z. Win and A. F. Molisch, The ultra-wide bandwidth indoor channel: from statistical model to simulations. *IEEE J. Select. Areas Commun. (JSAC)*, **20**: 6, (2002), 1247–1257.
122. K. Siwiak, H. Bertoni and S. M. Yano, Relation between multipath and wave propagation attenuation. *IEEE Commun. Lett.*, **39**: 1, (2003), 142–143.
123. K. Siwiak, UWB channel measurements and model for below 1 GHz (VHF, UHF). In *IEEE 802.15.4a standard*, Doc. no. 15-04-0505-04-004a, Nov. 2004.
124. Federal Communications Commission, Part 15 - Radio Frequency Devices, cfr 15.255: Operation within the band 57-64 GHz. Oct. 2006. [Online]. Available: http://www.access.gpo.gov/nara/cfr/waisidx_06/47cfr15_06.html
125. IEEE 802.15 WPAN millimeter wave alternative PHY Task Group 3c (TG3c). [Online]. Available: http://www.ieee802.org/15/pub/TG3c.html
126. J. Kunisch, E. Zollinger, J. Pamp and A. Winkelmann, MEDIAN 60 GHz wideband indoor radio channel measurements and model. In *Proc. IEEE Veh. Technol. Conf. (VTC)*, vol. 4, Sep. 1997, pp. 2393–2397.
127. C. Liu, E. Skafidas, T. Pollock, *et al.*, NICTA indoor 60 GHz channel measurements and analysis update. Mar. 2006, IEEE 802.15-06-0112-00-003c.
128. P. Pagani, N. Malhouroux, I. Siaud and V. Guillet, Characterization and modeling of the 60 GHz indoor channel in the office and residential environments. Jan. 2006, IEEE 802.15-06-0027-02-003c.
129. K. Sato, T. Manabe, T. Ihara, *et al.*, Measurements of reflection and transmission of office building in the 60 GHz band. *IEEE Trans. Antennas Propag.*, **45**: 12, (1997), 1783–1792.
130. T. Zwick, T. J. Beukema and H. Nam, Wideband channel sounder with measurements and model for the 60 GHz indoor radio channel. *IEEE Trans. Veh. Technol.*, **54**: 4, (2005), 1266–1277.
131. P. Pagani, I. Siaud, N. Malhouroux and W. Li, Adaptation of the France Telecom 60 GHz channel model to the TG3c framework. Apr. 2006, IEEE 802.15-06-0218-00-003c.
132. M. Fiacco, M. Parks, H. Radi and S. R. Saunders, Final report: Indoor propagation factors at 17 and 60 GHz. Technical Report, University of Surrey, Aug. 1998.
133. S. K. Yong, C. C. Chong and S. S. Lee, Generalization and parameterization of the mmWave channel models. May 2006, IEEE 802.15-05-0261-01-003c.
134. H. B. Yang, M. H. A. J. Herben and P. F. M. Smulders, Impact of antenna pattern and reflective environment on 60 GHz indoor radio channel characteristics. *IEEE Ant. Wireless Prop. Lett.*, **4**, (2005), 300–303.

135. S. Gezici, A Survey on wireless position estimation. *Wireless Personal Commun.*, **44**: 3, (2008), 263–282.
136. J. J. Caffery, *Wireless Location in CDMA Cellular Radio Systems*. (Boston: Kluwer Academic Publishers, 2000).
137. A. J. Weiss, Direct position determination of narrowband radio frequency transmitters. *IEEE Sig. Processing Lett.*, **11**: 5, (2004), 513–516.
138. Y. Qi, H. Kobayashi and H. Suda, Analysis of wireless geolocation in a non-line-of-sight environment. *IEEE Trans. Wireless Commun.*, **5**: 3, (2006), 672–681.
139. Y. Qi, Wireless geolocation in a non-line-of-sight environment. Ph.D. Dissertation, Princeton University, Dec. 2004.
140. H. V. Poor, *An Introduction to Signal Detection and Estimation*. (New York: Springer-Verlag, 1994).
141. A. Mallat, J. Louveaux and L. Vandendorpe, UWB based positioning in multipath channels: CRBs for AOA and for hybrid TOA-AOA based methods. In *Proc. IEEE Int. Conf. on Commun. (ICC)*, Glasgow, Scotland, June 2007.
142. G. L. Turin, An introduction to matched filters. *IRE Trans. Information Theory*, **IT-6**: 3, (1960), 311–329.
143. C. E. Cook and M. Bernfeld, *Radar Signals: An Introduction to Theory and Applications*. (Norwood, MA: Academic Press, 1970).
144. C. Knapp and G. Carter, The generalized correlation method for estimation of time delay. *IEEE Trans. Acoust., Speech, and Sig. Processing (ICASSP)*, **24**, (1976), 320–327.
145. B. Champagne, M. Eizenman and S. Pasupathy, Exact maximum likelihood time delay estimation. In *Proc. IEEE Int. Conf. on Acoust., Speech, and Sig. Processing (ICASSP)*, vol. 4, Glasgow, Scotland, May 1989, pp. 23–26.
146. S. P. Belanger, Multisensor TDOA estimation in a multipath propagation environment using the em algorithm. In *Proc. Asilomar Conf. on Signals, Syst. and Computers*, vol. 2, Pacific Grove, CA, Nov. 1995, pp. 1096–1100.
147. M. Aatique, Evaluation of TDOA techniques for position location in CDMA. Master's Thesis, Virginia Polytechnic Institute and State University, Sep. 1997.
148. C. Nerguizian, C. Despins and S. Affes, Framework for indoor geolocation using an intelligent system. In *Proc. IEEE Workshop on Wireless LANs*, Newton, MA, Sep. 2001.
149. M. Triki, D. T. M. Slock, V. Rigal and P. Francois, Mobile terminal positioning via power delay profile fingerprinting: Reproducible validation simulations. In *Proc. IEEE Veh. Technol. Conf. (VTC)*, Montreal, Canada, Sep. 2006.
150. F. Althaus, F. Troesch and A. Wittneben, UWB geo-regioning in rich multipath environment. In *Proc. IEEE Veh. Technol. Conf. (VTC)*, vol. 2, Dallas, TX, Sep. 2005, pp. 1001–1005.
151. T. Nypan, K. Gade and T. Maseng, Location using estimated impulse responses in a mobile communication system. In *Proc. Norwegian Sig. Processing Symp.*, Trondheim, Norway, Oct. 2001.
152. C. Steiner, F. Althaus and A. Wittneben, On the performance of UWB geo-regioning. In *Proc. IEEE Workshop on Sig. Processing Advances in Wireless Commun. (SPAWC)*, July 2006, pp. 1–5.
153. C. Nerguizian, C. Despins and S. Affes, Geolocation in mines with an impulse response fingerprinting technique and neural networks. *IEEE Trans. Wireless Commun.*, **5**: 3, (2006), 603–611.

154. H. Lim, L.-C. Kung, J. C. Hou and H. Luo, Zero-configuration, robust indoor localization: Theory and experimentation. In *Proc. IEEE Int. Conf. on Computer Commun. (INFOCOM)*, Barcelona, Spain, Apr. 2006, pp. 1–12.
155. M. McGuire, K. N. Plataniotis and A. N. Venetsanopoulos, Location of mobile terminals using time measurements and survey points. *IEEE Trans. Veh. Technol.*, **52**: 4, (2003), 999–1011.
156. S. Gezici, H. Kobayashi and H. V. Poor, A new approach to mobile position tracking. In *Proc. IEEE Sarnoff Symp. On Advances in Wired and Wireless Commun.*, Ewing, NJ, March 2003, pp. 204–207.
157. T.-N. Lin and P.-C. Lin, Performance comparison of indoor positioning techniques based on location fingerprinting in wireless networks. In *Proc. Int. Conf. on Wireless Networks, Commun. and Mobile Computing*, vol. 2, June 2005, pp. 1569–1574.
158. J. Kwon, B. Dundar and P. Varaiya, Hybrid algorithm for indoor positioning using wireless LAN. In *Proc. IEEE Veh. Technol. Conf. (VTC)*, vol. 7, Los Angeles, CA, Sep. 2004, pp. 4625–4629.
159. R. O. Duda, P. E. Hart and D. G. Stork, *Pattern Classification*, 2nd ed. (New York: Wiley-Interscience, 2000).
160. A. H. Sayed, A. Tarighat and N. Khajehnouri, Network-based wireless location. *IEEE Sig. Processing Mag.*, **22**: 4, (2005), 24–40.
161. L. Cong and W. Zhuang, Hybrid TOA/AOA mobile user location for wideband CDMA cellular systems. *IEEE Trans. Wireless Commun.*, **1**: 3, (2002), 439–447.
162. R. I. Reza, Data fusion for improved TOA/TDOA position determination in wireless systems. Ph.D. Dissertation, Virginia Tech., 2000.
163. L. Cong and W. Zhuang, Non-line-of-sight error mitigation in mobile location. *IEEE Trans. Wireless Commun.*, **4**, (2005), 560–573.
164. R. Casas, A. Marco, J. J. Guerrero and J. Falco, Robust estimator for non-line-of-sight error mitigation in indoor localization. *EURASIP J. Applied Sig. Processing*, Article ID 43 429, 8 pp., 2006, doi:10.1155/ASP/2006/43429.
165. S. Al-Jazzar and J. J. Caffery, ML and Bayesian TOA location estimators for NLOS environments. In *Proc. IEEE Veh. Technol. Conf. (VTC)*, vol. 2, Vancouver, BC, 2002, pp. 1178–1181.
166. J. Borras, P. Hatrack and N. B. Mandayam, Decision theoretic framework for NLOS identification. In *Proc. IEEE Veh. Technol. Conf. (VTC)*, vol. 2, Ontario, Canada, May 1998, pp. 1583–1587.
167. S. Venkatraman and J. Caffery, A statistical approach to non-line-of-sight BS identification. In *Proc. Int. Symp. on Wireless Personal Multimedia Commun.*, Honolulu, HI, Oct. 2002, pp. 296–300.
168. S. Gezici, H. Kobayashi and H. V. Poor, Non-parametric non-line-of-sight identification. In *Proc. IEEE Veh. Technol. Conf. (VTC)*, vol. 4, Orlando, FL, Oct. 2003, pp. 2544–2548.
169. S. Al-Jazzar, J. J. Caffery and H.-R. You, A scattering model based approach to NLOS mitigation in TOA location systems. In *Proc. IEEE Veh. Technol. Conf. (VTC)*, Birmingham, AL, May 2002, pp. 861–865.
170. Y. Qi, H. Kobayashi and H. Suda, On time-of-arrival positioning in a multipath environment. *IEEE Trans. Veh. Technol.*, **55**: 5, (2006), 1516–1526.
171. Y. Qi and H. Kobayashi, On relation among time delay and signal strength based geolocation methods. In *Proc. IEEE Global Commun. Conf. (GLOBECOM)*, vol. 7, San Francisco, CA, Dec. 2003, pp. 4079–4083.

172. A. Catovic and Z. Sahinoglu, The Cramer–Rao bounds of hybrid TOA/RSS and TDOA/RSS location estimation schemes, *IEEE Commun. Lett.*, **8**, (2004), 626–628.
173. P. Masatt and K. Rudick, Geometric formulas for dilution of precision calculation. *J. Inst. Navigation*, **37**, (1991), 379–391.
174. D. J. Torrieri, Statistical theory of passive location systems, *IEEE Trans. Aerosp. Electron. Syst.*, **2 (AES-20)**, (1984), 183–198.
175. N. Levanon, Lowest GDOP in 2-D scenarios. *IEE Proc.-Radar Sonar Navig.*, **147**: 3, (2000), 149–155.
176. W. H. Foy, Position-location solutions by Taylor-series estimation. *IEEE Trans. Aerosp. Electron. Syst.*, **AES-12**, (1976), 187–194.
177. S. Arulampalam, S. Maskell, N. Gordon and T. Clapp, A tutorial on particle filters for on-line non-linear/non-Gaussian Bayesian tracking. *IEEE Trans. Sig. Processing*, **50**: 2, (2002), 174–188.
178. D. Fox, J. Hightower, L. Liao, D. Schulz and G. Borriello, Bayesian filtering for location estimation. *IEEE Pervasive Computing*, **2**: 3, (2003), 24–33.
179. R. E. Kalman and R. S. Bucy, New results in linear filtering and prediction theory. *Trans. of the ASME – J. Basic Engng*, **83**, (1961), 95–107.
180. A. H. Jazwinski, *Stochastic Process and Filtering Theory*. (New York: Academic Press, 1970).
181. S. Julier, A skewed approach to filtering. *Signal Data Proc. Small Targets*, **3373**, (1998), 271–282.
182. E. A. Wan and R. V. der Merwe, The unscented Kalman filter for non-linear estimation. In *Proc. Symp. 2000 on Adaptive Syst. for Sig. Processing, Commun. and Control*, Lake Louise, Alberta, Canada, Oct. 2000.
183. Y. Bar-Shalom and X.-R. Li, *Multitarget-Multisensor Tracking: Principles and Techniques*. (Yaakov Bar-Shalom, 1995).
184. L. R. Rabiner and B. H. Juang, An introduction to hidden Markov models. *IEEE Advances in Speech Sig. Processing (ASSP) Mag.*, **3**, (1996), 4–16.
185. R. L. Streit and R. F. Barrett, Frequency line tracking using hidden Markov models. *Proc. IEEE Trans. Accoust., Speech and Sig. Proc.*, **38**: 4, (1990), 586–598.
186. L. R. Rabiner, A tutorial on hidden Markov models and selected applications in speech recognition. *Proc. IEEE*, **77**: 2, (1989), 257–285.
187. A. Doucet, N. D. Freitas and N. Gordon (eds.), *Sequential Monte Carlo Methods in Practice*. (New York: Springer, 2005).
188. F. Gustafsson, F. Gunnarsson, N. Bergman, *et al.*, Particle filters for positioning, navigation, and tracking. *IEEE Trans. Sig. Processing*, **50**, (2002), 425–435.
189. J. Vermaak, N. Ikoma and S. J. Godsill, Sequential Monte Carlo framework for extended object tracking. *IEE Proc.-Radar Sonar Navig.*, **152**: 5, (2005), 353–363.
190. A. J. Haug, A tutorial on Bayesian estimation and tracking techniques applicable to nonlinear and non-gaussian processes. *MITRE Technical Report MTR 05W0000004*, Sep. 2005.
191. S. J. Godsill and J. Vermaak, Variable rate particle filters for tracking applications. In *Proc. IEEE/SP Workshop on Statistical Sig. Processing*, Bordeaux, France, July 2005, pp. 1280–1285.
192. H. Sidenbladh and S. L. Wirkander, Tracking random sets of vehicles in terrain. In *Proc. IEEE Workshop on Multi-Object Tracking*, Madison, WI, June 2003.

193. S. Sarkka, A. Vehtari and J. Lampinen, Rao-Blackwellized Monte Carlo data association for multiple target tracking. In *Proc. Int. Conf. on Information Fusion (FUSION)*, Stockholm, Sweden, June 2004.

194. D. B. Jourdan, J. J. Deyst Jr., M. Z. Win and N. Roy, Monte Carlo localization in dense multipath environments using UWB ranging. In *Proc. IEEE Int. Conf. UWB (ICU)*, Zurich, Switzerland, Sep. 2005, pp. 314–319.

195. B. Denis, L. Ouvry, B. Uguen and F. Tchoffo-Talom, Advanced Bayesian filtering techniques for UWB tracking systems in indoor environments. In *Proc. IEEE Int. Conf. UWB (ICU)*, Zurich, Switzerland, Sep. 2005, pp. 638–643.

196. S. Gezici, Z. Sahinoglu, H. Kobayashi and H. V. Poor, *Ultra Wideband Geolocation*. In *Ultra-wideband Wireless Communications*. (John Wiley & Sons, Inc., 2005).

197. S. Venkatesh and R. M. Buehrer, A linear programming approach to NLOS error mitigation in sensor networks. In *Proc. IEEE Int. Symp. Information Processing in Sensor Networks (IPSN)*, Nashville, Tennessee, Apr. 2006.

198. B. Allen, M. Dohler, E. E. Okon, W. Q. Malik, A. K. Brown and D. J. Edwards, *Ultra-wideband Antennas and Propagation for Communications, Radar, and Imaging*. (West Sussex, England: John Wiley & Sons, Ltd., 2007), p. 393.

199. F. Chin, Impulse radio signaling for communication and ranging, April, 2004, doc.: IEEE 802.15-05-0231-00-004a. [Online]. Available: http://802wirelessworld.com

200. I. Guvenc, Z. Sahinoglu, A. F. Molisch and P. Orlik, Non-coherent TOA estimation in IR-UWB systems with different signal waveforms. In *Proc. IEEE Int. Workshop on Ultra-wideband Networks (UWBNETS)*, Boston, MA, October 2005, pp. 245–251, (invited paper).

201. Z. Sahinoglu and I. Guvenc, Multiuser interference mitigation in noncoherent UWB ranging via nonlinear filtering. *EURASIP J. Wireless Commun. Networking*, Article ID 56 849, 10 pages, (2006).

202. S. Gezici and Z. Sahinoglu, UWB geolocation techniques for IEEE 802.15.4a personal area networks. MERL technical report, Cambridge, MA, Aug. 2004.

203. D. B. Jourdan and N. Roy, Optimal sensor placement for agent localization. In *Proc. IEEE Position, Location, and Navigation Symp. (PLANS)*, San Diego, CA, Apr. 2006, pp. 128–139.

204. V. Dizdarevic and K. Witrisal, On impact of topology and cost function on LSE position determination in wireless networks. In *Proc. Workshop on Positioning, Navigation, and Commun. (WPNC)*, Hannover, Germany, Mar. 2006, pp. 129–138.

205. J. Riba and A. Urruela, A non-line-of-sight mitigation technique based on ML-detection. In *Proc. IEEE Int. Conf. Acoust., Speech, and Sig. Processing (ICASSP)*, vol. 2, Quebec, Canada, May 2004, pp. 153–156.

206. B. Denis, J. B. Pierrot and C. Abou-Rjeily, Joint distributed synchronization and positioning in UWB ad hoc networks using TOA. *IEEE Trans. Microw. Theory and Techniques*, **54**: 4, (2006), 1896–1911.

207. D. B. Jourdan, D. Dardari and M. Z. Win, Position error bound for UWB localization in dense cluttered environments. In *Proc. IEEE Int. Conf. Commun. (ICC)*, vol. 8, Istanbul, Turkey, June 2006, pp. 3705–3710.

208. M. P. Wylie and J. Holtzman, The non-line of sight problem in mobile location estimation. In *Proc. IEEE Int. Conf. Universal Personal Commun.*, Cambridge, MA, Sep. 1996, pp. 827–831.

209. X. Diao and F. Guo, A method distinguishing line of sight (LOS) from non-line-of-sight (NLOS) in CDMA mobile communication system. Europe Patent No. 02706591.1, Mar. 2002.
210. A. Rabbachin, I. Oppermann and B. Denis, ML time-of-arrival estimation based on low complexity UWB energy detection. In *Proc. IEEE Int. Conf. Ultra-Wideband (ICUWB)*, Waltham, MA, Sep. 2006, pp. 598–604.
211. Y. Shimizu and Y. Sanada, Accuracy of relative distance measurement with ultra wideband system. In *Proc. IEEE Conf. Ultra-Wideband Syst. Technol. (UWBST)*, Reston, VA, Nov. 2003, pp. 374–378.
212. F. Coppens, First arrival picking on common-offset trace collections for automatic estimation of static corrections. *Geophys. Prospecting*, **33**: 8, (1985), 1212–1231.
213. R. E. Jativa and J. Vidal, Coarse first arriving path detection for subscriber location in mobile communication systems. In *Proc. European Sig. Processing Conf. (EUSIPCO)*, Toulouse, France, Sep. 2002, pp. 2733–2736.
214. J. Vidal and R. E. Jativa, First arriving path detection for subscriber location in mobile communication systems. In *Proc. IEEE Conf. on Acoust., Speech, and Sig. Processing (ICASSP)*, vol. 3, Orlando, FL, May 2002, pp. 2733–2736.
215. J. Vidal, M. Najar and R. E. Jativa, High resolution time-of-arrival detection for wireless positioning systems. In *Proc. IEEE Veh. Technol. Conf. (VTC)*, vol. 4, Vancouver, Canada, Sep. 2002, pp. 2283–2287.
216. J. J. van de Beek, P. O. Brjesson, H. Eriksson, J.-O. Gustavsson and L. Olsson, MMSE estimation of arrival time with application to ultrasonic signals. Research Report, Apr. 1993. [Online]. Available: citeseer.ist.psu.edu/vandebeek93mmse.html
217. T. G. Manickam, R. J. Vaccaro and D. W. Tufts, A least-squares algorithm for multipath time-delay estimation. *IEEE Trans. Sig. Processing*, **42**: 11, (1994), 3229–3233.
218. H. Saarnisaari, ML time delay estimation in a multipath channel. In *Proc. IEEE Int. Symp. Spread Spectrum Techniques and Applications*, Mainz, Germany, Sep. 1996, pp. 1007–1011.
219. L. Reggiani and G. M. Maggio, Rapid search algorithms for code acquisition in UWB impulse radio communications. *IEEE J. Select. Areas Commun.*, **23**: 5, (2005), 898–908.
220. Z. Tian and V. Lottici, Efficient timing acquisition in dense multipath for UWB communications. In *Proc. IEEE Veh. Technol. Conf. (VTC)*, vol. 2, Orlando, FL, Oct. 2003, pp. 1318–1322.
221. Z. Tian and L. Wu, Timing acquisition with noisy template for ultra-wideband communications in dense multipath. *EURASIP J. Applied Sig. Processing*, **2005**: 3, (2005), 439–454.
222. Z. Tian and G. B. Giannakis, A GLRT approach to data-aided timing acquisition in UWB radios – Part I: Algorithms. *IEEE Trans. Wireless Commun.*, **4**: 6, (2005), 2956–2967.
223. Z. Tian and G. B. Giannakis, A GLRT approach to data-aided timing acquisition in UWB radios – Part II: Training sequence design. *IEEE Trans. Wireless Commun.*, **4**: 6, (2005), 2994–3004.
224. L. Yang and G. B. Giannakis, Blind UWB timing with dirty templates. In *Proc. IEEE Conf. on Acoust., Speech, Sig. Processing (ICASSP)*, vol. 4, Quebec, Canada, May 2004, pp. 509–512.
225. J. Yu and Y. Yao, Detection performance of time-hopping ultra-wideband LPI waveforms. In *Proc. IEEE Sarnoff Symp.*, Princeton, NJ, April 2005, pp. 137–140.
226. A. Rabbachin and I. Oppermann, Synchronization analysis for UWB systems with a low-complexity energy collection receiver. In *Proc. IEEE Ultra-wideband Syst. Technol. (UWBST)*, Kyoto, Japan, May 2004, pp. 288–292.

227. C. Falsi, D. Dardari, L. Mucchi and M. Z. Win, Time of arrival estimation for UWB localizers in realistic environments. *EURASIP J. Applied Sig. Processing*, vol. 2006, Article ID32082, 13 pp. 2006. doi:10,1155/ASP/2006/32082.
228. D. Dardari and M. Z. Win, Threshold-based time-of-arrival estimators in UWB dense multipath channels. In *Proc. IEEE Int. Conf. Commun. (ICC)*, vol. 10, Istanbul, Turkey, June 2006, pp. 4723–4728.
229. I. Guvenc, Z. Sahinoglu and P. Orlik, TOA estimation for IR-UWB systems with different transceiver types. *IEEE Trans. Microw. Theory and Techniques (Special Issue on Ultrawideband)*, **54**: 4, (2006), 1876–1886.
230. D. Dardari, C. C. Chong and M. Z. Win, Analysis of threshold-based TOA estimator in UWB channels. In *Proc. European Sig. Processing Conf. (EUSIPCO)*, Florence, Italy, Sep. 2006.
231. D. Dardari, C. C. Chong and M. Win, Threshold-based time-of-arrival estimators in UWB dense multipath channels. *to appear in IEEE Trans. Commun.*, (2007).
232. I. Guvenc and Z. Sahinoglu, Threshold-based TOA estimation for impulse radio UWB systems. In *Proc. IEEE Int. Conf. UWB (ICU)*, Zurich, Switzerland, Sep. 2005, pp. 420–425.
233. J. G. Proakis, *Digital Communications*, 4th ed. (New York: McGraw-Hill, 2001).
234. P. A. Humblet and M. Azizoglu, On the bit error rate of lightwave system with optical amplifiers. *J. of Lightwave Technol.*, **9**: 11, (1991), 1576–1582.
235. A. J. Weiss and E. Weinstein, Fundamental limitations in passive time delay estimation – part I: Narrow-band systems, *IEEE Trans. Acoust., Speech, Sig. Processing*, **31**: 2, (1983), 472–486.
236. E. Weinstein and A. J. Weiss, Fundamental limitations in passive time delay estimation – part II: Wide-band systems. *IEEE Trans. Acoust., Speech, Sig. Processing*, **32**: 5, (1984), 1064–1078.
237. L. Huang, Z. Xu and B. M. Sadler, Simplified Ziv-Zakai time delay estimation bound for ultra-wideband signals. In *Proc. Workshop on Short Range Ultra-Wideband Radio Syst.*, Santa Monica, CA, Apr. 2006.
238. D. Chazan, M. Zakai and J. Ziv, Improved lower bounds on signal parameter estimation. *IEEE Trans. Inform. Theory*, **21**: 1, (1975), 90–93.
239. E. Weinstein and A. J. Weiss, A general class of lower bounds in parameter estimation. *IEEE Trans. Inform. Theory*, **34**: 2, (1988), 338–342.
240. A. Zeira and P. M. Schultheiss, Realizable lower bounds for time delay estimation. *IEEE Trans. Sig. Processing*, **41**: 11, (1993), 3102–3113.
241. A. Zeira and P. M. Schultheiss, Realizable lower bounds for time delay estimation: Part 2 – threshold phenomena. *IEEE Trans. Sig. Processing*, **32**: 5, (1994), 1001–1007.
242. K. L. Bell, Y. Steinberg, Y. Ephraim and H. L. V. Trees, Extended Ziv-Zakai lower bound for vector parameter estimation. *IEEE Trans. Inform. Theory*, **43**: 2, (1997), 624–637.
243. S. Gezici, Z. Tian, G. B. Giannakis, *et al.*, Localization via ultra-wideband radios: A look at positioning aspects for future sensor networks. *IEEE Sig. Processing Mag.*, **22**: 4, (2005), 70–84.
244. D. Dardari, C. C. Chong and M. Z. Win, Improved lower bounds on time-of-arrival estimation error in realistic UWB channels. In *Proc. IEEE Int. Conf. on Ultra-wideband (ICUWB)*, Weltham, MA, Sep. 2006, pp. 531–537.
245. V. Lottici, A. D. Andrea and U. Mengali, Channel estimation for ultra-wideband communications. *IEEE J. Select. Areas Commun. (JSAC)*, **20**: 12, (2002), 1638–1645.
246. M. Moeneclaey, A fundamental lower bound on the performance of practical joint carrier and bit synchronizers. *IEEE Trans. Commun.*, **32**: 9, (1984), 1007–1012.

247. J. Zhang, R. A. Kennedy and T. D. Abhayapala, Cramer–Rao lower bounds for the time delay estimation of UWB signals. In *Proc. IEEE Int. Conf. Commun. (ICC)*, Paris, France, May 2004, pp. 3424–3428.
248. B. T. Sieskul and T. Kaiser, Cramer–Rao bound for TOA estimation in UWB positioning systems. In *Proc. IEEE Int. Conf. on Ultra-wideband (ICUWB)*, Weltham, MA, Sep. 2006, pp. 408–413.
249. H. Celebi and H. Arslan, Ranging accuracy in dynamic spectrum access networks. *IEEE Commun. Lett.*, **11**: 5, (2007), 405–407.
250. H. L. V. Trees, *Detection, Estimation, and Modulation Theory: Part I*, 2nd ed. (New York, NY: John Wiley & Sons, Inc., 2001).
251. H. Anouar, A. M. Hayar, R. Knopp and C. Bonnet, Ziv-Zakai lower bound on the time delay estimation of UWB signals. In *Proc. Int. Symp. on Commun., Control, and Sig. Processing (ISCCSP)*, Marrakech, Morocco, Mar. 2006.
252. M. Z. Win and R. A. Scholtz, Characterization of ultra-wideband width wireless indoor channels: A communication theoretic view. *IEEE J. Select. Areas. Commun.*, **20**: 9, (2002), 1613–1627.
253. J.-Y. Lee and R. A. Scholtz, Ranging in a dense multipath environment using an UWB radio link. *IEEE J. Select. Areas Commun.*, **20**: 9, (2002), 1677–1683.
254. R. F. Nau, Statistical forecasting (course notes), May 2005. [Online]. Available: http://www.duke.edu/rnau/compare.htm
255. S. M. Kay, *Fundamentals of Statistical Signal Processing: Detection Theory*. (Upper Saddle River, NJ: Prentice Hall, Inc., 1998).
256. I. Guvenc and Z. Sahinoglu, TOA estimation with different IR-UWB transceiver types. In *Proc. IEEE Int. Conf. UWB (ICU)*, Zurich, Switzerland, Sep. 2005, pp. 426–431.
257. S. Verdu, *Multiuser Detection*. 1st ed. (Cambridge, UK: Cambridge University Press, 1998).
258. S. Gezici, Z. Sahinoglu, H. Kobayashi, H. V. Poor and A. F. Molisch, A two-step time of arrival estimation algorithm for impulse radio ultra-wideband systems. In *Proc. European Sig. Processing Conf. (EUSIPCO)*, Antalya, Turkey, Sep. 2005.
259. L. Yang and G. B. Giannakis, Ultra-wideband communications: An idea whose time has come. *IEEE Sig. Processing Mag.*, **21**: 6, (2004), 26–54.
260. L. Yang and G. B. Giannakis, Timing ultra-wideband signals with dirty templates. *IEEE Trans. Commun.*, **53**: 11, (2005), 1952–1963.
261. R. A. Scholtz and J. Y. Lee, Problems in modeling UWB channels. In *Proc. IEEE Asilomar Conf. Signals, Syst. Computers*, vol. 1, Pacific Grove, CA, Nov. 2002, pp. 706–711.
262. A. S. Tanenbaum, *Computer Networks*, 3rd ed. (Upper Saddle River, NJ: Prentice Hall, 1996).
263. IEEE 802.15.4b standard specification. Jan. 2007. [Online]. Available: http://www.ieee802.org/15/pub/TG4b.html
264. R. Fontana and S. J. Gunderson, Ultra-wideband precision asset location system. In *Proc. IEEE Conf. on Ultra-wideband Syst. and Technol. (UWBST)*, Baltimore, MD, May 2002, pp. 147–150.
265. R. Hach, Symmetric double sided two-way ranging. Jun. 2005, doc: IEEE 15-05-0334-00-004a. [Online]. Available: http://www.ieee802.org/15/pub/TG4a.html
266. R. J. Fontana, Experimental results from an ultra-wideband precision geolocation system. In *Multispectral Solutions, Inc. EuroEM Geolocation*, May 2000, pp. 1–6.
267. D. Kelly, G. Shreve and D. Langford, Fusing communications and positioning - Ultra-wideband offers exciting possibilities. (Time Domain Corporation, 1998).

268. D. Bertsekas and R. Gallager, *Data Networks*, 2nd ed. (Upper Saddle River, NJ: Prentice Hall, 1992).
269. I. Guvenc and H. Arslan, UWB channel estimation with various sampling rate options. In *Proc. IEEE Sarnoff Symp.*, Apr. 2005, pp. 1–6.
270. R. Qiu, A study of the ultra-wideband wireless propagation channel and optimum UWB receiver design. *IEEE J. Select. Areas Commun.*, **20**: 9, (2002), 1628–1637.
271. F. Chin, Impulse radio signalings, communication and ranging, Jul. 2005, doc: IEEE 15-05-0231-07-004a. [Online]. Available: http://www.ieee802.org/15/pub/TG4a.html
272. M. Negreiros, L. Carlo and A. A. Susin, An improved RF loopback for test time reduction. In *Proc. Design, Automation and Test in Europe*, Munich, Germany, Mar. 2006, pp. 1–6.
273. J. S. Yoon and W. R. Eisenstadt, Embedded loopback test for RF ICs. *IEEE Trans. Instr. Meas.*, **54**: 5, (2006), 1715–1720.
274. G. Srinivasan, A. Chatterjee and F. Taenzler, Alternate loop-back diagnostic tests for wafer-level diagnosis of modern wireless transceivers using spectral signatures. In *Proc. IEEE VLSI Test Symp.*, Berkeley, CA, May 2006, pp. 68–73.
275. S. Bhattacharya and A. Chatterjee, A built-in loopback test methodology for RF transceiver circuits using embedded sensor circuits. In *Proc. Asian Test Symp. (ATS)*, Taiwan, 2004, pp. 68–73.
276. Y. Qi, B. Zhen, H. B. Li, S. Hara and R. Kohno, Optional MAC protocol: TDMA type multiplexed preamble, Nov. 2005, doc.: IEEE 15-05-0698-00-004a. [Online]. Available: http://www.ieee802.org/15/pub/TG4a.html
277. X. Chu and R. D. Murch, The effect of NBI on UWB time hopping systems. *IEEE Trans. Wireless Commun.*, **3**: 5, (2004), 1431–1436.
278. Z. Lou, H. Gao, Y. Liu and J. Gao, A new UWB pulse design method for narrowband interference suppression. In *Proc. IEEE Global Telecommun. Conf. (GLOBECOM)*, Dallas, TX, Dec. 2004, pp. 3488–3492.
279. K. Taniguchi and R. Kohno, Design and analysis of synthesized template waveform for receiving UWB signals. *IEICE Trans. Fundamentals*, **E88**: 9, (2005), 2299–2309.
280. K. Ohno and T. Ikegami, Interference mitigation study for UWB radio using template waveform processing. *IEEE Trans. Microw. Theory and Techniques*, **54**: 4, (2006), 1782–1792.
281. Y. Wang, X. Dong and I. J. Fair, A method for spectrum shaping and NBI suppression in UWB communications. In *Proc. IEEE Int. Conf. Commun. (ICC)*, Istanbul, Turkey, Jun. 2006, pp. 1476–1481.
282. N. C. Beaulieu and B. Hu, A soft-limiting receiver structure for time-hopping UWB in multiple access interference. In *Proc. Int. Symp. on Spread Spectrum Techniques and Applications (ISSSTA)*, Manaus, Brazil, Aug. 2006, pp. 417–421.
283. A. Taha and K. M. Chugg, A theoretical study on the effects of interference on UWB multiple access impulse radio. In *Proc. Asilomar Conf. Signals, Syst. and Computers*, Pacific Grove, CA, Nov. 2002, pp. 728–732.
284. M. Win, Ultra-wideband bandwidth time-hopping spread-spectrum impulse radio for wireless multiple-access communications. *IEEE Trans. Commun.*, **48**: 4, (2000), 679–691.
285. G. Durisi and S. Benedetto, Performance evaluation of TH-PPM UWB systems in the presence of multiuser interference. *IEEE Commun. Lett.*, **7**: 5, (2003), 224–226.
286. Y. Dhibi and T. Kaiser, On the impulsiveness of multiuser intereferences in TH-PPM UWB systems. *IEEE Trans. Sig. Processing*, **54**: 7, (2006), 2853–2857.

287. D. Middleton, Statistical models of electromagnetic interference. *IEEE Trans. Electromagn. Compat.*, **EMC-19**: 3, (1977), 106–127.
288. H. G. Senel, R. A. Peters and B. Dawant, Topological median filters. *IEEE Trans. Image Proc.*, **11**: 2, (2002), 89–104.
289. N. C. Gallagher and G. L. Wise, A theoretical analysis of the properties of median filters. *IEEE Trans. Acous. Speech Sig. Proc.*, **29**: 6, (1981), 1136–1141.
290. Z. Sahinoglu and I. Guvenc, Interference suppression in non-coherent time-hopping IR-UWB ranging. In *IEEE Int. Conf. Ultra-wideband (ICUWB)*, Waltham, MA, Sep. 2006, pp. 507–511.
291. D. Dardari, A. Giorgetti and M. Win, Time-of-arrival estimation in the presence of narrow and wide bandwidth interference in UWB channels. To appear in *IEEE Int. Conf. on Ultra-wideband (ICUWB)*, Marina Bay, Singapore, Sep. 2007.
292. I. Ramachandran and S. Roy, Clear channel assessment in energy constrained wideband wireless networks. *IEEE Wireless Commun. Mag.*, **14**: 3, (2007), 70–78.
293. Y. Qi, B. Zhen, H. Li, S. Hara and R. Kohno, The TDMA type multiplexed preamble. Nov. 2005, doc:IEEE 15-05-0698-00-004a. [Online]. Available: http://www.ieee802.org/15/pub/TG4a.html
294. S. Capkun and J. Hubaux, Secure positioning of wireless devices with applications to sensor networks. In *Proc. IEEE Int. Conf. on Computer Commun. (INFOCOM)*, Miami, FL, Mar 2005, pp. 1917–1928.
295. L. Lazos and R. Poovendran, SeRLoc: Secure range-independent localization for wireless sensor networks. In *Proc. Int. Conf. on Web Information Syst. Engineering (WiSe)*, Brisbane, Australia, Oct 2004, pp. 21–30.
296. Y. Zhang, W. Liu, Y. Fang and D. Wu, Secure localization and authentication in ultra-wideband sensor networks. *IEEE J. Select. Areas Commun.*, **24**: 4, (2006), 829–835.
297. I. Lakkis, Pulse compression for TG4a. Doc.: IEEE 15-05-0456-01-004a, July, 2005. [Online]. Available: http://www.ieee802.org/15/pub/2005/15-05-0456-00-004a-pulse-compression.ppt
298. Y.-S. Kwok, F. Chin and X. Peng, Ranging mechanism, preamble generation, and performance with IEEE 802.15.4a low-rate low-power UWB systems. In *Proc. IEEE Int. Conf. Ultra-wideband (ICUWB)*, Waltham, MA, Sep. 2006, pp. 525–530.
299. D. Wu, P. Spasojevic and I. Seskar, Ternary complementary sets for multiple channel DS-UWB with reduced peak to average power ratio. In *Proc. IEEE Global Telecommun. Conf. (GLOBECOM)*, vol. 5, Dec. 2004, pp. 3230–3234.
300. D. Wu, P. Spasojevic and I. Seskar, Ternary complementary sets for orthogonal pulse based UWB. In *Proc. Asilomar Conf. on Signals, Syst., and Computers*, vol. 2, Pacific Grove, CA, Nov. 2003, pp. 1776–1780.
301. I. Lakkis and S. Safavi, Band plan, PRF, preamble & modulation for TG4a, Doc.: IEEE 15-05-0250-03-004a, Apr., 2005. [Online]. Available: http://grouper.ieee.org/groups/802/15/pub/2005/15-05-0250-01-004a-revised-frequency-plan-and-prf-proposal-tg4a.ppt
302. S. R. Saunders, *Antennas and Propagation for Wireless Communication Systems*. (Chichester, West Sussex: Wiley; 1999).
303. J. O. Nielsen, G. F. Pedersen, K. Olesen and I. Z. Kovacs, Statistics of measured body loss for mobile phones. *IEEE Trans. on Antennas and Propag.*, **49**, (2001), 1351–1353.
304. H. Kim, Y. Joo and S. Jung, A tunable CMOS UWB pulse generator. In *Proc. IEEE Int. Conf. on Ultra-Wideband (ICUWB)*, Waltham, MA, Sep. 2006, pp. 109–112.

305. H. Kim and D. P. Y. Joo, All-digital low-power CMOS pulse generator for UWB system. *IEE Electronics Letters*, **40**: (2004), 1534–1535.
306. Y. Shimizu, Y. Takeuchi and Y. Sanada, Novel positioning system with extremely low power radio. In *Proc. IEEE Conf. Ultra-wideband Syst. Technol. (UWBST)*, Kyoto, Japan, May 2004, pp. 391–394.
307. W. Yao and Y. Wang, Direct antenna modulation – a promise for ultra-wideband (UWB) transmitting. In *Proc. IEEE MTT-S Int. Microw. Symp.*, vol. 2, Fort Worth, TX, June 2004, pp. 1273–1276.
308. S. D. Keller, W. D. Palmer and W. T. Joines, Direct antenna modulation: analysis, design, and experiment. In *Proc. IEEE Antennas and Propag. Society Int. Symp.*, Albuquerque, NM, July 2006, pp. 909–912.
309. W. Yao and Y. E. Wang, Radiating beyond the bandwidth using direct antenna modulation. In *Proc. IEEE Antennas and Propag. Society Int. Symp.*, vol. 1, Monterey, CA, June 2004, pp. 791–794.
310. L. Smaini, C. Tinella, D. Helal, *et al.*, Single-chip CMOS pulse generator for UWB systems. *IEEE J. of Solid-State Circuits*, **41**, (2006), 1551–1561.
311. Y. Jeong, S. Jung and J. Liu, A CMOS impulse generator for UWB wireless communication systems. In *Proc. IEEE Int. Symp. on Circuits and Syst. (ISCAS)*, vol. 4, Vancouver, Canada, May 2004, pp. 129–132.
312. K. Li, UWB bandpass filter: structure, performance and application to UWB pulse generation. In *Proc. Asia-Pacific Microw. Conf. (APMC)*, vol. 1, Suzhou, China, Dec. 2005.
313. K. Li, Experimental study on UWB pulse generation using UWB bandpass filters. In *Proc. IEEE Int. Conf. on Ultra-Wideband (ICUWB)*, Waltham, MA, Sep. 2006, pp. 103–108.
314. B. Jung, Y.-H. Tseng, J. Harvey and R. Harjani, Pulse generator design for UWB IR communication systems. In *Proc. IEEE Int. Symp. on Circuits and Syst. (ISCAS)*, vol. 5, Kobe, Japan, May 2005, pp. 4381–4384.
315. X. Zhang, S. Ghosh and M. Bayoumi, A low power CMOS UWB pulse generator. In *Proc. Midwest Symp. on Circuits and Syst.*, vol. 2, Cincinnati, OH, Aug. 2005, pp. 1410–1413.
316. A. Azakkour, M. Regis, F. Pourchet and G. Alquie, A new integrated monocycle generator and transmitter for ultra-wideband communications. In *Proc. IEEE Radio Frequency Integrated Circuits (RFIC) Symp.*, Long Beach, CA, June 2005, pp. 79–82.
317. S. Iida, K. Tanaka, H. Suzuki, *et al.*, A 3.1 to 5 GHz CMOS DSSS UWB transceiver for WPANs. In *Proc. IEEE Int. Solid-State Circuits Conf. (ISSCC)*, vol. 1, San Francisco, CA, Feb. 2005, pp. 214–215.
318. D. D. Wentzloff and A. P. Chandrakasan, Gaussian pulse generators for sub-banded ultra-wideband transmitters. *IEEE Trans. Microw. Theory Techniques*, **54**, (2006), 1647–1655.
319. T. Terada, S. Yoshizumi, M. Muqsith, Y. Sanada and T. Kuroda, A CMOS ultra-wideband impulse radio transceiver for 1-Mb/s data communications and ±2.5-cm range finding. *IEEE J. of Solid-State Circuits*, **41**, (2006), 891–898.
320. R. Harjani, J. Harvey and R. Sainati, Analog/RF physical layer issues for UWB systems. In *Proc. Int. Conf. on VLSI Design*, Mumbai, India, Jan. 2004, pp. 941–948.
321. H. C. Hsu, Z. W. Wang and G. K. Ma, A low power CMOS full-band UWB power amplifier using wideband RLC matching method. In *Proc. IEEE Int. Conf. on Electron Devices and Solid-State Circuits*, Washington, DC, Dec. 2005, pp. 233–236.
322. G. L. Puma, A. Wiesbauer and C. Sander, Fully integrated power amplifier in CMOS technology, optimized for UWB transmitters. In *Proc. IEEE Radio Frequency Integrated Circuits (RFIC) Symp.*, Fort Worth, TX, June 2004, pp. 87–90.

323. X. Guan and C. Nguyen, A 0.25-μm CMOS ultra-wideband amplifier for time-domain UWB applications. In *Proc. IEEE Radio Frequency Integrated Circuits (RFIC) Symp.*, Long Beach, CA, June 2005, pp. 339–342.
324. B. M. Ballweber, R. Gupta and D. J. Allstot, A fully integrated 0.5-5.5 GHz CMOS distributed amplifier. *IEEE Trans. Solid-State Circuits*, **35**: 2, (2000), 231–239.
325. R.-C. Liu, K.-L. Deng and H. Wang, A 0.6-22 GHz broadband CMOS distributed amplifier. In *Proc. IEEE Radio Frequency Integrated Circuits (RFIC) Symp.*, Philadelphia, PN, June 8–10 2003, pp. 103–106.
326. S. Andersson, C. Svensson and O. Drugge, Wideband LNA for a multistandard wireless receiver in 0.18 μm CMOS. In *Proc. European Solid-State Circuits Conf. (ESSCIRC)*, Lisbon, Portugal, Sep. 16–18 2003, pp. 655–658.
327. F. Bruccoleri, E. A. M. Klumperink and B. Nauta, Noise canceling in wideband CMOS LNAs. In *Proc. Int. Solid-State Circuits Conf.*, vol. 45, San Francisco, CA, Feb. 2002, pp. 406–407.
328. A. Bevilacqua and A. M. Niknejad, An ultra-wideband CMOS LNA for 3.1 to 10.6 GHz wireless receiver. In *Proc. Int. Solid-State Circuits Conf.*, vol. 45, San Francisco, CA, Feb. 2004, pp. 382–383.
329. C. Lu, A. V. Pham and M. Shaw, A CMOS power amplifier for full-band UWB transmitters. In *Proc. IEEE Radio Frequency Integrated Circuits (RFIC) Symp.*, Honolulu, HI, June 11–13 2006.
330. A. Ismail and A. Abidi, 10 GHz LNA using a wideband LC-ladder matching network. In *Proc. Int. Solid-State Circuits Conf.*, vol. 45, San Francisco, CA, Feb. 2004, pp. 384–385.
331. Z. N. Chen, M. J. Ammann, X. Qing, X. H. Wu, T. S. P. See and A. Cai, Planar antennas. *IEEE Microw. Mag.*, **7**: 6, (2006), 63–73.
332. J. Powell, Antenna design for ultra wideband radio. M.S. thesis, Massachusetts Institute of Technology, May 2004.
333. Z. N. Chen and M. Y. W. Chia, *Broadband Planar Antennas: Design and Applications*. (West Sussex, UK: Wiley, 2006).
334. D. Lamensdorf and L. Susman, Baseband-pulse-antenna techniques. *IEEE Antennas Propag. Mag.*, **36**: 1, (1994), 20–30.
335. X. H. Wu, Z. N. Chen and N. Yang, Optimization of planar diamond antenna for single/multi-band UWB wireless communications. *Microw. Optical Technol. Letters*, **42**: 6, (2004), 451–455.
336. M. J. Ammann and Z. N. Chen, A wideband shorted planar monopole with bevel. *IEEE Trans. Antennas Propag.*, **51**: 4, (2003), 901–903.
337. D. Ghosh, A. De, M. C. Taylor, T. K. Sarkar, M. C. Wicks and E. L. Mokole, Transmission and reception by ultra-wideband (UWB) antennas. *IEEE Antennas Propag. Mag.*, **48**: 5, (2006), 67–99.
338. Z. N. Chen, M. J. Ammann and M. Y. W. Chia, Broadband square annular planar monopoles. *Microw. Optical Technol. Lett.*, **36**: 6, (2003), 449–454.
339. D. Valderas, J. Melendez and I. Sancho, Some design criteria for UWB planar monopole antennas: Application to a slotted rectangular monopole. *Microw. Optical Technol. Lett.*, **46**: 1, (2005), 6–11.
340. M. J. Ammann and Z. N. Chen, An asymmetrical feed arrangement for improved impedance bandwidth of planar monopole antennas. *Microw. Optical Technol. Lett.*, **40**: 2, (2004), 156–158.
341. E. Antonino-Daviu, M. Cabedo-Fabres, M. Ferrando-Bataller and A. Valero-Nogueira, Wideband double-fed planar monopole antennas. *Electron. Lett.*, **39**: 23, (2003), 1635–1636.

342. Z. N. Chen, M. J. Ammann, M. Y. W. Chia and T. S. P. See, Circular annular planar monopoles with EM coupling. *Proc. IEE Microw. Antennas, Propag.*, **150**, (2003), 269–273.
343. R. T. Lee and G. S. Smith, On the characteristic impedance of the TEM horn antenna. *IEEE Trans. Antennas Propag.*, **52**, (2004), 315–318.
344. K. L. Shlager, G. S. Smith and J. Maloney, Accurate analysis of TEM horn antennas for pulse radiation. *IEEE Trans. Electromagn. Compat.*, 38: 3, (1996), 414–423.
345. J. R. Andrews, UWB signal sources, antennas and propagation. In *Picosecond Pulse Labs, Application Note AN-14a*, Aug. 2003, pp. 1–11.
346. L. T. Chang and W. D. Burnside, An ultrawide-bandwidth tapered resistive TEM horn antenna. *IEEE Trans. Antennas Propag.*, **48**: 12, (2000), 1848–1857.
347. B. Scheers, M. Acheroy and A. V. Vorst, Time-domain simulation and characterisation of TEM horns using a normalised impulse response. *Proc. IEE Microw., Antennas, Propag.*, **147**, (2000), 463–468.
348. S. S. Sandler and R. W. P. King, Compact conical antennas for wide-band coverage. *IEEE Trans. Antennas Propag.*, **42**: 3, (1994), 436–439.
349. S. N. Samaddar and E. L. Mokole, Biconical antennas with unequal cone angles. *IEEE Trans. Antennas Propag.*, 46: 2, (1998), 181–193.
350. T. T. Wu and R. W. P. King, The cylindrical antenna with nonreflecting resistive loading. *IEEE Trans. Antennas Propag.*, **13**: 3, (1965), 369–373.
351. J. G. Maloney and G. S. Smith, A study of transient radiation from the Wu-King resistive monopole–FDTD analysis and experimental measurements. *IEEE Trans. Antennas Propag.*, **41**: 5, (1993), 668–676.
352. K. Noguchi, S. I. Betsudan, T. Katagi and M. Mizusawa, A compact broad-band helical antenna with two-wire helix. *IEEE Trans. Antennas Propag.*, **51**: 9, (2003), 2176–2181.
353. M. A. Peyrot-Solis, G. M. Galvan-Tejada and H. Jardon-Aguilar, State of the art in ultra-wideband antennas. In *Proc. Int. Conf. on Electrical and Electronics Engineering*, Sep. 2005, pp. 101–105.
354. E. Guillanton, J. Y. Dauvignac, C. Pichot and J. Cashman, A new design tapered slot antenna for ultra-wideband applications. *Microw. Optical Technol. Lett.*, **19**, (1998), 286–289.
355. Z. N. Chen, Broadband roll monopole. *IEEE Trans. Antennas Propag.*, **51**: 11, (2003), 3175–3177.
356. Z. N. Chen, A new bi-arm roll antenna for UWB applications. *IEEE Trans. Antennas Propag.*, **53**: 2, (2005), 672–677.
357. W. Namgoong, A channelized digital ultra-wideband receiver. *IEEE Trans. Wireless Commun.*, **2**: 3, (2003), 502–510.
358. P. P. Newaskar, R. Blazquez and A. P. Chandrakasan, A/D precision requirements for an ultra-wideband radio receiver. In *Proc. IEEE Workshop on Sig. Processing Syst. (SIPS)*, San Diego, CA, Oct. 16–18 2002, pp. 270–275.
359. R. Blazquez, F. S. Lee, D. D. Wentzloff, P. P. Newaskar, J. D. Powell and A. P. Chandrakasan, Digital architecture for an ultra-wideband radio receiver. In *Proc. IEEE Veh. Technol. Conf. (VTC)*, vol. 2, Orlando, FL, Oct. 6-9 2003, pp. 1303–1307.
360. Y. Vanderperren, G. Leus and W. Dehaene, An approach for specifying the ADC and AGC requirements for UWB digital receivers. In *Proc. IEE Seminar on Ultra Wideband Syst., Technol. and Applications*, London, UK, Apr. 2006.
361. M. Verhelst, W. Vereecken, M. Steyaert and W. Dehaene, Architectures for low power ultra-wideband radio receivers in the 3.1–5 GHz band for data rates <10 Mbps. In *Proc. IEEE Int. Symp. on Low Power Electronics and Design (ISLPED)*, Aug. 11–19 2004, pp. 280–285.

362. A. F. Molisch, Y. P. Nakache, P. Orlik, *et al.*, An efficient low-cost time-hopping impulse radio for high data rate transmission. *EURASIP J. Applied Sig. Processing (Special Issue on UWB - State of the Art)*, **2005**: 3, (2005), 397–412.
363. H. Xie, S. Fan, X. Wang, *et al.*, A pulse-based non-carrier 7.5GHz UWB transceiver SoC with on-chip ADC. In *Proc. Int. Conf. on Solid-State and Integrated Circuit Technol. (ICSICT)*, Shanghai, China, Oct. 23–26 2006, pp. 1804–1807.
364. J. Lerdworatawee and W. Namgoong, Low-noise amplifier design for ultra-wideband radio. *IEEE Trans. Circuits Systems–I: Regular Papers*, **51**: 6, (2004), 1075–1087.
365. H. A. Haus, IRE standards on methods of measuring noise in linear two ports. *Proc. IRE*, **48**, (1960), 60–68.
366. Q. He and M. Feng, Low-power, high-gain, and high-linearity SiGe BiCMOS wide-band low-noise amplifier. *IEEE J. Solid-State Circuits*, **39**, (2004), 956–959.
367. D. Barras, F. Ellinger, H. Jackel and W. Hirt, A low supply voltage SiGe LNA for ultra-wideband frontends. *IEEE Microw. Wireless Compon. Letters*, **14**, (2004), 469–471.
368. C.-W. Kim, M.-S. Kang, P. T. Anh, H.-T. Kim and S.-G. Lee, An ultra-wideband CMOS low-noise amplifier for 3–5-GHz UWB system. *IEEE J. Solid-State Circuits*, **40**, (2005), 544–547.
369. B. Shi and M. Y. W. Chia, Design of a SiGe low-noise amplifier for 3.1-10.6 GHz ultra-wideband radio. In *Proc. IEEE Int. Symp. Circuits Syst. (ISCAS)*, vol. 1, Vancouver, Canada, May 2004, pp. 101–104.
370. Q. Li, Y. P. Zhang and G. X. Zheng, Gain-flatness considerations on the ultra-wideband low-noise amplifier design. In *Proc. Asia-Pacific Microw. Conf. (APMC)*, vol. 4, Suzhou, China, Dec. 4–7 2005.
371. Y.-J. Lin, S. S. H. Hsu, J.-D. Jin and C. Y. Chan, A 3.1–10.6 GHz ultra-wideband CMOS low noise amplifier with current-reused technique. *IEEE Microw. Wireless Compon. Letters*, **17**, (2007), 232–234.
372. S. Shekhar, X. Li and D. J. Allstot, A CMOS 3.1–10.6 GHz UWB LNA employing stagger-compensated series peaking. In *Proc. IEEE Radio Frequency Integrated Circuits (RFIC) Symp.*, San Francisco, CA, June 2007, pp. 63–66.
373. C.-T. Fu and C.-N. Kuo, 3–11 GHz CMOS UWB LNA using dual feedback for broadband matching. In *Proc. IEEE Radio Frequency Integrated Circuits (RFIC) Symp.*, San Francisco, CA, June 2007, pp. 67–70.
374. M. T. Reiha, J. R. Long and J. J. Pekarik, A 1.2 V reactive-feedback 3.1–10.6 GHz ultra-wideband low-noise amplifier in 0.13 μm CMOS. In *Proc. IEEE Radio Frequency Integrated Circuits (RFIC) Symp.*, San Francisco, CA, June 2007, pp. 55–58.
375. J. Lee and J. D. Cressler, Analysis and design of an ultra-wideband low-noise amplifier using resistive feedback in SiGe HBT technology. *IEEE Trans. Microw. Theory Techniques*, **54**: 3, (2006), 1262–1268.
376. I. S.-C. Lu, N. Weste and S. Parameswaran, ADC precision requirement for digital ultra-wideband receivers with sublinear front-ends: a power and performance perspective. In *Proc. Int. Conf. on VLSI Design (VLSID)*, Hyderabad, India, Jan. 2006.
377. R. Blazquez, F. Lee, D. Wentzloff, B. Ginsburg, L. Powell and A. Chandrakasan, Direct conversion pulsed UWB transceiver architecture. In *Proc. Design, Automation and Test in Europe*, vol. 3, Munich, Germany, Mar. 2005, pp. 94–95.
378. R. Thirugnanam, D. S. Ha and S. S. Choi, Design of a 4-bit 1.4 Gsamples/s low power folding ADC for DS-CDMA UWB transceivers. In *Proc. IEEE Int. Conf. on Ultra-Wideband (ICU)*, Zurich, Switzerland, Sep. 2005, pp. 536–541.

379. L. Smaini and D. Helal, RF digital transceiver for impulse radio ultra wide band communications. In *Proc. European Solid-State Circuits Conf. Workshop*, Leuven, Belgium, Sep. 2004.
380. I. D. ODonnell, M. S. W. Chen, S. B. T. Wang and R. W. Brodersen, An integrated, low power, ultra-wideband transceiver architecture for low-rate, indoor wireless systems. In *Proc. IEEE CAS Workshop Wireless Commun. and Networking*, Pasadena, CA, Sep. 2002.
381. S. Hoyos, B. M. Sadler and G. R. Arce, Monobit digital receivers for ultra-wideband communications. *IEEE Trans. Wireless Commun.*, **4**: 4, (2005), 1337–1344.
382. R. Blazquez, P. P. Newaskar, F. S. Lee and A. P. Chandrakasan, A baseband processor for pulsed ultra-wideband signals. In *Proc. IEEE Custom Integrated Circuits Conf.*, Oct. 2004, pp. 587–590.
383. A. Petraglia and S. Mitra, Analysis of mismatch effects among A/D converters in a time-interleaved waveform digitizer. *IEEE Trans. Instrum. Meas.*, **40**, (1991), 831–835.
384. S. Hoyos and B. M. Sadler, Ultra-wideband analog-to-digital conversion via signal expansion. *IEEE Trans. Veh. Technol.*, **54**: 5, (2005), 1609–1622.
385. H. Lee, D. Ha and H. Lee, Toward digital UWB radios: Part I – Frequency domain UWB receiver with 1 bit ADCs. In *Proc. IEEE Conf. Ultra-wideband Syst. Technol. (UWBST)*, Kyoto, Japan, May 2004, pp. 248–252.
386. M. S.-W. Chen and R. W. Brodersen, A subsampling UWB radio architecture by analytic signaling. In *Proc. IEEE Int. Conf. Acoust., Speech, Sig. Processing (ICASSP)*, Montreal, Canada.
387. Y. Vanderperren, W. Dehaene and G.Leus, A flexible low power subsampling UWB receiver based on line spectrum estimation methods. In *Proc. IEEE Int. Conf. on Commun. (ICC)*, vol. 10, Istanbul, Turkey, June 2006, pp. 4694–4699.
388. J. H. Reed, *Software Radio: A Modern Approach to Radio Engineering*. (Upper Saddle River, NJ: Pearson Education, Inc., 2002).
389. B. Le, T. W. Rondeau, J. H. Reed and C. W. Bostian, Analog-to-digital converters. *IEEE Sig. Processing Mag.*, **22**: 6, (2005), 69–77.
390. M. B. Romdhane and P. Loumeau, Analog to digital conversion specifications for ultra wide band reception. In *Proc. IEEE Int. Symp. on Sig. Processing and Information Technol. (ISSPIT)*, Rome, Italy, Dec. 2004, pp. 157–160.
391. B. P. Ginsburg and A. P. Chandrakasan, A 500MS/s 5b ADC in 65nm CMOS. In *Proc. Symp. on VLSI Circuits*, Honolulu, HI, June 2006, pp. 140–141.
392. B. P. Ginsburg and A. P. Chandrakasan, Dual time-interleaved successive approximation register ADCs for an ultra-wideband receiver. *IEEE J. Solid-State Circuits*, **42**, (2007), 247–257.
393. Y. Rui and D. Florencio, New direct approaches to robust sound source localization. In *Proc. IEEE Int. Conf. on Multimedia and Expo (ICME)*, vol. 1, Baltimore, MD, July 2003, pp. 737–740.
394. I. Guvenc, C. C. Chong and F. Watanabe, Joint TOA estimation and localization technique for UWB sensor network applications. In *Proc. IEEE Veh. Technol. Conf. (VTC)*, Baltimore, MD, Apr. 2007, pp. 1574–1578.
395. L. Reggiani, M. Rydstrom, E. G. Strom and A. Svensson, Adapting the ranging algorithm to the positioning technique in UWB sensor networks. In *Proc. 4th Cost 289 Workshop*, Goteborg, Sweden, Apr. 2007.
396. D. L. Goeckel and Q. Zhang, Slightly frequency-shifted transmitted-reference ultrawideband radio. *To appear in IEEE Trans. Commun.*, 2007.

397. D. L. Goeckel and Q. Zhang, Slightly frequency-shifted reference ultra-wideband (UWB) radio: TR-UWB without the delay element. In *Proc. IEEE Military Commun. Conf. (MILCOM)*, Atlantic City, NJ, Oct. 2005, pp. 3029–3035.
398. H. Xu, L. Yang and D. Goeckel, Digital multi-carrier differential signaling for ultra-wideband radios. In *Proc. IEEE Global Telecommun. Conf. (GLOBECOM)*, San Francisco, CA, Nov. 2006.
399. H. Xu and L. Yang, Differential UWB communications with digital multi-carrier modulation. In *Proc. IEEE Int. Conf. Ultra-Wideband (ICUWB)*, Waltham, MA, Sep. 2006, pp. 49–54.
400. D. B. Jourdan, J. J. Deyst, M. Z. Win and N. Roy, Monte Carlo localization in dense multipath environments using UWB ranging. In *Proc. IEEE Int. Conf. UWB (ICU)*, Zurich, Switzerland, Sep. 2005, pp. 314–319.
401. K. Takizawa, H. B. Li and R. Kohno, Precise leading edge detection using a forward error correction coding. In *Proc. IEEE Int. Symp. Wireless Commun. Syst. (ISWCS)*, Valencia, Spain, Sep. 2006.
402. K. Whitehouse, C. Karlof, A. Woo, F. Jiang and D. Culler, The effects of ranging noise on multihop localization: an empirical study. In *Proc. IEEE Int. Symp. Information Processing in Sensor Networks (IPSN)*, Los Angeles, CA, Apr. 2005, pp. 73–80.
403. B. Alavi and K. Pahlavan, Modeling of the TOA-based distance measurement error using UWB indoor radio measurements. *IEEE Commun. Lett.*, **10**: 4, (2006), 275–277.
404. Staccato Communications. [Online]. Available: http://www.staccatocommunications.com
405. I. Guvenc, C. Abdallah, R. Jordan and O. Dedeoglu, Enhancements to RSS-based indoor tracking systems using Kalman filters. In *Proc. Global Sig. Processing Conf. (GSPx)*, Dallas, TX, Apr. 2003.
406. F. Althaus, C. Steiner and A. Wittneben, UWB geo-regioning - algorithm and performance. In *Proc. Workshop on Positioning, Navigation, and Commun. (WPNC)*, Hannover, Germany, Mar. 2006, p. 6.
407. Y. H. Jo, J. Y. Lee, D. H. Ha and S. H. Kang, Accuracy enhancement of UWB indoor positioning using ray tracing. In *Proc. IEEE Position, Location, and Navigation Symp. (PLANS)*, San Diego, CA, Apr. 2006, pp. 565–568.
408. J. Y. Lee, Y. H. Jo, S. H. Kang, D. H. Ha and S. J. Yoon, Determination of the existence of LoS blockage and its application to UWB localization. In *Proc. IEEE Military Commun. Conf. (MILCOM)*, Washington, DC, Oct. 2006, pp. 1–4.
409. B. Li, A. Dempster, C. Rizos and H. K. Lee, A database method to mitigate NLOS error in mobile phone positioning. In *Proc. IEEE Position, Location, and Navigation Symp. (PLANS)*, San Diego, CA, Apr. 2006, pp. 173–178.
410. D. Zimmerman, C. Pavlik, A. Ruggles and M. P. Armstrong, An experimental comparison of ordinary and universal kriging and inverse distance weighting. *Springer J. Math. Geol.*, **31**: 4, (1999), 1576–1582.
411. C. Morelli, M. Nicoli, V. Rampa and U. Spagnolini, Hidden Markov models for radio localization in mixed LOS/NLOS conditions. *IEEE Trans. Sig. Proc.*, **55**: 4, (2007), 1525–1542.
412. S. Venkatesh and R. M. Buehrer, Multiple-access design for ad hoc UWB position location networks. In *Proc. IEEE Wireless Commun. Networking Conf. (WCNC)*, vol. 4, Las Vegas, NV, Apr. 2006, pp. 1866–1873.
413. S. Kim, A. P. Brown, T. Pals, R. A. Iltis and H. Lee, Geolocation in ad hoc networks using DS-CDMA and generalized successive interference cancellation. *IEEE J. Select. Areas Commun.*, **23**: 5, (2005), 984–998.

414. S. Venkatesh and R. M. Buehrer, Power control in UWB position-location networks. In *Proc. IEEE Int. Conf. Commun. (ICC)*, vol. 9, Istanbul, Turkey, June 2006, pp. 3953–3959.
415. S. Haykin, Cognitive radio: Brain-empowered wireless communications. *IEEE J. Select Areas Commun.*, **23**: 2, (2005), 201–220.
416. S. Haykin, Cognitive radar. *IEEE Sig. Processing Mag.*, **23**: 1, (2006), 30–40. Jan. 2006.
417. S. Haykin, Cognitive radar networks. In *Proc. IEEE Int. Workshop on Computational Advances in Multi-Sensor Adaptive Processing*, Puerto Vallarta, Mexico, Dec. 2005, pp. 1–3.
418. H. Celebi and H. Arslan, Adaptive positioning systems for cognitive radios. In *Proc. IEEE Symp. on New Frontiers in Dynamic Spectrum Access Networks (DySpan)*, Dublin, Ireland, Apr. 2007.
419. H. Celebi and H. Arslan, Cognitive positioning systems. *to appear in IEEE Trans. Wireless Commun.*, 2007.
420. H. Celebi and H. Arslan, Utilization of location information in cognitive wireless networks. *to appear in IEEE Wireless Commun. Mag. - Special Issue on Cognitive Wireless Networks*, Aug. 2007.
421. B. Alavi and K. Pahlavan, Studying the effect of bandwidth on performance of UWB positioning systems. In *Proc. IEEE Wireless Commun. Networking Conf. (WCNC)*, vol. 2, Las Vegas, NV, Apr. 2006, pp. 884–889.
422. C. C. Chong, F. Watanabe and M. Z. Win, Effect of bandwidth on UWB ranging error. In *Proc. IEEE Wireless Commun. Networking Conf. (WCNC)*, Hong Kong, Mar. 2007, pp. 1559–1564.
423. J. S. Abel, Optimal sensor placement for passive source localization. In *Proc. IEEE Int. Conf. Acoust., Speech, and Sig. Processing (ICASSP)*, vol. 5, Albuquerque, NM, Apr. 1990, pp. 2927–2930.
424. J. Schroeder, S. Galler, K. Kyamakya and K. Jobmann, Practical considerations for optimal three dimensional localization. In *Proc. IEEE Int. Conf. Multisensor Fusion and Integration for Intelligent Syst. (MFI)*, Heidelberg, Germany, Sep. 2006, pp. 439–443.
425. B. Yang and J. Scheuing, Cramer–Rao bound and optimum sensor array for source localization from time differences of arrival. In *Proc. IEEE Int. Conf. Acoust., Speech, Sig. Processing (ICASSP)*, vol. 4, Philadelphia, PN, Mar. 2005, pp. 961–964.
426. E. M. Staderini, UWB radars in medicine. *IEEE Aerospace and Electron. Syst. Mag.*, **17**: 1, (2002), 13–18.
427. S. Venkatesh, C. R. Anderson, N. V. Rivera and R. M. Buehrer, Implementation and analysis of respiration-rate estimation using impulse-based UWB. In *Proc. IEEE Military Commun. Conf. (MILCOM)*, vol. 5, Atlantic City, NJ, Oct. 2005, pp. 3314–3320.
428. A. G. Yarovoy, L. P. Ligthart, J. Matuzas and B. Levitas, UWB radar for human being detection. *IEEE Aerospace and Electron. Syst. Mag.*, **21**: 11, (2006), 22–26.
429. S. Gezici and Z. Sahinoglu, Theoretical limits for estimation of vital signal parameters using impulse radio UWB. In *Proc. IEEE Int. Conf. Commun. (ICC)*, Glasgow, Scotland, June 2007.
430. M. W. M. G. Dissanayake, P. Newman, S. Clark, H. F. D. Whyte and M. Csorba, A solution to the simultaneous localization and map building (slam) problem. *IEEE Trans. Robotics Automation*, **17**: 3, (2001), 229–241.
431. R. Sim and N. Roy, Global A-optimal robot exploration in SLAM. In *Proc. IEEE Int. Conf. Robotics and Automation (ICRA)*, Barcelona, Spain, Apr. 2005, pp. 661–666.
432. W. Guo, N. P. Filer and S. K. Barton, 2D indoor mapping and positioning using an impulse radio network. In *Proc. IEEE Int. Conf. Ultra-Wideband (ICU)*, Zurich, Switzerland, Sep. 2005, pp. 296–301.

433. W. Guo, N. P. Filer and R. Zetik, Indoor mapping and positioning using impulse radios. In *Proc. IEEE Position, Location, and Navigation Symp. (PLANS)*, San Diego, CA, Apr. 2006, pp. 153–163.
434. W. Guo and N. P. Filer, 2.5D indoor mapping and location-sensing using an impulse radio network. In *Proc. IET Seminar on Ultra-wideband Syst., Technol., and Applications*, London, UK, Apr. 2006, pp. 211–215.
435. Emergency ultra-wideband radio for positioning and communications. [Online]. Available: http://www.ist-europcom.org
436. V. Pammer and K. Witrisal, Ultra wideband communication system as sensor technology for 3D mapping. In *Proc. IEEE Int. Symp. Personal, Indoor, Mobile Commun. (PIMRC)*, Helsinki, Finland, Sep. 2006, pp. 1–5.
437. L. Lazos and R. Poovendran, HiRLoc: High-resolution robust localization for wireless sensor networks. *IEEE J. Select Areas Commun.*, **24**: 2, (2006), 233–246.
438. S. Capkun and J. P. Hubaux, Secure positioning in wireless networks. *IEEE J. Select Areas Commun.*, **24**: 2, (2006), 221–232.
439. Y. Zhang, W. Liu, W. Lou and Y. Fang, Location-based compromise-tolerant security mechanisms for wireless sensor networks. *IEEE J. Select Areas Commun.*, **24**: 2, (2006), 247–260.
440. Y. Zhang, W. Liu and Y. Fang, Secure localization in wireless sensor networks. In *Proc. IEEE Military Commun. Conf. (MILCOM)*, Atlantic City, NJ, Oct. 2005, pp. 1–7.
441. F. Anjum, S. Pandey and P. Agrawal, Secure localization in sensor networks using transmission range variation. In *Proc. IEEE Mobile Adhoc and Sensor Syst. Conf. (MASS)*, Washington, DC, Nov. 2005, pp. 195–203.

Index

Aetherwire Inc., 18
ALOHA protocol, 159
amplifiers
 BPF-based input matching, 218
 distributed, 218
 shunt feedback, 218
analog-to-digital conversion, 223
angle of arrival, *see* AOA
antenna efficiency, 46
antennas
 impedance bandwidth, 218
 radiation efficiency, 219
 radiation pattern, 219
anti-aliasing filter, 36
AOA, 7, 8, 17, 60, 67, 69, 79, 101
AOA based positioning, 89
artificial neural network, 229

bandwidth
 absolute, 20, 234
 effective, 69, 71, 86, 151, 233
 fractional, 20
Baranakin bound, 119
Bayesian estimator, 80, 133
Bessel function of the first kind, 114
binary hypothesis testing, 123
binary phase shift keying, 40
BPM-BPSK, 40, 189
BPSK-PPM modulation, 164
burst position modulation, 40

carrier sense multiple access, 232
carrier sensing methods, 181
CCA, 182, 196, 198
CDMA, 7, 134, 183
centralized chi-square distribution, 113
channel impulse response, *see* CIR
channel modeling techniques, 44
channel sounding, 44, 159, 160, 177
channel sounding request, 177
chirp spread spectrum, 39, 149
CIR, 56, 74, 125, 164, 229
circular error probability, 91

clear channel assessment, *see* CCA
clock frequency offsets, 153, 171
cluster arrival rate, 51
cluster arrival times, 51
coarse synchronization, 108
code division multiple-access, *see* CDMA
coded payload modulation, 182
cognitive radar, 233
coherent, 106, 108, 163, 183, 210
colored Gaussian noise, 112
complementary CDF, 113
complementary cumulative distribution function,
 see complementary CDF
complex constellation, 35
confidence interval scaling factor, 170
context-aware services, 2
convolutional encoder, 33
cooperative localization techniques, 228
correlated Gaussian noise model, 83
CRLB, 68, 71, 85, 89, 117, 119, 120, 125, 151,
 205, 227, 235
cross power spectral density, 73
cross-correlation function, 73
cumulative distribution function, 204

decay time constant
 cluster, 52
 intra-cluster, 51
degrees of freedom, *see* DOF
delay-and-correlate receiver, 112, 114
detect-and-avoid, 31
deterministic modeling, 49
dielectric constant, 48, 49
 relative, 48
dielectric layer, 47
differential TOA protocol, 148, 199
digital-to-analog converter, 36
dilution of precision
 geometric, *see* GDOP
 horizontal, 235
 vertical, 235

Index

direct positioning, 63
direct sampling receiver, 111
directional antennas, 60
dirty-template scheme, 135, 137
DOF, 113, 114, 130
DS-CDMA, 23
dual-carrier modulation, 36
DV-hop algorithm, 10
dynamic preamble selection, 200
dynamic spectrum access networks, 234

EC, 30
ECC, 29
Ecma International, 32, 38
Ecma standard, 33, 34, 36, 37
EIRP, 29, 32
Electronic Communications Committee, 29
Emergency-911 services, 2
emission limits, 25
    ECC, 29
    FCC, 25
    MIC, 30
emission point, 20
energy detection receiver, 53, 112
energy matrix with non-linear filtering, 193
equivalent isotropically-radiated power, see EIRP
EUROPCOM, 237
European Commission, 29
    Radio Spectrum Committee of, 29
excess delay
    maximum, 52, 103, 129
    mean, 52, 57, 231
exponential decay, 51

fading
    large-scale, 50, 60
    multipath, 64
    small-scale, 49, 52, 58, 60, 214
FCC, 1, 20, 25, 44, 59
FCC limits, 26
FDS, 37, 38
Federal Communications Commissions, see FCC
FFI, 33
figure of merit, see FoM
FIM, 85, 121
fingerprinting technique, 74, 230
finite state machine, 33
Fisher information matrix, see FIM
FoM, 160, 169, 174, 178
forward error correction, 33
frame synchronization error, 164
free-space propagation, 46
frequency dependency, 50, 55, 210
frequency-shifted reference, see FSR
FSR, 228
Fujitsu, 18

Gabor transform, 185
Gaussian mono-cycle, 22
GDOP, 90, 92, 235
generalized cross-correlation techniques, 73
generalized likelihood ratio test, see GLRT
generalized maximum likelihood technique, see GML
geometric positioning technique, 76
GLRT, 108, 127
GML, 127, 129, 130, 139
GPS, 25
Great Duck Island project, 11

heavy-tailed distribution, 57
hidden Markov model filters, 96
hybrid positioning systems, 89

IDFT, 37
IEEE 802.11, 14
IEEE 802.15.3, 32
IEEE 802.15.3a, 32
IEEE 802.15.3c, 44, 53, 61
IEEE 802.15.4, 32, 39
IEEE 802.15.4a, 1, 10, 42, 44, 53, 148, 158, 169, 175, 181, 186
IEEE 802.15.4b, 157
IEEE802.15.4a, 44
IF processing, 24
IFI, 53
imaging systems, 27
IMEC, 18
impulse radio, 21, 110
infrared signals, 7
interference, 24, 104
    fluorescent, 15
    inter-frame, 44, see IFI
    inter-pulse, 121
    multiuser, see MUI, 233
    multiple-access, see MAI
    narrowband, see NBI, 193
    side-lobe, see SLI
interference mitigation, 30, 181
interleaving, 35
    fixed-frequency, see FFI
    intra-symbol, 35
    time-frequency, see TFI
IR-UWB, 39, 40, 44, 109, 148, 158, 186
IR-UWB systems, 53
IR-UWB transmitter, 40

jump back and search forward algorithm, 139

k-nearest-neighbor, 74
Kalman filter, 93, 94
Kalman gain matrix, 93
Kolmogorov-Smirnov (K–S) hypothesis test, 57

Laplacian distribution, 187
link budget calculations, 50
LNA, 111, 222, 224
localization systems, 9
   active, 9
   centralized, 9
   distributed, 9
   hardware based, 9
   passive, 9
   software based, 9
log-normal distribution, 50
log-normal random variable, 52
loop-back test, 167
low Earth orbiting satellites, 27
low noise amplifier, *see* LNA

M-ary ternary orthogonal keying, *see* MTOK
MAC layer acknowledgment, 152
MAC layer management entity, 177
MAC sublayer, 150, 152, 153, 175
MAE, 131
MAI, 107
MAP estimator, 80
mapping techniques, 74
Marcum-Q function, 114
matched filter receiver, 112
maximum *a posteriori* estimator, *see* MAP estimator
maximum likelihood estimation, 80
MB-OFDM, 33
mean absolute error, *see* MAE
mean cluster energy, 51
mean number of clusters, 60
mean square error, 86, 117–119, 122, 145
median filter, 191, 194, 199, 233
MIC, 30
millimeter wave channel, 60
minimum variance estimators, 108
Ministry of Internal Affairs and Communications, *see* MIC
mixture probability, 51
ML estimation, 80, 84, 121, 125–128
ML estimator, 117, 134
MMSE, 94, 108
modified Hermite polynomials, 22
MSDU, 175
MTOK, 105, 106, 146, 238
MUI, 183, 186, 187
multiband OFDM, 32, 33
multicluster model, 54
multihypothesis tracking approach, 94
multiple access via carrier sensing, 181
multiplexed preamble scheme
   CDMA type, 197
   TDMA type, 196
Multispectral Solutions, Inc., 13, 18
multivariate Gaussian random variable, 83

Nakagami, 52, 58, 61, 62
narrowband systems, 44, 45, 108, 183, 203, 219
NBI, 183, 184, 224
neural networks, 74, 80, 229
next higher layer, *see* NHL
NHL, 150, 177, 178
NLOS, 54, 57, 66, 82, 86
non-coherent, 107, 108, 160, 163–165, 183, 191, 208, 210, 233
non-linear least squares technique, 102
Nyquist rate, 108, 111, 113, 117, 122, 128, 139, 222, 224, 235

OFDM, 7
open systems interconnection (OSI) model, 149
optical signals, 7
optimum geometries, 236
orthogonal frequency division multiplexing (OFDM), 7
out-of-band noise, 111

PAL 650, 18
particle filters, 96
passive sensing systems, 27
path loss, 49
pattern matching, 8, 14, 74
PBTS, 189, 190, 200, 201, 215
PDP, 73
peak power, 32
peak-to-lead delay, 52, 53
perfectly balanced ternary sequences, *see* PBTS
PHR, 159, 165, 169, 172, 181, 182
PHY layer management entity, 165, 177
PHY management service, 150
PHY PIB attributes, 165
PHY protocol data units, *see* PPDU
PHY service data unit, *see* PSDU
physical layer header, *see* PHR
PL exponent, 50, 54, 59, 65, 66, 89, 97
platonic solids, 236
polynomial fitting, 56
power amplifier, 215, 218
power control, 233
power delay profile, *see* PDP, 50, 52, 54
PPDU, 150, 172, 175
PRF, 41, 206, 208, 209
   mean, 41, 42, 208, 226
   minimum, 208
   peak, 208, 226
PRI, 189, 205, 206, 226
private ranging, 198
probability density function, 53
PSDU, 36, 159, 175
pulse generator, 209, 215
pulse repetition frequency, *see* PRF
pulse repetition interval, *see* PRI
PulseON350, 18

QPSK, 35

radar cross section, 47
radio frequency, 7
Radio Spectrum Committee, 29
range reply frame, 151
range request frame, 151
ranging capable device, 151
ranging protocol
    differential two way, 153
    symmetric double sided, *see* SDS
    two-way time-of-arrival, *see* TW-TOA
    type-II differential two-way, 154
ranging reference point, 39
ray arrival rates, 51, 60
ray arrival times, 51, 60
ray tracing, 44
real time location systems, *see* RTLS, 13
received signal strength, *see* RSS
rectangular lattices, 68
Reed–Solomon encoder, *see* RS encoder
reflection, 46
regulations, 25
    FCC, 25, 26, 28, 59, 215
remote-positioning, 8
residual error, 102, 156, 157, 227
residual weighting algorithm, 80
RF loop-back tests, 153
RFID, 7, 13, 18
RMS delay spread, 52, 53, 57, 59, 108, 146, 229, 231
RMSE, 90, 96, 131, 204, 205, 225
root-mean-square delay spread, 52
root-mean-square error, *see* RMSE
RS encoder, 40
RSS, 7, 8, 63, 64, 66, 67, 70, 76, 101
RSS based positioning, 88
RTLS, 6, 12, 13, 17

Sapphire DART, 13
scalability, 10
SDS, 155–157
seamless positioning, 234
search-back algorithm, 53
self-organization, 10
sequential Monte Carlo methods, 96
serial backward search algorithm, 141
service access point
    MAC common part sublayer, 150
    MAC sublayer management entity, 150
    PHY data, 149
    PHY layer management entity, 150
SFD, 159, 162, 169, 181, 182
SFD waveform coefficient vector, 162
shadowing, 49, 50, 59, 65, 214
Shannon capacity, 24
SHR, 159
SHR preamble waveform, 162

signal-to-noise ratio, *see* SNR
simultaneously operating piconets, *see* SOP
SLI, 105
SNR, 24, 56, 69, 72, 86, 113, 151, 162, 181, 192
SOP, 107, 187, 232
spectral mask, 25
    FCC, 25
spreading
    frequency-domain, *see* FDS
    time-domain, *see* TDS
standard
    ECMA-368, 33
    ECMA-369, 33
    IEEE 802.15.4a, 18, 39–42, 57, 58, 165–167, 178, 182, 187, 189, 190, 200, 209, 210
start of frame delimiter, *see* SFD
statistical modeling, 49
statistical positioning techniques, 79
sub-Nyquist, 108, 128, 183, 229
support vector regression, *see* SVR
surveillance systems, 27
SVR, 74, 75
symmetric double-sided ranging protocol, *see* SDS
synchronization header preamble, *see* SHR
synchronization preamble, 36

TDOA, 7, 8, 17, 72, 79, 84, 101, 157
TDOA-based positioning, 88
TDOA protocol, 148, 157, 158
TDS, 37, 38
TFC, 33, 36
TFI, 33
threshold-based acquisition scheme, 138
threshold-based ranging algorithm, 139
time difference of arrival, *see* TDOA
Time Domain Corporation, 18
time of flight, 148
time-based ranging protocols, 148
time-frequency code, *see* TFC
time-hopping code, 21, 109
time-of-arrival, *see* TOA
time-stamp report, 152, 157, 168, 174
timing jitter, 108
TOA, 7, 8, 61, 63, 70, 76, 79, 101, 133, 135, 138, 148
TOA-based positioning, 88
TR, 105, 115, 116, 183, 227
transmission, 48
transmitted reference, *see* TR
truncated Gaussian, 58
turn-around time, 151, 171
TW-TOA, 148, 151, 153–155, 157, 159, 178, 199, 200
two-dimensional energy matrix, 193
two-step TOA estimators, 135
two-way ranging protocol, 70
two-way TOA protocol, 148

Ubisense, 17
UCA, 68
ULA, 67, 69
ultrasound signals, 7
uniform circular arrays, *see* UCA
uniform linear array, *see* ULA
universal kriging technique, 231
UWB antennas, 218
   radiation efficiency of, 219
UWB Forum, 32

variance weighted least squares technique, 80
vehicular radar systems, 27
verifiable multilateration algorithm, 239
vital signal parameters, 236

wideband systems, 45, 196, 218
WiMedia Alliance, 32
wireless local area networks, *see* WLAN
wireless personal area networks,
   see WPAN
wireless sensor networks, *see* WSN
wireless USB, 26
WLAN, 1, 3, 9, 183
WPAN, 1, 20, 26, 156
WSN, 1, 10, 11, 13

ZCZ, 105, 107
zero correlation zone, *see* ZCZ
Ziv–Zakai lower bound, *see* ZZLB
ZZLB, 122, 124, 125, 227